B/N A

USMG   NATO ASI

LES HOUCHES
Session XXXVII
1981

THÉORIES DE JAUGE
EN PHYSIQUE DES HAUTES ÉNERGIES

GAUGE THEORIES
IN HIGH ENERGY PHYSICS

CONFÉRENCIERS

Richard C. BROWER
Sidney COLEMAN
John ELLIS
Leon M. LEDERMAN
Chris QUIGG
Chris T.C. SACHRAJDA
Julius WESS
Björn WIIK

Gordon L. Kane, David N. Schramm,
Larry R. Sulak, Michael S. Turner.

LES HOUCHES

SESSION XXXVII

3 Août–11 Septembre 1981

THÉORIES DE JAUGE
EN PHYSIQUE DES HAUTES ÉNERGIES

GAUGE THEORIES
IN HIGH ENERGY PHYSICS

édité par

MARY K. GAILLARD *et* RAYMOND STORA

Université de Grenoble, Ecole d'Ete de Physique Theorique, Les Houches, Publications.

PART II

1983

NORTH-HOLLAND PUBLISHING COMPANY

AMSTERDAM · NEW YORK · OXFORD

© NORTH-HOLLAND PUBLISHING COMPANY, 1983

ISBN 0 444 86543 8 Set
    0 444 86722 8 Part I
    0 444 86723 6 Part II

Published by:
NORTH-HOLLAND PUBLISHING COMPANY
AMSTERDAM·NEW YORK·OXFORD

Sole distributors for the USA and Canada:
ELSEVIER SCIENCE PUBLISHING COMPANY, INC.
52 VANDERBILT AVENUE
NEW YORK, N.Y. 10017

Library of Congress Cataloging in Publication Data
Main entry under title:

Théories de jauge en physique des hautes énergies
   Gauge theories in high energy physics.

   English and French.
   At head of title: USMG, NATO ASI.
   Lectures delivered at Les Houches, École d'été de
physique théorique, session XXXVII.
   Includes bibliographies.
   1. Gauge fields (Physics)--Addresses, essays,
lectures. 2. Particles (Nuclear physics)--Addresses,
essays, lectures. 3. Quantum chromodynamics--
Addresses, essays, lectures. I. Gaillard, Mary K.
II. Stora, Raymond, 1930-   . III. NATO Advanced
Study Institute. IV. Université seientifique et
médicale de Grenoble. École d'été de physique
théorique. V. Gauge theories in high energy physics.
QC793.3.F5T46 1983   539.7'6'01   83-13083
ISBN 0-444-86543-8 (Elsevier Science)

Printed in The Netherlands

LES HOUCHES
ÉCOLE D'ÉTÉ DE PHYSIQUE THÉORIQUE

ORGANISME D'INTÉRÊT COMMUN DE L'UNIVERSITÉ
SCIENTIFIQUE ET MÉDICALE DE GRENOBLE ET DE
L'INSTITUT NATIONAL POLYTECHNIQUE DE
GRENOBLE
AIDÉ PAR LE COMMISSARIAT A L'ÉNERGIE
ATOMIQUE

SESSION XXXVII
INSTITUT D'ÉTUDES AVANCÉES DE L'OTAN
NATO ADVANCED STUDY INSTITUTE

3 Aout–11 Septembre 1981

Directeur scientifique de la session: Mary K. Gaillard, LAPP, Chemin de
Bellevue, BP no. 909, F 74019 Annecy-le-Vieux (nouvelle adresse: Physics
Department, University of California, Berkeley, CA 94720, USA)

SESSIONS PRÉCÉDENTES

*Sessions ayant reçu l'appui du Comité Scientifique de l'OTAN.

LECTURERS

BROWER, *Richard*, Physics Department, Natural Sciences 2, University of California, Santa Cruz, CA 95064, USA.

COLEMAN, *Sidney*, Theoretical Physics Group, Lyman Laboratory, Harvard University, Cambridge, MA 02138, USA.

ELLIS, *John*, Theory Division, CERN, CH-1211 Genève 23, Switzerland.

KANE, *Gordon*, Randall Laboratory of Physics, University of Michigan, Ann Arbor, MI 46109, USA.

LEDERMAN, *Leon*, Fermi National Accelerator Laboratory, PO Box 500, Batavia, IL 60510, USA.

QUIGG, *Chris*, Fermi National Accelerator Laboratory, PO Box 500, Batavia, IL 60510, USA.

SACHRAJDA, *Chris*, Department of Physics, University of Southampton, Southampton SO9 5NH, UK.

SCHRAMM, *David*, Astrophysical and Astronomical Center 100, University of Chicago, 5840 S. Ellis Avenue, Chicago, IL 60637, USA.

SULAK, *Larry*, Randall Laboratory of Physics, University of Michigan, Ann Arbor, MI 46109, USA.

TURNER, *Michael*, Astrophysical and Astronomical Center, University of Chicago, 5640 Ellis Avenue, Chicago, IL 60637, USA.

WESS, *Julius*, Institut für Theoretische Physik, Universität Karlsruhe, Kaiserstrasse 12, Postfach 6380, D-7500 Karlsruhe 1, FRG.

WIIK, *Björn*, II Institut für Experimentalphysik, Universität Hamburg, Luruper Chaussee 149, 2000 Hamburg 50, FRG.

PARTICIPANTS

Alvarez, Enrique, Departamento de Fisica Teórica, CXI, Universidad Autónoma de Madrid, Canto Blanco, Madrid 34, Spain.

Bagger, Jonathan A., Department of Physics, Princeton University, Princeton, NJ 08544, USA.

Bailey, David, High Energy Physics, Rutherford Physics Bldg., McGill University, 3600 University, Montreal, Québec, Canada.

Bandelloni, Giuseppe, Istituto di Scienze Fisiche, Viale Benedetto XV-5, I-16132 Genova, Italy.

Barate, Robert, DPhPE/SEE, CEN Saclay, BP 2, 91191 Gif-sur Yvette, France.

Carpenter, David B., Physics Department, Westfield College, Kidderpore Ave, London NW3, England.

Chiu, Ting-Wai, Physics Department, University of California, Irvine, CA 92717, USA.

Cohen, Eyal, Department of Nuclear Physics, Weizmann Institute of Science, Rehovot, Israel.

Coquereaux, Robert, CPT-CNRS, Case 907, Luminy, 13288 Marseille cedex 2, France.

Dawson, Sara, Theory Group, Box 500, Fermilab, Batavia, IL 60510, USA.

Dhar, Avinash, Theory Group, T.I.F.R., Bombay 400005, India.

DiLieto, Christine, Blackett Laboratory, Imperial College, Prince Consort Road, London SW7, England. From Oct.: SLAC Theory Division, Stanford, CA, USA.

El Hassouni, Abdellah, Laboratoire de Physique Théorique, Département de Physique, Faculté des Sciences de Rabat, BP 1014, Rabat, Maroc.

Fayard, Louis, Laboratoire de l'Accélérateur Linéaire, Université Paris 11, 91405 Orsay, France.

Gavai, Rajiv V., Fakultät für Physik, Universität Bielefeld, Postfach 8640, D-4800 Bielefeld 1, FRG.

Gavela Legazpi, Maria Belén, Laboratoire de Physique des Particules, Chemin de Bellevue, BP 909, 74019 Annecy-le-Vieux cedex, France.

Gelmini, Graciela, Sektion Physik der Universität München, Theoretische Physik, Theresienstrasse 37, 8000 München 2, FRG.

Goossens, Michel, Division EP, CERN, CH-1211 Genève 23, Switzerland.

Govaerts, Jan, Université Catholique de Louvain, Institut de Physique Théorique, Chemin du Cyclotron 2, 1348 Louvain-la-Neuve, Belgium.

Haggerty, John, Physics Department, Fermilab, PO Box 500, Batavia, IL 60510, USA

Hasenfratz, Anna, CRIP, PO Box 114, Budapest, Hungary.

Hornaes, Arne, Fysisk Institutt Allégt. 55, 5014-U, Bergen, Norway.

Jonsson, Thordur, Raunvisindastofnun Haskolans, Dunhaga 3, 107 Reykjavik, Iceland.

Kaplunovsky, Vadim, Department of Physics and Astronomy, Tel-Aviv University, Ramat-Aviv, Tel-Aviv, Israel.

Karasinski, Piotr, Theory group, Gibbs Research Laboratory, Yale University, New Haven, CT 06520, USA

Kleiss, Ronald, Lorentz Instituut, Niewsteeg 18, 2311 BS Leiden, The Netherlands.

Klinkhamer, Frans, Astronomisch Instituut, PO Box 9513, 2300 Leiden, The Netherlands.

Lauwers, Paul, Niels Bohr Institutet, Blegdamsvej 17, DK-2100 København Ø, Denmark.

Leung, Chung Ngoc, School of Physics and Astronomy, University of Minnesota, 116 Church Street SE, Minneapolis, MN 55455, USA.

Machet, Bruno, CNRS, Centre de Physique Théorique, Section II, Luminy, Case 907, 13288 Marseille cedex 2, France.

Maciel, Arthur, Rua viúva Lacerda 433/401, Humaitá, Rio de Janeiro, Brasil.

McBride, Patricia, 5th floor JWG, Department of Physics, Box 6666, 260 Whitney Ave, New Haven, CT 06511, USA

Mukhi, Sunil, ICTP, PO Box 586, Miramare, 34100 Trieste, Italy.

Nadkarni, Sudhir, Yale University, Department of Physics, 217 Prospect Street, PO Box 666, New Haven, CT 06511, USA.

Napoly, Olivier, Service de Physique Théorique, CEN Saclay, BP 2, 91191 Gif-sur-Yvette, France.

Neufeld, Helmut, Institut für Theoretische Physik der Universität Wien, Boltzmanngasse 5, A-1090 Wien, Austria.

Oudrhiri Safiani, El Ghali, Laboratoire de Physique Théorique, Département de Physique, Faculté des Sciences de Rabat, BP 1014, Rabat, Maroc.

Petcov, Sergey, Institute of Nuclear Research and Nuclear Energy, Bulgarian Academy of Sciences, Boul. Lenin 72, 1184 Sofia, Bulgaria.

Pitman, Dale, Physics Department, University of Toronto, Toronto, Ontario M5S 1A7, Canada.

Polley, Lutz, Institut f. Kernphysik, Technische Hochschule Darmstadt, Schlossgarten Str. 9, D-6100 Darmstadt, FRG.

Roudeau, Patrick, L.A.L., Université Paris-Sud, Bât. 200, 91405 Orsay cedex, France.

Sever, Ramazan, Middle East Technical University, Physics Department, Ankara, Turkey.

Sjöstrand, Torbjörn, Department of Theoretical Physics, University of Lund, Sölvegatan 14 A, S-223 62 Lund, Sweden.

Sodano, Pasquale, Istituto di Fisica, Università di Salerno, 84100 Salerno, Italy.

Van Ramshorst, Jan Gerrit, Instituut voor Theoretische Natuurkunde der R.U.U., PO Box 80006, 3508 TA Utrecht, The Netherlands.

Verschelde, Henri, Seminarie voor Theoretische Vaste Stof, en lage Energie Kern-Fysica R.U.G., Krijgslaan 271,S9, 9000 Gent, Belgium.

Wheater, John, Department of Theoretical Physics, University of Oxford, 1 Keble Road, Oxford, UK.

Wrigley, James C., HEP Group, Cavendish Laboratory, Madingley Road, Cambridge, UK.

Zaks, Alexander, Tel-Aviv University, Ramat-Aviv, Israel.

Zhang, Yi-Cheng, SISSA (I.C.T.P.), Trieste, Italy.

Zoupanos, Georges, Theoretical Physics, National Technical University of Athens, Zografou Campus, Athens 624, Greece.

PRÉFACE

Depuis la session d'été de 1976 à l'Ecole des Houches sur la physique des particules à haute énergie, les théories de jauge se sont imposées comme le cadre commun à l'intérieur duquel les théoriciens tentent de décrire et de prévoir les phénomènes, et avec lequel les expérimentateurs confrontent leurs données.

Les prédictions du modèle standard "SU(2)×U(1)" des interactions électromagnétiques et faibles ont maintenant trouvé un succès incontesté, la chromodynamique quantique est devenue le language communément accepté pour l'étude des hadrons et de leurs interactions fortes, et, même les théories de jauge qui tentent d'unifier les trois forces fondamentales de la physique des particules à haute énergie ont reçu un soutien phénoménologique encourageant. Il a donc semblé approprié de présenter une revue exhaustive des théories de jauge: leur motivation et les principes sous-jacents, l'évolution rapide des développements techniques des traitements aussi bien perturbatifs que non perturbatifs de ces théories, leur large spectre d'applications à la physique des particules aussi bien qu'à l'astrophysique et à la cosmologie, leur succès sur le plan expérimental et les possibilités de futurs tests de leur structure qui seront offertes par les nouveaux équipements expérimentaux.

Le volume que nous proposons inclut la rédaction de douze articles écrits à partir de huit séries de cours et de quatre séries de séminaires.

Les théories de jauge ont été introduites dans les conférences de Julius Wess, avec une focalisation particulière sur la construction des théories de jauge supersymétriques qui récemment ont été largement reconnues comme susceptibles d'offrir une approche prometteuse à la résolution de certaines des difficultés rencontrées dans les modèles réalistes. Ces conférences comprennent aussi une introduction à la théorie de la gravitation et à sa formulation supersymétrique.

Dans une série de cours simultanément proposés aux étudiants, Chris Sachrajda a décrit l'approche perturbative à la chromodynamique quantique, sujet qui a connu un considérable développement technique au cours des cinq dernières années avec un élargissement du domaine d'applicabilité permettant d'inclure certains phénomènes exclusifs en plus du régime standard des processus inclusifs profondément inélastiques.

John Ellis a passé en revue les applications des théories de jauge aux interactions électromagnétiques et faibles, commençant par la théorie électrofaible maintenant standard, développant ensuite les idées qui cherchent à unifier cette théorie avec la chromodynamique quantique, et offrant pour terminer certaines spéculations sur une ultime unification avec la gravité.

Ces conférences ont été complètées par des revues de l'interpénétration entre la physique des particules et la cosmologie par David Schramm et Michael Turner, une discussion par Gordon Kane des particules scalaires insaisissables que l'on doit introduire dans la théorie pour comprendre la brisure spontanée des théories de jauge, et une revue par Larry Sulak des expériences à venir destinées à mesurer la vie moyenne du proton.

Il y a eu aussi d'importants développements dans le traitement non perturbatif des théories de jauge. Sidney Coleman a décrit des configurations classiques appelées monopôles et les solutions correspondantes qui doivent apparaître dans les théories de jauge spontanément brisées utilisées dans les descriptions unifiées des interactions entre particules. Un autre sujet auquel s'est appliquée une activité d'importance majeure de la part des théoriciens des particules, avec l'espoir de pénétrer les mécanismes du phénomène de confinement, est l'étude des théories de jauge sur réseau qui a été le sujet des conférences de Richard Brower.

Tandis que la plupart des théoriciens travaillent aujourd'hui dans le language des quarks et des gluons, les propriétés de ces objets se manifestent par l'intermédiaire de l'étude expérimentale des hadrons et l'une des questions fondamentales est la description quantitative des hadrons eux-mêmes à partir de la théorie quantique des champs sousjacente. L'état actuel de cet art a été décrit dans les conférences de Chris Quigg.

Finalement, le sort ultime de toute théorie repose sur sa vérification expérimentale. Leon Lederman a donné une description personnalisée des pièges, mais aussi des heures de gloire de la physique expérimentale et décrit le travail qui reste à faire pour étudier plus avant la structure des interactions entre particules aussi bien à l'aide des accélérateurs de protons sur cibles fixes que des anneaux de collision proton–anti-proton. Björn Wiik a décrit les accélérateurs actuels en insistant sur les anneaux de collision électron–positron et sur la physique qu'ils ont produit jusqu'à maintenant. Il a aussi passé en revue les résultats sur la lepto-production et conclu avec une perspective des équipements à venir pour les interactions de très haute énergie induites par des leptons.

Outre les conférences reproduites dans ce volume, des séminaires ont été présentés sur des aspects plus spécialisés de chacun des sujets, aussi bien par des étudiants que par des orateurs invités. Comme ils traitent de travaux déjà publiés, à paraître, ou dans un état préliminaire, ces séminaires ne sont pas reproduits, mais répertoriés à la place qui leur est appropriée.

Cependant, les qualités dynamiques de l'Ecole ne peuvent être reproduites par l'imprimerie. Elles doivent aussi bien au dévouement des conférenciers qui ont encouragé le dialogue et fréquemment participé aux sessions de discussion, qu'à l'enthousiasme des étudiants qui ont organisé les discussions non seulement dans un but de clarification et d'approfondissement des sujets discutés dans les conférences, mais aussi sur de nombreux autres sujets connexes qui les intéressaient spécialement. Nous désirons remercier ici tous les participants, conférenciers, orateurs de séminaires pour leur contribution à cette magnifique session de six semaines à l'Ecole de Physique Théorique des Houches, la 37ème depuis la création de l'Ecole, qui a regroupé 51 étudiants appartenant à plus de 20 nations, et 27 conférenciers et spécialistes.

Remerciements

La réalisation de la session XXXVII de l'Ecole des Houches et la publication de ce volume sont le résultat de nombreuses contributions:

–le soutien financier de l'Université Scientifique et Médicale de Grenoble, et les subventions de la Division des Affaires Scientifiques de l'OTAN, qui a inclus cette session dans son programme d'Instituts d'Etudes Avancées, et du Commissariat à l'Energie Atomique;

–l'orientation et le soutien effectif du Conseil de l'Ecole;

–le soin apporté par Valérie Lecuyer dans la préparation et la frappe des manuscrits;

–la coopération de tous les participants qui ont contribué de façon inestimable à la richesse du programme scientifique, par leur aide active au cours de l'élaboration des notes de cours et les compléments sous forme de séminaires;

–l'aide d'Henri et Nicole Coiffier, d'Anny Battendier, et de toute l'équipe qui a rendu la vie de tous les jours aussi confortable que possible;

–la direction enthousiaste et communicative des opérations photo-graphiques assurée par Piotr Karasinski.

Nous tenons finalement à remercier les conférenciers pour avoir donné leur temps à la préparation et à la rédaction des cours.

Mary K. Gaillard
Raymond Stora

PREFACE

Since the 1976 Les Houches summer school on high energy particle physics, gauge theories have asserted themselves as the standard framework within which theorists attempt to describe and predict phenomena, and against which experimenters confront their data.

The predictions of the standard $SU(2) \times U(1)$ model of electromagnetic and weak interactions have by now met with uncontested success, quantum chromodynamics has become the accepted language for the study of hadrons and their strong interactions, and even those gauge theories which attempt to unify the three basic forces of high energy particle physics have received encouraging phenomenological support. It therefore appeared appropriate to present a comprehensive survey of gauge theories: their motivation and underlying principles, the rapidly evolving technical developments in both perturbative and non-perturbative treatments of these theories, their wide range of applications to particle physics as well as to astrophysics and cosmology, their experimental successes, and the possibilities for future tests and probes of their structure, which will be offered by new experimental facilities.

The present volume includes twelve articles, written from eight sets of lectures and four series of seminars.

Gauge theories were introduced in the lectures of Julius Wess, with particular emphasis on the construction of supersymmetric gauge theories which have recently become widely recognized as providing a promising approach to the resolution of some of the difficulties confronting realistic models. These lectures also include an introduction to the theory of gravitation and to its supersymmetric formulation.

In a concurrent series of lectures, Chris Sachrajda described the perturbative approach to quantum chromodynamics, a field which has undergone considerable technical development over the past five years, widening its domain of applicability to include certain exclusive phenomena in addition to the standard regime of deep inelastic inclusive processes.

John Ellis reviewed the applications of gauge theories to non-strong interactions, starting with the by now "standard" electroweak theory, developing the ideas which seek to unify this theory with quantum chromodynamics, and finally offering some speculations on an ultimate unification with gravity. These lectures were supplemented with reviews on the interplay between particle physics and cosmology by David Schramm and Michael Turner, a discussion by Gordon Kane of the properties of the elusive scalar particles which must be introduced into the theory to understand the spontaneous breaking of gauge theories, and an overview by Larry Sulak of the forthcoming experiments designed to measure the proton lifetime.

There have also been important developments in the non-perturbative treatment of gauge theories. Sidney Coleman described those classical configurations known as monopoles and the corresponding solutions which are expected to appear in the spontaneously broken quantum gauge theories used in unified descriptions of particle interactions. Another field which has become a major activity among particle theorists, with the hope of gaining insight into the phenomenon of confinement, is the study of lattice gauge theories which was the subject of the lecture series by Richard Brower.

While most theorists today work in the language of quarks and gluons, the properties of these objects are discerned through the experimental study of hadrons, and one of the outstanding issues is the quantitative description of the hadrons themselves in terms of the underlying quantum field theory. The present state of this art was described in the lectures of Chris Quigg.

Finally, the ultimate fate of any theory rests on its experimental verification. Leon Lederman gave a personalized account of the pitfalls and rewards of experimental physics and described the work still to be done by both fixed-target proton accelerators and proton–antiproton colliding rings in probing further the structure of particle interactions. Björn Wiik described existing accelerators with emphasis on electron–positron colliding rings and the physics they have so far produced. He also reviewed leptoproduction results and concluded with a perspective of future facilities for very high energy lepton interactions.

In addition to the lectures reproduced in this volume seminars were presented by both students and outside speakers on more specialized aspects of each topic. As they cover work that is published, soon to be published or still in a preliminary stage, these seminars are not reproduced, but are listed at the end of the appropriate section. However, the dynamic quality of the School cannot be reproduced in print. This was due both to the dedication of the lecturers who encouraged a dialogue and participated frequently in discussion sessions, and to the enthusiasm of the students who organized on-going discussions, not only for clarification of and further insight into subjects covered by the lectures, but also on many other related topics of special interest to themselves. We wish to thank here all the participants, lecturers and seminar speakers for their contribution to a rewarding six weeks session at the Les Houches summer school, the 37th since its creation, which gathered 51 students from more than 20 nations, and 27 lecturers and seminar speakers.

Acknowledgements

The XXXVIIth Session of the Les Houches summer school and the publication of this volume of lecture notes would not have been possible without:
   –the financial support from the Université Scientifique et Médicale de Grenoble, the NATO Scientific Affairs Division (which included this session in its Advanced Studies Institute Programme) and the Commissariat à l'Energie Atomique;
   –the guidance of the school board;
   –the careful preparation and typing of the manuscripts by Valérie Lecuyer;
   –the cooperation of all the participants who contributed in an invaluable manner to the whole scientific programme, including substantial help in the preparation of many of the lecture notes and providing complements to the courses in the form of carefully chosen seminars;
   –the help of Henri and Nicole Coiffier, Anny Battendier and the whole team, in making everyday life as comfortable as possible;
   –Piotr Karasinski and his enthusiastic and communicative direction of the photographic operations.
   Special thanks are due to the lecturers for giving so much of their time in preparing the lectures, and writing them down.

Mary K. Gaillard
Raymond Stora

CONTENTS

Seminar 3. *Generalized Higgs physics and technicolor, by Gordon L. Kane* 415

Seminar 4. *Waiting for the proton to decay: a comparison of the new experiments, by Larry R. Sulak* — *441*

Course 4. *Le monopôle magnétique cinquante ans après, par Sidney Coleman (trad. Raymond Stora)* — *461*

PART II

COURSE 6

MODELS FOR HADRONS

Chris QUIGG

Fermi National Accelerator Laboratory*
Batavia, IL 60510, USA
and
Laboratoire de Physique Théorique de l'Ecole Normale Supérieure**
75231 Paris-Cédex 05, France

*Permanent address. Fermilab is operated by Universities Research Association, Inc., under contract with the United States Department of Energy.
**Laboratoire propre du Centre National de la Recherche Scientifique, associé à l'Ecole Normale Supérieure et à l'Université de Paris-Sud.

M.K. Gaillard and R. Stora, eds.
Les Houches, Session XXXVII, 1981
Théories de jauge en physique des hautes énergies / Gauge theories in high energy physics
© *North-Holland Publishing Company, 1983*

Contents

1. Introduction

In the course of the past fifteen years, high-energy physicists have developed a strong collective conviction that the smallest subunits of matter are the quarks and leptons. "Smallest" is of course to be interpreted as the smallest we have yet observed. Truly elementary particles must be indivisible and structureless. What is known so far about the quarks and leptons is that they are structureless down to a scale of a few times 10^{-16} cm. Nothing more complex than quarks and leptons qualifies as elementary. Nothing smaller is yet indicated by experiment. Thus, the possibility that quarks and leptons themselves have constituents is entertained largely by tradition, in hopes of bringing order to the burgeoning spectrum of apparently fundamental fermions.

We have also achieved a certain understanding of the weak and electromagnetic interactions of quarks and leptons. The Weinberg–Salam theory is calculable, incorporates all observational systematics, and agrees with experiment insofar as it has been tested. The theory as developed until now has some remaining arbitrariness, embodied in the weak mixing angle θ_W, the mass of the Higgs boson, and fermion masses. It is also incompletely motivated, in that the left-handedness of the charged currents is not explained. Progress may come in the form of a deeper understanding of the mechanism of spontaneous symmetry breaking.

If, however, we are willing to regard the fermion spectrum as given, then the electroweak interactions of quarks and leptons may be considered as understood, at least at an engineering level. This means that apart from rare, lepton-number violating processes (and of course gravity) the lepton sector is completely disposed of.

What about the strongly-interacting particles represented by the quarks? We have a gauge theory, Quantum ChromoDynamics, which purports to describe the strong interactions among quarks. It incorporates—by construction—all observational systematics, it has no experimental embarrassments, and it is calculable in perturbation theory under restrictive circumstances. This is promising! But is it enough?

If you know the elementary particles and their interactions, and you call yourself a physicist, you ought to be able to calculate the consequences—or at least you should feel guilty if you can't! The desire to share my guilt, which is amplified by the fact that ordinary laboratory experience concerns hadrons rather than quarks, will then be a motive force for these lectures. Our specific aspirations should include these:

–to compute the properties of hadrons, explain the absence of unseen species, and predict the existence of new varieties of hadrons;

–to explain why quarks and the quanta of the color forces, gluons, are not observed;

–to derive the interactions among hadrons as a collective effect of the interactions among constituents.

These lectures will be concerned principally with the first task, understanding the hadron spectrum. The second and third will be discussed only in passing. My approach will be to review the basis of the quark hypothesis [Gell-Mann 1964, Zweig 1964a, b, c], and deduce the consequences of this picture. Because I do not know how to solve QCD in general, it will be necessary to proceed by iteration, using a progression of models motivated by simplicity, or QCD, or both.

The detailed plan of these lectures is to be found in the table of contents, but a short summary may be useful at this point. Following a brief review of basic concepts, the elementary quark model will be introduced as a classification tool. The quark model will then be extended to the consideration of static properties of hadrons. The ensuing description of magnetic moments, charge radii, and such, is often called the naïve quark model. What is meant is that detailed dynamical assumptions play little or no rôle in the predictions of the model. Many consequences of the quark model are shared with symmetry schemes. I have not been fastidious in pointing out these overlaps, because I regard the quark description as the more fundamental. The relative success of the predictions argues that the constituent (quark) picture be taken seriously, and that dynamical explanations for its validity be sought. These considerations lead us to the idea of QCD, and to the construction of that theory. With QCD then serving as inspiration, we shall consider a variety of idealized descriptions of hadrons and of the force between quarks, including strings, bags, nonrelativistic potential models, and the $1/N$ expansion. As applications of these ideas we shall examine, in addition to the conventional hadrons, novel configurations such as glueballs and baryonium. A few words will follow on future prospects, and on the possibilities of unconfined quarks.

Within the conceptual framework to which this Les Houches session has been devoted, it is easy to identify three Great Questions—in addition to the overlying issue of whether the entire structure is defective. The first question concerns the spectroscopy of fundamental fermions: how many are there, and why do they have the properties they have? The second question has to do with the spectroscopy of gauge bosons or, equivalently, with the identification of the gauge groups involved. A corollary to these basic questions, which apply at the level of field theories of the quarks and leptons, is more in the nature of an applied science problem. This third Great Question, which may in some ways be the most difficult to answer, concerns the problem of hadron structure: to understand why hadrons have the form they do, and why hadrons interact as they do.

These lectures are intended to be relatively self-contained, and to begin at a rather elementary level. Nevertheless there are some prerequisites, including a general knowledge of particle physics as conferred by any of the standard textbooks, such as Frauenfelder and Henley (1974), Frazer (1966), Gasiorowicz (1966), Perkins (1982), and Perl (1974). In addition to the specific references cited within, the reader will find much of interest in the following quark model reviews and monographs: Close (1979), Dalitz (1965, 1977), Feld (1969), Feynman (1972, 1974), Greenberg (1978), Hendry and Lichtenberg (1978), Kokkedee (1969), Lipkin (1973a), Rosner (1981a). It will also be useful to have at least a cultural appreciation of the logic of gauge theories, which may be gained from my St. Croix lectures [Quigg 1981], from my book [Quigg 1983], or from the lectures at this school by Wess (1983).

2. Quarks for SU(2)

In order to recall some basic concepts and to simplify arithmetic, let us open our exploration of constituent models by considering only the nonstrange hadrons. The generalization to SU(3) can be accomplished largely by transcription. That will be done in section 3, where the use of SU(2) subgroups will be found convenient. Here we shall introduce terminology and notation in the somewhat simplified context of flavor SU(2), or isospin. After brief reviews of the idea and the calculus of isospin symmetry, we shall construct the nonstrange baryons out of the fundamental representation of SU(2). This will lead us to consider the spin × flavor group SU(4). We shall next make use of the baryon wavefunctions to investigate some elementary static properties. Techniques developed there will be applied repeatedly throughout the course of these lectures. The section concludes with a short discussion of the nonstrange mesons.

2.1. The idea of isospin

The concept of isospin arose in nuclear physics during the 1930s. Heisenberg (1932) first suggested a relation between the neutron and proton because of the near equality of their masses [Particle Data Group 1980]:

$$M(n) = 939.5731 \pm 0.0027 \text{ MeV}/c^2,$$
$$M(p) = 938.2796 \pm 0.0027 \text{ MeV}/c^2, \tag{2.1}$$

which is to say $\Delta M/M \approx 1.4 \times 10^{-3}$. It was subsequently noticed by Breit et al. (1936), among others, that the neutron–proton and proton–proton forces in nuclei are exceedingly similar. This is made strikingly apparent by modern data on nuclear levels.

The binding energy of the ^{3}H ground state, a (pnn) composite, is 8.482 MeV, whereas that of the ^{3}He ground state, a (ppn) composite, is 7.718 MeV [Fiarman and Hanna 1975]. These are equal, up to the Coulomb repulsion in ^{3}He which we may estimate, using the mean charge radius measured in electron scattering, to be of order $\alpha/(1.88 \text{ fm}) \approx 0.77$ MeV.

The level structures of mirror nuclei further exhibit the charge-independence of nuclear forces. The energy levels of ^{7}Li (3p + 4n) and ^{7}Be (4p + 3n) are shown in fig. 1, and those of ^{11}B (5p + 6n) and ^{11}C (6p + 5n) are compared in fig. 2. In both cases the correspondence between energy levels of the mirror nuclei is precise.

Following these early observations, the use of isospin as a good quantum number of the nuclear interaction was discussed by Cassen and Condon (1936) and by Wigner (1937), who codified the idea of isospin symmetry as invariance under SU(2), the group of 2×2 unitary matrices with determinant $+1$. Elaboration of the concept occupied more than a decade of interplay between theory and experiment. The evolution may be traced in any of the standard textbooks on nuclear physics, such as Blatt and Weisskopf (1952) or Bohr and Mottelson (1969).

We regard the proton and neutron as an isospin doublet of nucleons

$$|p\rangle = |I = 1/2, I_3 = 1/2\rangle = \begin{pmatrix} 1 \\ 0 \end{pmatrix},$$

$$|n\rangle = |I = 1/2, I_3 = -1/2\rangle = \begin{pmatrix} 0 \\ 1 \end{pmatrix}. \tag{2.2}$$

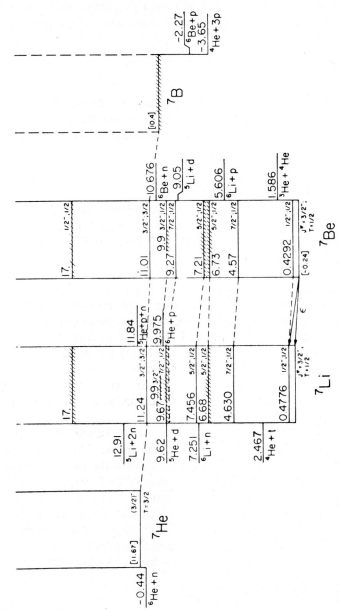

Fig. 1. Energy levels for $A = 7$ nuclei [from Ajzenberg-Selove, 1979]. The diagrams for individual isobars have been shifted vertically to eliminate the neutron–proton mass difference and the Coulomb energy, taken as $E_C = (0.6 \text{ MeV}) Z(Z-1)/A^{1/3}$. Energies in square brackets represent the approximate nuclear binding energy $E_N = M(Z, A) - Z M_p - (A - Z) M_n - E_C$, minus the corresponding quantity for ^{7}Li. Note the one-to-one correspondence between levels of the mirror nuclei ^{7}Li and ^{7}Be.

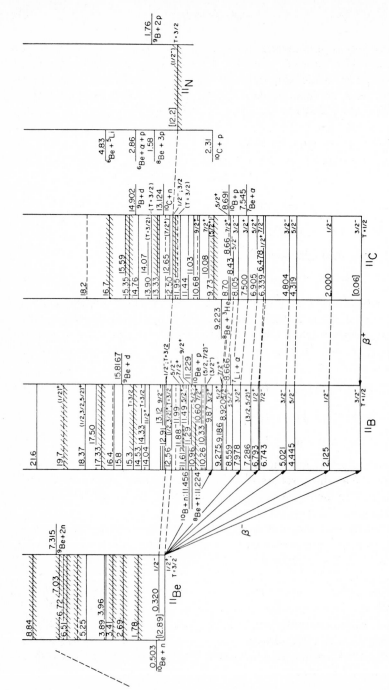

Fig. 2. Energy levels for $A = 11$ nuclei [from Ajzenberg-Selove, 1975]. Notation as in fig. 1, with binding energies referred to ^{11}B.

Evidence for isospin invariance in nuclei can also be found in the energy spectra, in the form of "isobaric analog states." Thus in the $A = 7$ and $A = 11$ systems depicted in figs. 1 and 2, we may note that the ^{7}He and ^{7}B ground states are identifiable as $I = 3/2$ partners of the ^{7}Li (11.24 MeV) and ^{7}Be (11.01 MeV) levels, while the ^{11}Be and ^{11}N spectra have analog levels in ^{11}B and ^{11}C. However, it is more revealing to consider two-nucleon states systematically.

From the basis states (2.2) we may construct isospin triplet states

$$|1,1\rangle = |p_1\rangle|p_2\rangle,$$

$$|1,0\rangle = (|p_1\rangle|n_2\rangle + |n_1\rangle|p_2\rangle)/\sqrt{2},$$

$$|1,-1\rangle = |n_1\rangle|n_2\rangle, \tag{2.3}$$

and an isospin singlet state

$$|0,0\rangle = (|p_1\rangle|n_2\rangle - |n_1\rangle|p_2\rangle)/\sqrt{2}. \tag{2.4}$$

Among two-nucleon states, only the isoscalar deuteron is bound. Thus, to look for evidence of isobaric analog levels we must consider two nucleons outside a core which is hoped to be inert. The textbook example is the $A = 14$ system of ^{12}C plus two nucleons. Neglecting the core, we classify the isobars as:

$$^{14}\text{O} = {}^{12}\text{C} + (\text{pp}), \quad I_3 = 1;$$

$$^{14}\text{N} = {}^{12}\text{C} + (\text{pn}), \quad I_3 = 0;$$

$$^{14}\text{C} = {}^{12}\text{C} + (\text{nn}), \quad I_3 = -1. \tag{2.5}$$

The energy levels are shown in fig. 3. In addition to the $I = 1$ levels common to all three isobars, ^{14}N has a large number of additional levels, which may be identified as isoscalar. The isospin assignments in ^{14}N are confirmed in ^{14}N(α, α)^{14}N* reactions, among others.

Data such as these, and measurements of nucleon–nucleon scattering provide abundant evidence that isospin is useful both as a classification symmetry, and as a dynamical symmetry of the nuclear interaction. The extension to elementary particle physics, motivated in the first instance by degeneracies in particle masses, is a central element in the description of hadrons. In combination with charge-conjugation invariance, isospin yields the useful discrete symmetry of G-parity. The relevant formalism is explained in the textbooks cited in section 1, and in the monograph by Sakurai (1964). Problems 1–3 provide some practice in the application of these elementary ideas.

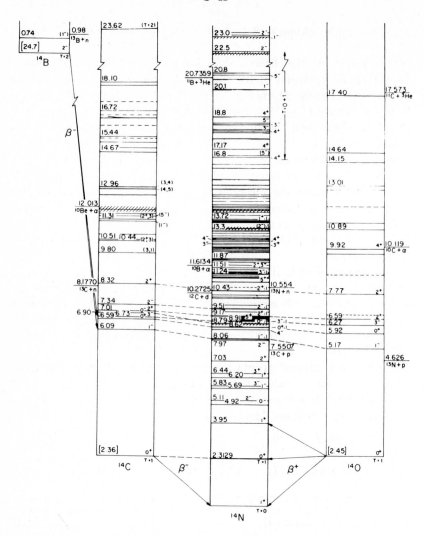

Fig. 3. Energy levels for $A = 14$ nuclei [from Ajzenberg-Selove, 1976]. Notation as in fig. 1, with binding energies referred to ^{14}N.

2.2. *The isospin calculus and the baryon wavefunctions*

We have just recalled how the similarity of the proton and neutron motivated the invention of the first flavor symmetry, isospin. The two nucleons are nearly degenerate in mass, experience the same nuclear forces, both have $J^P = 1/2^+$, and differ only in electric charge. Isospin of course proved to be a "good" symmetry of the strong interaction among hadrons.

With the proliferation of "elementary" particles in the early 1960s, it was natural to try to extend the idea of isospin symmetry and seek familial relations among the then known pseudoscalar mesons, vector mesons, and the baryons of $J^P = 1/2^+$ or $3/2^+$. Great effort went into finding the correct generalization, which was by no means obvious. Some of the steps toward SU(3) symmetry may be traced in the retrospective by Zweig (1980).

Before proceeding to flavor SU(3), let us review the calculus of isospin and introduce a notation that will match the conventional notation for larger groups. This will serve as preparation for the SU(N) calculations to follow. Accessible introductions to the group theory are given by Carruthers (1966), Close (1979), Hamermesh (1963), Lichtenberg (1978), and Lipkin (1966).

The lowest nontrivial isospin is $I = 1/2$. Two isospinors can be combined in two ways, leading to total isospin $I = 1$ (symmetric) or $I = 0$ (antisymmetric). It is convenient to label representations by the multiplicity of states they contain. Thus we write

Isospin I	0	1/2	1	$\cdots$	I
Dimension	**(1)**	**(2)**	**(3)**	$\cdots$	**($2I+1$)**

With this notation we can reëxpress the coupling of two isospinors as

$$(\mathbf{2}) \otimes (\mathbf{2}) = (\mathbf{1}) \oplus (\mathbf{3}), \tag{2.6}$$

which emphasizes the fact that the number of states which can be formed is the product of the numbers of states in the representations being combined. For a general representation $(\mathbf{n})$, it is easy to see that

$$(\mathbf{2}) \otimes (\mathbf{n}) = (\mathbf{n+1}) \oplus (\mathbf{n-1}). \tag{2.7}$$

We also note that

$$(\mathbf{2})^k \supset (\mathbf{k+1}) \oplus (\mathbf{k-1}) \oplus \cdots \oplus (\mathbf{2/1}). \tag{2.8}$$

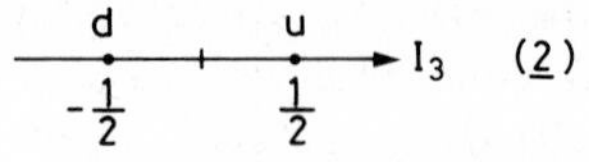

C. Quigg

$$\begin{array}{ccc} d & & u \\ \bullet & \;\;\;\dashv & \bullet \\ -\frac{1}{2} & & \frac{1}{2} \end{array} \longrightarrow I_3 \quad (\underline{2})$$

Fig. 4. The isospin quarks.

Any isospin state $|I, I_3\rangle$ can be constructed out of products of $|1/2, \pm 1/2\rangle$. In other words, products of (2) can lead to an arbitrary representation ($\boldsymbol{n}$). We may therefore imagine constructing the observed isospin multiplets out of fundamental objects which lie in a (2), as shown in fig. 4. The constituents are labeled u and d, for isospin up and down. This idea, like the alternative view of "nuclear democracy" [Chew and Frautschi 1961, Chew 1965], in which all hadrons are regarded as composites of each other, finds its modern roots in the work of Fermi and Yang (1949). The lowest-lying nonstrange baryons are the nucleons (2.2) and the quartet of nucleon resonances

$$\Delta^{++} = |3/2, 3/2\rangle,$$
$$\Delta^{+} = |3/2, 1/2\rangle,$$
$$\Delta^{0} = |3/2, -1/2\rangle,$$
$$\Delta^{-} = |3/2, -3/2\rangle, \tag{2.9}$$

depicted in fig. 5.

To build the quartet we require (at least) the product

$$(\boldsymbol{2}) \otimes (\boldsymbol{2}) \otimes (\boldsymbol{2}) = (\boldsymbol{2}) \oplus (\boldsymbol{2}) \oplus (\boldsymbol{4}) \tag{2.10}$$

of three fundamental objects. In the interest of simplicity, we choose the minimal configuration of three constituents. Here we face a choice: should we invent new fundamental constituents, or should we use the nucleons as our elementary states (in terms of which $\Delta^{++} = pp\bar{n}$, for

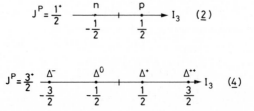

Fig. 5. Isospin assignments of the nucleons and nucleon resonances.

example)? We reject the nucleon alternative on two grounds. First, nucleons are manifestly composite and have a finite geometrical size. Secondly, when generalized to include strange particles the nucleon scheme becomes the Sakata (1956) model, which generates the wrong multiplet structure. We are thus led to quarks.

The nucleon resonances may now be constructed as

$$\Delta^{++} = uuu,$$
$$\Delta^{+} = u\{u,d\},$$
$$\Delta^{0} = d\{u,d\},$$
$$\Delta^{-} = ddd, \tag{2.11}$$

where the notation $\{a_1, a_2, \ldots, a_n\}$ has been introduced for the totally symmetrized product of n objects, which in this case denotes the $I=1$ configuration of $(ud + du)/\sqrt{2}$. If there are no further flavor distinctions among quarks, it is obvious that both up and down quarks have baryon number $B = 1/3$ and that the electric charge assignments must be $e_u = 2e/3, e_d = -e/3$ in agreement with the Gell-Mann (1953)–Nakano and Nishijima (1955) formula for displaced charge multiplets.

To construct explicit wave functions for the baryons, we first transcribe eqs. (2.3) and (2.4) to list the two-body states of definite isospin in the quark basis. These are

$$|1,1\rangle = uu,$$
$$|1,0\rangle = (ud + du)/\sqrt{2},$$
$$|1,-1\rangle = dd, \tag{2.12}$$

and

$$|0,0\rangle = (ud - du)/\sqrt{2}. \tag{2.13}$$

These two-body states all have definite permutation symmetries. The isovector states are symmetric under interchange of the constituents, whereas the isoscalar state is antisymmetric.

States with isospin $= 3/2$ can be obtained only by combining the third quark with an isovector pair, with the result

$$|3/2,3/2\rangle = u\underline{u}\underline{u},$$
$$|3/2,1/2\rangle = [u\underline{u}d + (ud + du)\underline{u}]/\sqrt{3},$$
$$|3/2,-1/2\rangle = [(ud + du)\underline{d} + dd\underline{u}]/\sqrt{3},$$
$$|3/2,-3/2\rangle = dd\underline{d}. \tag{2.14}$$

The last quark added has been underlined, to emphasize the structure of the wavefunctions. For $I = 3/2$, the wavefunctions are symmetric under the interchange of any pair of quarks, as expected.

There are two distinct paths to isospinor final states. First, the third quark can be combined with an isoscalar pair to give

$$\varphi_{M,A}^{p} \equiv |1/2, 1/2\rangle_0 = (ud - du)\underline{u}/\sqrt{2},$$

$$\varphi_{M,A}^{n} \equiv |1/2, -1/2\rangle_0 = (ud - du)\underline{d}/\sqrt{2}, \tag{2.15}$$

which are antisymmetric under interchange of quarks 1 and 2. Alternatively the final quark may be added to an isovector pair; this yields

$$\varphi_{M,S}^{p} \equiv |1/2, 1/2\rangle_1 = -[(ud + du)\underline{u} - 2uu\underline{d}]/\sqrt{6},$$

$$\varphi_{M,S}^{n} \equiv |1/2, -1/2\rangle_1 = [(ud + du)\underline{d} - 2dd\underline{u}]/\sqrt{6}, \tag{2.16}$$

which are symmetric under interchange of particles 1 and 2 [under (12)].

The isospinor wavefunctions (2.15) and (2.16) are said to be of mixed symmetry because permutations involving quark 3 mix the states with definite symmetry properties under (12). For example, interchange of 1 and 3 yields

$$(13)\varphi_{M,A}^{p} = (\varphi_{M,A}^{p} - \sqrt{3}\,\varphi_{M,S}^{p})/2. \tag{2.17}$$

Further elaboration of the meaning of mixed symmetry for these states is to be found in problem 4. Notice that the construction just undertaken has led to an explicit realization of all the three-quark states anticipated in (2.10), namely a quartet and two doublets.

To give a complete description of the baryons, it is necessary to specify the spin wavefunctions as well. The spin wavefunctions can be read off from the isospin wavefunctions (2.14–2.16), with the replacements

$$\text{flavor} \left.\begin{matrix} u \\ d \end{matrix}\right\} \rightarrow \left\{\begin{matrix} \uparrow \\ \downarrow \end{matrix}\right. \text{spin.} \tag{2.18}$$

The spin-3/2 wavefunctions are then

$$|s = 3/2, s_z = 3/2\rangle = \uparrow\uparrow\uparrow,$$

$$|s = 3/2, s_z = 1/2\rangle = [\uparrow\uparrow\downarrow + (\uparrow\downarrow + \downarrow\uparrow)\uparrow]/\sqrt{3},$$

$$|s = 3/2, s_z = -1/2\rangle = [(\uparrow\downarrow + \downarrow\uparrow)\downarrow + \downarrow\downarrow\uparrow]/\sqrt{3},$$

$$|s = 3/2, s_z = -3/2\rangle = \downarrow\downarrow\downarrow, \tag{2.19}$$

while for spin-1/2 the possibilities are

$$\chi_{M,A}^{+} = (\uparrow\downarrow - \downarrow\uparrow)\uparrow/\sqrt{2}\,,$$
$$\chi_{M,A}^{-} = (\uparrow\downarrow - \downarrow\uparrow)\downarrow/\sqrt{2}\,, \tag{2.20}$$

and

$$\chi_{M,S}^{+} = -[(\uparrow\downarrow + \downarrow\uparrow)\uparrow - 2\uparrow\uparrow\downarrow]/\sqrt{6}\,,$$
$$\chi_{M,S}^{-} = [(\uparrow\downarrow + \downarrow\uparrow)\downarrow - 2\downarrow\downarrow\uparrow]/\sqrt{6}\,. \tag{2.21}$$

Consequently we may write the wavefunctions for Δ^{++} in definite spin states as

$$|\Delta^{++};3/2\rangle = u_{\uparrow}u_{\uparrow}u_{\uparrow}\,,$$
$$|\Delta^{++};1/2\rangle = (u_{\uparrow}u_{\uparrow}u_{\downarrow} + u_{\uparrow}u_{\downarrow}u_{\uparrow} + u_{\downarrow}u_{\uparrow}u_{\uparrow})/\sqrt{3}\,, \tag{2.22}$$

etc.

These wavefunctions, and indeed all those for the Δ-resonances, are symmetric in spin $\times$ flavor (isospin). Because $\Delta(1232)$ is the $J^{P} = 3/2^{+}$ object with the lowest mass, it is reasonable to suppose that all the quarks are in relative s-waves. This assumption yields the correct parity for the Δ states at the price of fermion wavefunctions that apparently are in conflict with the generalized Pauli principle because the wavefunctions are symmetric in space $\times$ spin $\times$ flavor. This is a matter to which we shall have to return if we find that the model otherwise makes theoretical and experimental sense.

What has been accomplished so far? We have found that the nucleons and spin-3/2 nucleon resonances can be generated from three-body configurations of spin-1/2 fundamental particles. We may further notice that all known nonstrange baryons have isospin $= 1/2$ or 3/2, and that the model in which baryons are constructed of three isospinors limits them to precisely these values.

Now let us construct the nucleon wavefunctions explicitly. The result of problem 5 will show that the nucleons and Δ resonances are plausibly members of the same flavor $\times$ spin supermultiplet, a 20-dimensional representation of SU(4) (Young tableau ⊞⊞) which is decomposed under SU(2)$\otimes$SU(2) as $(2I+1,2s+1) = (4,4)\oplus(2,2)$. Thus the nucleon wavefunctions, like those for the Δ resonances, must be symmetric in flavor $\times$ spin. The fully symmetrized, normalized wavefunction for a proton with spin up is given in terms of the flavor and spin wavefunctions (2.15), (2.16) and (2.20), (2.21) by

$$|p\uparrow\rangle = (\varphi^P_{M,A}\chi^+_{M,A} + \varphi^P_{M,S}\chi^+_{M,S})/\sqrt{2}. \tag{2.23}$$

Thus the proton and neutron wavefunctions have the explicit forms

$$\begin{aligned}
|p\uparrow\rangle = (1/\sqrt{18})(2u_\uparrow d_\downarrow u_\uparrow &- u_\downarrow d_\uparrow u_\uparrow - u_\uparrow d_\uparrow u_\downarrow \\
&- d_\uparrow u_\downarrow u_\uparrow + 2d_\downarrow u_\uparrow u_\uparrow - d_\uparrow u_\uparrow u_\downarrow \\
&- u_\uparrow u_\downarrow d_\uparrow - u_\downarrow u_\uparrow d_\uparrow + 2u_\uparrow u_\uparrow d_\downarrow),
\end{aligned} \tag{2.24}$$

and

$$\begin{aligned}
|n\uparrow\rangle = -(1/\sqrt{18})(2d_\uparrow u_\downarrow d_\uparrow &- d_\downarrow u_\uparrow d_\uparrow - d_\uparrow u_\uparrow d_\downarrow \\
&- u_\uparrow d_\downarrow d_\uparrow + 2u_\downarrow d_\uparrow d_\uparrow - u_\uparrow d_\uparrow d_\downarrow \\
&- d_\uparrow d_\downarrow u_\uparrow - d_\downarrow d_\uparrow u_\uparrow + 2d_\uparrow d_\uparrow u_\downarrow).
\end{aligned} \tag{2.25}$$

Several remarks about the spin structure of these wavefunctions are in order. Note first that in the proton wavefunction the pair of up quarks is always in a symmetric $(I=1)$ flavor state. Therefore, because of the total symmetry of the wavefunction, the up quarks must always be in a symmetric spin state: $|s=1, s_z=1\rangle$ for the terms with coefficient 2, and $|s=1, s_z=0\rangle$ for the terms with coefficient $(-)1$. For pairs of quarks with total spin s, the expectation value of $\sigma_1 \cdot \sigma_2$ is easily evaluated as

$$\begin{aligned}
\langle \sigma_1 \cdot \sigma_2 \rangle = \langle 4s_1 \cdot s_2 \rangle &= 2\langle s^2 - s_1^2 - s_2^2 \rangle = 2[s(s+1) - 2\times 3/4] \\
&= \begin{cases} 1, & s=1; \\ -3, & s=0. \end{cases}
\end{aligned} \tag{2.26}$$

Thus, in the proton wavefunction (2.24) we have at once that

$$\langle \sigma_u \cdot \sigma_u \rangle = 1, \tag{2.27}$$

while in the neutron wavefunction (2.25),

$$\langle \sigma_d \cdot \sigma_d \rangle = 1. \tag{2.28}$$

For three quarks with total spin s, a similar trick applies:

$$\begin{aligned}
\left\langle \sum_{i<j} \sigma_i \cdot \sigma_j \right\rangle = 4\left\langle \sum_{i<j} s_i \cdot s_j \right\rangle &= 4\langle (s_1 \cdot s_2 + s_2 \cdot s_3 + s_1 \cdot s_3) \rangle \\
&= 2\langle (s^2 - s_1^2 - s_2^2 - s_3^2) \rangle = 2[s(s+1) - 3\times 3/4] \\
&= \begin{cases} 3, & s=3/2; \\ -3, & s=1/2. \end{cases}
\end{aligned} \tag{2.29}$$

In a proton, then,

$$\langle p\uparrow | \sum_{i<j} \sigma_i \cdot \sigma_j | p\uparrow \rangle = 2\langle p\uparrow | \sigma_u \cdot \sigma_d | p\uparrow \rangle + \langle p\uparrow | \sigma_u \cdot \sigma_u | p\uparrow \rangle = -3,$$

$$(2.30)$$

implies, in combination with (2.27), that

$$\langle \sigma_u \cdot \sigma_d \rangle_p = -2. \tag{2.31}$$

Similar arithmetic shows that

$$\langle \sigma_u \cdot \sigma_d \rangle_n = -2, \tag{2.32}$$

as well. We shall find numerous applications for these and related results.

2.3. Baryons: electromagnetic properties

As a first application and test of the wavefunctions we have constructed, let us consider some elementary electromagnetic properties of the baryons. The total charge carried by a hadron is

$$Q(h) = \langle h | Q | h \rangle, \tag{2.33}$$

where the charge operator Q is the sum of the charge operators for the constituents. For baryons containing three quarks, this is simply

$$Q = Q_{(1)} + Q_{(2)} + Q_{(3)}.$$

Because of the total symmetry of the wavefunction, the total charge of a baryon can be expressed as

$$Q(B) = \langle B | Q | B \rangle = \sum_{i=1}^{3} \langle B | Q_{(i)} | B \rangle = 3\langle B | Q_{(3)} | B \rangle, \tag{2.34}$$

for example. One immediately verifies that $Q(p) = +1$, $Q(n) = 0$, etc.

Although this success is a trivial one, having been assured by construction, the distribution of charge within a hadron presents a sterner challenge. As we shall see in greater detail in subsection 3.4.5, a convenient parameter of the charge distribution is the mean-squared charge radius, defined as

$$\langle r_{EM}^2 \rangle = \int d^3 r \rho(r) r^2, \tag{2.35}$$

where $\rho(r)$ is the charge density. In a nonrelativistic constituent picture,

$\langle r_{EM}^2 \rangle$ is therefore given by

$$\langle r_{EM}^2 \rangle = \left\langle \sum_{i=1}^{3} e_i(r_i - R)^2 \right\rangle, \tag{2.36}$$

where r_i is the coordinate of the i-th quark and R gives the position of the baryon center-of-mass which for equal-mass quarks is:

$$R = (r_1 + r_2 + r_3)/3. \tag{2.37}$$

In the present approximation the baryon wavefunctions are completely symmetric, so eq. (2.36) is equivalent to

$$\langle r_{EM}^2 \rangle = \langle (r_3 - R)^2 \rangle \sum_{i=1}^{3} e_i. \tag{2.38}$$

Because $\Sigma e_i = 0$ for the neutron, our model implies that

$$\langle r_{EM}^2 \rangle_n = 0. \tag{2.39}$$

or, equivalently, that the neutron is uniformly neutral. This is not in fact the case [Hofstadter 1963], and an understanding of the neutron's charge distribution will have to await a less ingenuous model of the nucleon. (See section 5.3.)

Let us next analyze the implications of the quark model wavefunctions for electromagnetic mass differences. Within a multiplet one may identify three sorts of contributions to electromagnetic mass differences [Dolgov et al. 1965, Gerasimov 1966; Thirring 1966]:

– a difference between the masses of the up- and down-quarks,
– pairwise Coulomb interactions among quarks,
– pairwise hyperfine or magnetic moment interactions among quarks.

A plausible expression for particle masses within a multiplet is then

$$M = M_0 + N_d(m_d - m_u)$$
$$+ \left\langle \frac{\alpha}{r} \right\rangle \sum_{i<j} e_i e_j - \frac{8\pi}{3} |\Psi_{ij}(0)|^2 \left\langle \sum_{i<j} \mu_i \mu_j \sigma_i \cdot \sigma_j \right\rangle. \tag{2.40}$$

In this equation, N_d is the number of downquarks in the hadron, e_i and μ_i are the electric charge and magnetic moment of the i-th quark, and $\Psi_{ij}(0)$ is the wavefunction at zero separation between the i-th and j-th quark. The symmetry of the wave function has been exploited in writing the Coulomb term. The Fermi (1930) hyperfine interaction will be derived for general values of the orbital angular momentum in problem 16.

A dynamical theory of the interactions among quarks would permit the calculation of quantities such as $\langle r^{-1} \rangle$ and $|\Psi(0)|^2$ from first principles.

In the temporary absence of such a theory, it is necessary to parametrize the unknown dynamical quantities in eq. (2.40). The baryon masses then take the form

$$M = M_0 + N_d(m_d - m_u) + \sum_{i<j} e_i e_j \delta M_C + \left\langle \sum_{i<j} e_i e_j \sigma_i \cdot \sigma_j \right\rangle \delta M_m,$$

$$(2.41)$$

where it has been assumed provisionally that $\mu_i = \text{const.} \times e_i$. This would be the case for Dirac particles, if $m_u \approx m_d$. The quantities M_0, δM_C, and δM_m evidently may vary from one multiplet to another. Those for the $\Delta(N^*)$ multiplet will be denoted by an asterisk. Contributions to the EM mass differences are gathered in table 1. In constructing table 1, use has been made of the spin averages (2.26–2.28) and (2.31), (2.32). The resulting baryon masses are:

$$M(p) = M_0 + (m_d - m_u) \qquad\qquad + \tfrac{4}{3}\delta M_m, \qquad\qquad (2.42)$$

$$M(n) = M_0 + 2(m_d - m_u) \quad - \tfrac{1}{3}\delta M_C + \delta M_m, \qquad\qquad (2.43)$$

$$M(\Delta^{++}) = M_0^* \qquad\qquad + \tfrac{4}{3}\delta M_C^* + \tfrac{4}{3}\delta M_m^*, \qquad (2.44)$$

$$M(\Delta^+) = M_0^* + (m_d - m_u), \qquad\qquad\qquad\qquad (2.45)$$

$$M(\Delta^0) = M_0^* + 2(m_d - m_u) \; - \tfrac{1}{3}\delta M_C^* - \tfrac{1}{3}\delta M_m^*, \qquad (2.46)$$

$$M(\Delta^-) = M_0^* + 3(m_d - m_u) + \tfrac{1}{3}\delta M_C^* + \tfrac{1}{3}\delta M_m^*. \qquad (2.47)$$

Table 1
Contributions to electromagnetic mass differences in nonstrange baryons

Particle	N_d	$\sum\limits_{i<j} e_i e_j$	$\left\langle \sum\limits_{i<j} e_i e_j \sigma_i \cdot \sigma_j \right\rangle$
p	1	0	4/3
n	2	$-1/3$	1
Δ^{++}	0	4/3	4/3
Δ^+	1	0	0
Δ^0	2	$-1/3$	$-1/3$
Δ^-	3	1/3	1/3

In section 5.3, where the strong hyperfine splitting is discussed, we shall attempt to relate $M_0^* - M_0$ to the magnetic moment term δM_m. For the moment we make the plausible approximations

$$\delta M_C = \delta M_C^* \tag{2.48}$$

and

$$\delta M_m = \delta M_m^*, \tag{2.49}$$

and note the following simple relations:

$$M(n) - M(p) = (m_d - m_u) - \tfrac{1}{3}\delta M_C - \tfrac{1}{3}\delta M_m$$
$$= 1.29343 \pm 0.00004 \text{ MeV}/c^2; \tag{2.50}$$

$$M(\Delta^-) - M(\Delta^{++}) = 3(m_d - m_u) - \delta M_C - \delta M_m$$
$$= 3[M(n) - M(p)]; \tag{2.51}$$

$$M(\Delta^0) - M(\Delta^+) = (m_d - m_u) - \tfrac{1}{3}\delta M_C - \tfrac{1}{3}\delta M_m$$
$$= M(n) - M(p). \tag{2.52}$$

Regrettably, these predictions have not been tested experimentally, in large part because of the breadth ($\Gamma \approx 115$ MeV) of the Δ. Upon making the generalization to flavor SU(3), however, we will immediately deduce similar mass differences for the strange particles, for which experimental comparisons are possible.

Of more immediate experimental interest are the magnetic moments of baryons. In terms of quark constituents, the magnetic moment of a hadron of spin s is defined to be

$$\mu_h = \sum_i \langle h; s_z = s / \mu_i \sigma_z^{(i)} / h; s_z = s \rangle. \tag{2.53}$$

If quarks are assumed to be Dirac particles, with

$$\mu_i = e_i \hbar / 2m_i c, \tag{2.54}$$

then

$$\mu_d = -\tfrac{1}{2}\mu_u \equiv -\tfrac{1}{3}\mu. \tag{2.55}$$

Again using the symmetry of the baryon wavefunctions, we may rewrite (2.53) as

$$\mu_h = 3\mu \langle h | e_i \sigma_z^{(i)} | h \rangle \quad \text{not summed,} \tag{2.56}$$

which leads to:

$$\mu_p = \mu, \tag{2.57}$$

and

$$\mu_n = -\tfrac{2}{3}\mu, \tag{2.58}$$

from which

$$\mu_n/\mu_p = -2/3. \tag{2.59}$$

Experimentally [Particle Data Group 1980], the nucleon magnetic moments are known to impressive accuracy*:

$$\mu_p = 2.7928456 \pm 0.0000011 \text{ n.m.}, \tag{2.60}$$

$$\mu_n = -1.91304184 \pm 0.00000088 \text{ n.m.}, \tag{2.61}$$

whence

$$\mu_n/\mu_p = -0.68497945 \pm 0.00000058. \tag{2.62}$$

The agreement with the simple quark model prediction is quite satisfying, even if theory and experiment do differ by $\pi \times 10^4$ standard deviations!

[A general description of nucleon magnetic moment measurements is given by Ramsey (1956), especially chapters VI and VII. The neutron moment is reported by Greene et al. (1979). Measurement of the fundamental properties of the neutron can be expected to advance greatly with the collection of ultracold neutrons, described by Golub et al. (1979).]

If μ_u and μ_d are treated as independent parameters, the nucleon magnetic moments become

$$\mu_p = (4\mu_u - \mu_d)/3, \tag{2.63}$$

$$\mu_n = (4\mu_d - \mu_u)/3, \tag{2.64}$$

from which one may extract

$$\mu_u = 1.85166812 \text{ n.m.} \tag{2.65}$$

and

$$\mu_d = -0.97186433 \text{ n.m.} \tag{2.66}$$

The ratio of quark magnetic moments,

$$\mu_d/\mu_u = -0.5248, \tag{2.67}$$

differs only slightly from the symmetry limit of eq. (2.55), but not in the direction expected if the down quark is more massive than the up quark,

---

*The standard unit for baryon magnetic moments is the nuclear magneton (n.m.) $\equiv eh/2m_pc \approx 3.15 \times 10^{-12}$ eV/G.

as suggested by the neutron–proton mass difference. Indeed, if the quarks have Dirac moments (2.54), then the effective quark masses may be determined from the quark moments (2.60), (2.61). The results are

$$m_u \approx 338 \text{ MeV}/c^2, \tag{2.68}$$

$$m_d \approx 322 \text{ MeV}/c^2. \tag{2.69}$$

What seems noteworthy to me is not the precise values, but the fact that the masses are so reasonable—approximately one third of a proton mass.

It is also straightforward, as problem 6 will demonstrate, to compute the magnetic moments of the Δ resonances. The lifetime of these resonances are too short to permit magnetic moment measurements by the standard precession techniques. The Δ^{++} moment has, however, recently been determined indirectly in measurements of the bremsstrahlung reaction $\pi^+ p \rightarrow \pi^+ p\gamma$ by Nefkens et al. (1978). Their determination,

$$\mu_{\Delta^{++}} = 5.7 \pm 1 \text{ n.m.}, \tag{2.70}$$

is in reasonable agreement with the quark model prediction

$$\mu_{\Delta^{++}} = 3\mu_u \approx 2\mu_p \approx 5.6 \text{ n.m.} \tag{2.71}$$

Before leaving the subject of magnetic moments, let us mention the transition magnetic moment that characterizes the M1 decay $\Delta^+ \rightarrow p\gamma$. It is straightforward to calculate that

$$\mu^* \equiv \langle \Delta^+; 1/2 | \sum_{i=1}^{3} \mu_i \sigma_z^{(i)} | p; 1/2 \rangle$$

$$= \tfrac{2}{3}\sqrt{2}\,(\mu_u - \mu_d) \approx \sqrt{2}\,\mu_u \approx \sqrt{\tfrac{8}{9}}\,\mu_p. \tag{2.72}$$

This is in fair agreement with experiment [Dalitz and Sutherland 1966, Gilman and Karliner 1974], but for higher-lying resonances it is essential to incorporate recoil corrections. A vast literature exists on the subject of resonance photoproduction. This may be traced from chapter 7 of the book by Close (1979), or from the lecture notes by Rosner (1981a).

2.4. Mesons

The introduction of isospin quarks has been motivated by an analysis of the baryon spectrum. Let us now understand the implications of the model for mesons. The simplest meson configuration is a quark–anti-quark pair $(q\bar{q})$. The quark basis can be written as $\binom{u}{d}$, which transforms as a doublet under SU(2). In this basis the G-parity operator may be represented as

$$G = Ce^{i\pi\tau_2/2} = Ci\tau_2 = C\begin{pmatrix} 0 & 1 \\ -1 & 0 \end{pmatrix}, \tag{2.73}$$

where C is the charge conjugation operator. The antiquark doublet

$$G\begin{pmatrix} u \\ d \end{pmatrix} = C\begin{pmatrix} d \\ -u \end{pmatrix} = \begin{pmatrix} \bar{d} \\ -\bar{u} \end{pmatrix} \tag{2.74}$$

also transforms as an isospin doublet. The minus sign conforms to the usual SU(2) phase convention [Carruthers (1966) p. 5, Close (1979) p. 26].

In the language of flavor $\times$ spin SU(4), the ($q\bar{q}$) mesons lie in the

$$4 \otimes 4^* = 1 \oplus 15 \tag{2.75}$$

representations. It is easy to verify that these SU(4) representations decompose under SU(2)$\otimes$SU(2) as

$$1 = (1,1) \tag{2.76}$$

$$15 = (3,1) \oplus (1,3) \oplus (3,3) \tag{2.77}$$

in the $(2I + 1, 2s + 1)$ notation we have employed before. The elements of the **15** correspond to the π, ω, and ρ respectively, and we may label the SU(4) singlet as "η" for the moment. With the results of problem 2 in hand, we readily conclude that for s-wave configurations the quantum numbers $I^G J^P$ of the ($q\bar{q}$) bound states justify these assignments.

Explicit wavefunctions for the mesons are formed by combining flavor and spin wavefunctions as before. The wavefunctions presented here are manifestly eigenstates of G-parity, but for many applications that is an unnecessary frill:

$$|"\eta"\rangle = \frac{(u\bar{u} + d\bar{d}) + (\bar{u}u + \bar{d}d)}{2} \cdot \frac{(\uparrow\downarrow - \downarrow\uparrow)}{\sqrt{2}}, \tag{2.78}$$

$$|\pi^+\rangle = \left(\frac{u\bar{d} + \bar{d}u}{\sqrt{2}}\right)\left(\frac{\uparrow\downarrow - \downarrow\uparrow}{\sqrt{2}}\right)$$

$$= \left(u_\uparrow \bar{d}_\downarrow - u_\downarrow \bar{d}_\uparrow + \bar{d}_\uparrow u_\downarrow - \bar{d}_\downarrow u_\uparrow\right)/2, \tag{2.79}$$

$$|\pi^0\rangle = \left(\frac{(-u\bar{u} + d\bar{d}) + (-\bar{u}u + \bar{d}d)}{2}\right)\left(\frac{\uparrow\downarrow - \downarrow\uparrow}{\sqrt{2}}\right), \tag{2.80}$$

$$|\pi^-\rangle = -\left(\frac{d\bar{u} + \bar{u}d}{\sqrt{2}}\right)\left(\frac{\uparrow\downarrow - \downarrow\uparrow}{\sqrt{2}}\right), \tag{2.81}$$

$$|\omega\rangle = \left(\frac{(u\bar{u} + d\bar{d}) - (\bar{u}u + \bar{d}d)}{2}\right) \times \text{spin}, \tag{2.82}$$

$$|\rho^+\rangle = \left(\frac{u\bar{d} - \bar{d}u}{\sqrt{2}}\right) \times \text{spin}, \tag{2.83}$$

$$|\rho^0\rangle = \left(\frac{(-u\bar{u} + d\bar{d}) - (-\bar{u}u + \bar{d}d)}{2}\right) \times \text{spin}, \tag{2.84}$$

$$|\rho^-\rangle = \left(\frac{-d\bar{u} + \bar{u}d}{\sqrt{2}}\right) \times \text{spin}. \tag{2.85}$$

The neutral states are also seen to be eigenstates of charge conjugation, as required.

The $(q\bar{q})$ construction has satisfactorily reproduced the spectrum of what may be called the ground state nonstrange mesons. As for the baryons, it is interesting to explore whether the quark model provides more than a classification symmetry. Again we being by considering electromagnetic mass differences.

In analogy with eqs. (2.40), (2.41) we write the meson masses within a multiplet as

$$M = M_0 + N_d(m_d - m_u) + \langle e_q e_{\bar{q}}\rangle \delta M_C + \langle e_q e_{\bar{q}}\sigma_q \cdot \sigma_{\bar{q}}\rangle \delta M_m. \tag{2.86}$$

Contributions to the meson masses are summarized in table 2. The

Table 2
Contributions to electromagnetic mass differences in nonstrange mesons

Particle	N_d	$\langle e_q e_{\bar{q}}\rangle$	$\langle e_q e_{\bar{q}}\sigma_q \cdot \sigma_{\bar{q}}\rangle$
"η"	1	$-5/18$	$5/6$
π^+	1	$2/9$	$-2/3$
π^0	1	$-5/18$	$5/6$
π^-	1	$2/9$	$-2/3$
ω	1	$-5/18$	$-5/18$
ρ^+	1	$2/9$	$2/9$
ρ^0	1	$-5/18$	$-5/18$
ρ^-	1	$2/9$	$2/9$
$\langle\rho^0|M|\omega\rangle$	1	$1/6$	$1/6$

resulting meson masses are:

$$M(\text{``}\eta\text{''}) = M_0 + (m_d - m_u) - \tfrac{5}{18}\delta M_C + \tfrac{5}{6}\delta M_m = M(\pi^0), \quad (2.87)$$

$$M(\pi^+) = M_0 + (m_d - m_u) + \tfrac{2}{9}\delta M_C - \tfrac{2}{3}\delta M_m = M(\pi^-), \quad (2.88)$$

$$M(\omega) = M_0' + (m_d - m_u) - \tfrac{5}{18}\delta M_C' - \tfrac{5}{18}\delta M_m' = M(\rho^0), \quad (2.89)$$

$$M(\rho^+) = M_0' + (m_d - m_u) + \tfrac{2}{9}\delta M_C' + \tfrac{2}{9}\delta M_m' = M(\rho^-). \quad (2.90)$$

The equalities of the $\pi^{\pm}$ masses and $\rho^{\pm}$ masses are ensured by the construction of particle–antiparticle wavefunctions. The equalities of the ρ^0 and ω^0 masses (which is desirable), and of the π^0 and "η" masses (which is not) follow from the common quark compositions of these particles. We may hope, with some justification, that it is incorrect to identify "η" with the physical $\eta(549)$. Only one nonstrange meson electromagnetic mass difference is precisely known [Particle Data Group 1980]; it is recorded here for future reference:

$$M(\pi^+) - M(\pi^0) = (\delta M_C - 3\delta M_m)/2$$
$$= 4.6043 \pm 0.0037 \text{ MeV}/c^2. \quad (2.91)$$

It is worth noting that, as indicated in table 2, the electromagnetic Hamiltonian mediates transitions between ρ^0 and ω^0. It was observed long ago by Glashow (1961) that because ρ and ω are nearly degenerate the $\rho\omega$ mixing could give rise to substantial interference effects in the $\pi^+\pi^-$ mass spectrum. Such effects were observed in quasi two body final states [G. Goldhaber 1970] and were shown to provide useful information on the ρ and ω production mechanisms [A.S. Goldhaber et al. 1969]. Neglecting the $\rho - \omega$ mass difference, one may write the branching ratio for the isospin-violating decay $\omega \to \pi^+\pi^-$ as

$$\frac{\Gamma(\omega \to \pi^+\pi^-)}{\Gamma(\omega \to \text{all})} \approx \frac{4|\langle\rho^0|M|\omega\rangle|^2}{\Gamma_\rho\Gamma_\omega}. \quad (2.92)$$

Approximating the matrix element for mixing as

$$\langle\rho^0|M|\omega\rangle \approx [M(\pi^+) - M(\pi^0)]/3, \quad (2.93)$$

which entails neglecting the hyperfine interaction and $m_d - m_u$ and suppressing the distinction between vector meson and pseudoscalar meson wavefunctions, we estimate:

$$\langle\rho^0|M|\omega\rangle \approx 1.5 \text{ MeV}. \quad (2.94)$$

This implies a branching ratio on the order of a percent, in rough agreement with the current experimental average. We shall see below how to improve upon these gross approximations. It is of some interest that the decay $\omega \to \pi^+\pi^-$ has now been observed directly, without benefit of interference with $\rho^0 \to \pi^+\pi^-$, in the final state $\psi \to \pi^+\pi^- + \text{anything}$ [Gidal et al. 1981], in a data sample corresponding to about 1.3 million produced ψ's. In much of this sample, G-parity conservation ensures the incoherence of semifinal states containing ρ^0 and ω. This is but a single illustration of how the enormous data samples becoming available in electron–positron annihilations may contribute to the study of light-meson spectroscopy.

The static magnetic moments of the mesons are not of particular interest, because they are zero or are unmeasurable or both. Nevertheless, let us verify that the quark model predictions do no violence to common sense. The meson magnetic moments may be evaluated as

$$\mu_h = \langle h; s_z = s | \left(\mu_q \sigma_z^{(q)} + \mu_{\bar{q}} \sigma_z^{(\bar{q})} \right) | h; s_z = s \rangle. \tag{2.95}$$

The magnetic moments of the spinless mesons and of the neutral mesons vanish. The charged ρ-meson moments are

$$\mu_{\rho^+} = \left(\mu_u - \mu_d \right) = -\mu_{\rho^-}. \tag{2.96}$$

If quark magnetic moments have the same value in mesons as in baryons, we expect

$$\mu_{\rho^+} \approx \mu_p. \tag{2.97}$$

The possibility of relating observables for mesons and for baryons distinguishes the explicit quark model from the related symmetry schemes.

Of considerably more experimental interest are the transition magnetic moments which mediate radiative decays of vector mesons. For transitions between s-wave states, the magnetic dipole transition rate is

$$\Gamma(i \to f + \gamma) = \frac{\omega^3}{3\pi} | \langle f | \mu_q \sigma^{(q)} + \mu_{\bar{q}} \sigma^{(\bar{q})} | i \rangle |^2, \tag{2.98}$$

where ω is the energy of the emitted photon. The decay rate is independent of the polarization of the initial state, so we may choose the polarization to make the calculation as short as possible. For a pseudo-scalar final state it is apt to choose $s_z = 0$ in the initial state, so that only σ_z contributes.

The matrix elements of interest are

$$\langle \pi | M1 | \rho \rangle = \mu_u + \mu_d = 0.88 \text{ n.m.}, \tag{2.99}$$

and

$$\langle \pi^0 | M1 | \omega \rangle = \mu_u - \mu_d = 2.82 \text{ n.m.,} \tag{2.100}$$

where the numerical values derive from (2.65), (2.66), determined by the nucleon magnetic moments. Neglecting the $\rho - \omega$ mass difference, we expect

$$r = \frac{\Gamma(\omega \to \pi^0 \gamma)}{\Gamma(\rho \to \pi \gamma)} = \frac{(\mu_u - \mu_d)^2}{(\mu_u + \mu_d)^2} = 10.3, \tag{2.101}$$

or approximately $9:1$ in the SU(3) limit. Current experimental values,

$$\Gamma(\omega \to \pi^0 \gamma) = \begin{cases} (889 \pm 50) \text{ keV,} & \text{Particle Data Group (1980)} \\ (789 \pm 92) \text{ keV,} & \text{Ohshima (1980)} \end{cases}$$

$$\tag{2.102}$$

and

$$\Gamma(\rho \to \pi \gamma) = (67 \pm 7) \text{ keV,} \quad \text{Berg et al. (1980ab),} \tag{2.103}$$

imply a ratio

$$r = \begin{cases} 13.3 \pm 2.1 \\ 11.7 \pm 2.6 \end{cases}, \tag{2.104}$$

that is consistent with the theoretical expectation. A more critical discussion, with attention to the absolute rates, will be given in subsection 3.4.4.

With respect to experimental technique, note that whereas $\Gamma(\omega \to \pi^0 \gamma)/\Gamma(\omega \to \text{all}) \approx 8\%$, the ρ branching ratio for radiative decay is approximatively 4×10^{-4}. It is therefore relatively straightforward to measure the rate for radiative decay of ω by measuring the total width and branching ratio, although the measurement is still uncertain at the 10% level. This procedure would be unthinkable for the ρ. What must be done instead [Jensen 1980, 1981] is to measure the cross section for the Primakoff (1951) effect—the excitation of the pion in the Coulomb field of a heavy nucleus [Halprin et al. 1966].

3. SU(3) and light-quark spectroscopy

We now broaden our interest to include all of the "light" mesons and baryons by incorporating strange particles. The observed multiplet structure is that of the low-dimensionality representations of the flavor group

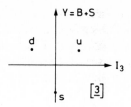

Fig. 6. The weight diagram for the fundamental [3] representation of SU(3).

SU(3): the 3×3 unitary matrices with determinant $+1$ [Gell-Mann 1961, 1962, Ne'eman 1961, Gell-Mann and Ne'eman 1964]. Thorough discussions of SU(3) and its applications are to be found in the books by Carruthers (1966), Gasiorowicz (1966), Gourdin (1967), Lichtenberg (1978), and Lipkin (1966). Although the pure symmetry aspects of SU(3) and its SU(2) subgroups will be discussed to a limited extent in this section, the main emphasis will be upon the SU(3) quark model as conceived by Gell-Mann (1964) and Zweig (1964 a, b, c, 1980). The applications will principally be those of the preceding section, with two important additions: a first look at formulas for strong-interaction mass differences, and a discussion of meson electromagnetic form factors.

3.1. The SU(3) quark model

Up, down, and strange quarks are assigned to the fundamental [3] representation as shown in fig. 6. The properties of the quarks are summarized in table 3. It is instructive to decompose the triplet under $SU(2)_{\text{isospin}} \otimes$ strangeness as

$$[3] = (2)_0 \oplus (1)_{-1}, \tag{3.1}$$

Table 3
Properties of the quarks

Quark	I	I_3	S	B	$Y = B + S$	$Q = I_3 + Y/2$
u	1/2	1/2	0	1/3	1/3	2/3
d	1/2	−1/2	0	1/3	1/3	−1/3
s	0	0	−1	1/3	−2/3	−1/3

where square brackets denote SU(3) representations, parentheses denote SU(2) representations, and the subscripts label the strangeness S. Using the isospin calculus reviewed in section 2, we may now build up mesons and baryons as $(q\bar{q})$ and (qqq) composites, respectively. Because the wavefunction factors into flavor × spin, the spin analysis of section 2 can be transplanted intact.

3.1.1. Mesons
As before, mesons are composed of a quark and antiquark, so they lie in the

$$[3] \otimes [3^*] = [1] \oplus [8] \tag{3.2}$$

dimensional representations of SU(3). In the notation of Young tableaux, the arithmetic is simply

$$\square \otimes \begin{array}{c}\square\\\square\end{array} = \begin{array}{c}\square\\\square\end{array} \oplus \begin{array}{cc}\square&\square\\\square&\end{array}. \tag{3.3}$$

Expanding (3.2) by inserting (3.1) and its conjugate we find

$$
\begin{aligned}
[3] \otimes [3^*] &= \{(2)_0 \oplus (1)_{-1}\} \otimes \{(2)_0 \oplus (1)_1\}\\
&= (1)_0 \oplus (3)_0 \oplus (2)_1 \oplus (2)_{-1} \oplus (1)_0. \tag{3.4}
\end{aligned}
$$

The first two terms are precisely the nonstrange mesons discussed in section 2. In the vector meson nonet they correspond to ω and ρ^+, ρ^0, ρ^-. The doublet with strangeness $S = +1$ corresponds to the $K^{*+}(u\bar{s})$ and $K^{*0}(d\bar{s})$. Their charge conjugates lie in the $S = -1$ doublet. The remaining singlet arises in the product of $(1)_1 \otimes (1)_{-1}$. It is to be identified with the $\phi(s\bar{s})$. The weight diagram for the vector nonet is presented in fig. 7.

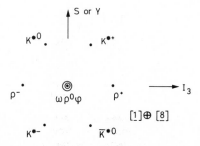

Fig. 7. Weight diagram for the vector meson nonet = $[1] \oplus [8]$.

It is often useful to represent the flavor content of the mesons in matrix form. For the vector states, the flavor matrix is

$$\mathcal{V} = \begin{array}{c} \bar{u} \\ \bar{d} \\ \bar{s} \end{array} \begin{pmatrix} (\omega + \rho^0)/\sqrt{2} & \rho^- & K^{*-} \\ \rho^+ & (\omega - \rho^0)/\sqrt{2} & \overline{K}^{*0} \\ K^{*+} & K^{*0} & \varphi \end{pmatrix}. \tag{3.5}$$

The SU(3) single state, in which all flavors receive equal weights, is then

$$[\mathbf{1}] = (1/\sqrt{3}) \operatorname{Tr} \mathcal{V} = (\varphi + \sqrt{2}\,\omega)/\sqrt{3} \equiv V_1. \tag{3.6}$$

The orthogonal isoscalar combination, which corresponds to the SU(3) generator

$$\lambda_8 = \frac{1}{\sqrt{3}} \begin{pmatrix} 1 & 0 & 0 \\ 0 & 1 & 0 \\ 0 & 0 & -2 \end{pmatrix}, \tag{3.7}$$

is

$$V_8 = (\omega - \sqrt{2}\,\varphi)/\sqrt{3}. \tag{3.8}$$

The physical states ω and ϕ thus are mixtures of the SU(3) singlet and octet states. The idea of singlet–octet mixing and the flavor assignments of ω and ϕ will be reviewed in subsection 3.3.2.

Ths decomposition of SU(3) into SU(2)$_{\text{isospin}} \otimes$ (strangeness or hypercharge) was convenient but not obligatory. For other purposes it may be preferable to single out other additive quantum numbers, such as electric charge, and other SU(2) subgroups. The remaining possibilities are illustrated in fig. 8. The virtues of the SU(2) subgroups have been emphasized by Levinson et al. (1962) and by Lipkin (1966).

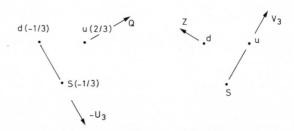

Fig. 8. Decompositions of the fundamental quark triplets with respect to the SU(2) subgroups U-spin and V-spin.

3.1.2. Baryons

The baryons are three-quark states, which lie in the

$$[3] \otimes [3] \otimes [3] = [1] \oplus [8] \oplus [8] \oplus [10] \tag{3.9}$$

representations of SU(3). Again in terms of Young tableaux, the product (3.9) is computed as

$$\Box \otimes \Box = \Box \oplus \Box\Box , \tag{3.10}$$

i.e. $[3] \otimes [3] = [3^*] \oplus [6]$, and thus that

$$\Box \otimes \Box \otimes \Box = \Box \oplus \Box\!\!\Box \oplus \Box\!\!\Box \oplus \Box\Box\Box . \tag{3.11}$$

The SU(3) representations that occur in the product have the following decompositions with respect to isospin and strangeness:

$$[10] = (4)_0 \oplus (3)_{-1} \oplus (2)_{-2} \oplus (1)_{-3}, \tag{3.12}$$

$$[8] = (2)_0 \oplus (3)_{-1} \oplus (1)_{-1} \oplus (2)_{-2}, \tag{3.13}$$

$$[1] = (1)_{-1}. \tag{3.14}$$

For the s-wave ground state, symmetric spin $\times$ flavor wavefunctions can be constructed for a spin-3/2 decimet and for one spin-1/2 octet. These are to be identified with the familiar $J^P = 3/2^+$ and $1/2^+$ baryons, as indicated in figs. 9 and 10.

3.1.3. Explicit wavefunctions

For explicit construction of the hadron wavefunctions, it is efficient to make use of the SU(2) subgroups of SU(3): I-spin, U-spin, and V-spin. The action of the flavor-changing SU(2) operators upon the quarks and

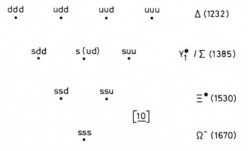

Fig. 9. The $J^P = 3/2^+$ baryon decimet.

d [ud]   u[ud]
  •   •   N (940)

sdd   sud   suu   Σ (1190)
  •   •   •
           Λ (1115)
                           [8]
ssd   ssu   Ξ (1315)
  •   •

Fig. 10. The $J^P = 1/2^+$ baryon octet.

antiquarks is summarized in fig. 11. The minus signs which appear in the antiquark figure reflect the fact that the proper SU(2) doublets of antiquarks are

$$\left(\begin{array}{c} \bar{d} \\ -\bar{u} \end{array} \right)_I, \left(\begin{array}{c} \bar{s} \\ -\bar{d} \end{array} \right)_U, \left(\begin{array}{c} \bar{s} \\ -\bar{u} \end{array} \right)_V, \tag{3.15}$$

so that, for example,

$$I_+|\bar{u}\rangle = -|\bar{d}\rangle. \tag{3.16}$$

As an illustration of how wavefunctions may be constructed, let us build up the vector meson flavor wavefunctions already summarized in the matrix (3.5). Apart from the detailed identification of the isoscalars with physical particles, the same procedure applies for the pseudoscalar mesons as well. We begin with any convenient known wavefunction, for example

$$|\rho^+\rangle = u\bar{d}, \tag{3.17}$$

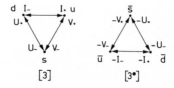

Fig. 11. Action of the *I*-spin, *U*-spin, and *V*-spin raising and lowering operators on the fundamental triplets of quarks [3] and antiquarks [3*].

and compute as follows:

$$I_-|\rho^+\rangle = d\bar{d} - u\bar{u} = \sqrt{2}|\rho^0\rangle; \tag{3.18}$$

$$I_-|\rho^0\rangle = \sqrt{2}\, d\bar{u} = \sqrt{2}|\rho^-\rangle; \tag{3.19}$$

$$U_+|\rho^+\rangle = -u\bar{s} = -|K^{*+}\rangle; \tag{3.20}$$

$$I_-|K^{*+}\rangle = d\bar{s} = |K^{*0}\rangle. \tag{3.21}$$

The action of

$$U_-|K^{*0}\rangle = s\bar{s} - d\bar{d} = \sqrt{2}|U = 1, U_3 = 0\rangle \tag{3.22}$$

yields a mixture of ρ^0 and V_8 [compare eq. (3.8)] which may be recognized as

$$|U = 1, U_3 = 0\rangle = -\tfrac{1}{2}\big(|\rho^0\rangle + \sqrt{3}|V_8\rangle\big). \tag{3.23}$$

The octet is completed by the operations

$$V_-|\rho^+\rangle = s\bar{d} = |\bar{K}^{*0}\rangle, \tag{3.24}$$

$$-I_-|\bar{K}^{*0}\rangle = s\bar{u} = |K^{*-}\rangle. \tag{3.25}$$

An analogous procedure can be devised for any SU(3) multiplet. The slightly more onerous task of constructing wavefunctions for the baryon octet is posed as problem 7.

With meson wavefunctions now in hand, we may return to the simple description of masses begun in section 2. Contributions to the masses of mesons containing strange quarks are listed in table 4, which may be viewed as an extension of table 2. There N_s denotes the number of strange quarks in a hadron, and it has been assumed that $\mu_s = \mu_d$, the SU(3)-symmetric relation. Symmetry breaking is easily incorporated. In analogy with (2.82–2.85) we may write (compare the early work by Zel'dovich and Sakharov 1966, Sakharov 1980):

$$M(\mathrm{K}^+) = M_0 + (m_s - m_u) + \tfrac{2}{9}\delta M_C - \tfrac{2}{3}\delta M_m, \tag{3.26}$$

$$M(\mathrm{K}^0) = M_0 + (m_d - m_u) + (m_s - m_u) - \tfrac{1}{9}\delta M_C + \tfrac{1}{3}\delta M_m, \tag{3.27}$$

for the pseudoscalars, and for the vectors:

$$M(\mathrm{K}^{*+}) = M_0' + (m_s - m_u) + \tfrac{2}{9}\delta M_C + \tfrac{2}{9}\delta M_m, \tag{3.28}$$

$$M(\mathrm{K}^{*0}) = M_0' + (m_d - m_u) + (m_s - m_u) - \tfrac{1}{9}\delta M_C - \tfrac{1}{9}\delta M_m, \tag{3.29}$$

$$M(\varphi) = M_0' + 2(m_s - m_u) - \tfrac{1}{9}\delta M_C - \tfrac{1}{9}\delta M_m. \tag{3.30}$$

Table 4
Contributions to meson masses for particles containing strange quarks

Particle	N_s	N_d	$\langle e_q e_{\bar{q}} \rangle$	$\langle e_q e_{\bar{q}} \sigma_q \cdot \sigma_{\bar{q}} \rangle$
K^+	1	0	2/9	$-2/3$
K^0	1	1	$-1/9$	1/3
$\overline{K}^0$	1	1	$-1/9$	1/3
K^-	1	0	2/9	$-2/3$
K^{*+}	1	0	2/9	2/9
K^{*0}	1	1	$-1/9$	$-1/9$
$\overline{K}^{*0}$	1	1	$-1/9$	$-1/9$
K^{*-}	1	0	2/9	2/9
φ	2	0	$-1/9$	$-1/9$

Among the pseudoscalar mesons, let us form the electromagnetic mass difference

$$M(K^+) - M(K^0) = \tfrac{1}{3}(\delta M_C - 3\delta M_m) - (m_d - m_u)$$
$$= -4.01 \pm 0.13 \text{ MeV}/c^2. \tag{3.31}$$

The first term is expected to be positive. Indeed from (2.86) we may estimate it as

$$\tfrac{1}{3}(\delta M_C - 3\delta M_m) = \tfrac{2}{3}[M(\pi^+) - M(\pi^0)] = 3.07 \text{ MeV}/c^2, \tag{3.32}$$

which implies that the down quark is more massive than the up quark:

$$m_d - m_u \approx 7 \text{ MeV}/c^2. \tag{3.33}$$

The sign of the quark mass difference is compatible with the observation that $M(n) > M(p)$.

For the vector mesons, we may also write

$$M(\rho^+) - M(\rho^0) = \tfrac{1}{2}(\delta M_C + \delta M_m), \tag{3.34}$$

$$M(K^{*+}) = \tfrac{1}{3}(\delta M_C + \delta M_m) - (m_d - m_u)$$
$$= -6.7 \pm 1.3 \text{ MeV}/c^2, \tag{3.35}$$

and

$$\langle \rho | M | \omega \rangle = M(\rho^+) - M(\rho^0) - M(K^{*+}) + M(K^{*0})$$
$$= (m_d - m_u) + \tfrac{1}{6}\delta M_C + \tfrac{1}{6}\delta M_m. \tag{3.36}$$

Notice that the $\rho^\pm - \rho^0$ mass difference, like the $\pi^\pm - \pi^0$ mass difference, has an explicit electromagnetic origin. In contrast, the $K^{*+}-K^{*0}$ mass difference and the off-diagonal $\rho\omega$ coupling are sensitive to the u–d quark mass difference, which may be identified with the tadpole term of Coleman and Glashow (1964).

Although mass differences have received only a superficial treatment in these introductory sections, they are of more than passing interest. More attention should be devoted to the experimental determination of mass differences for the ρ, K^*, and other meson families. The new tool of quarkonium decay makes possible measurements which are free from the biases of earlier experiments. We shall take up the question of quark masses again in section 5.3. Until then, a survey of traditional methods for calculating electromagnetic mass difference has been given by Zee (1972).

As a preview of our discussion of strong-interaction mass formulas, let us observe that eqs. (2.90) and (3.28–3.30) imply that

$$M(\varphi) - M(K^{*0}) = M(K^{*+}) - M(\rho^+) = m_s - m_d. \qquad (3.37)$$

The tabulated masses, which do not distinguish between different charges, yield

$$128 \text{ MeV}/c^2 \approx 116 \text{ MeV}/c^2, \qquad (3.38)$$

which is not bad for a beginning.

3.2. Some applications of U spin

The utility of the SU(2) subgroups of SU(3) for constructing wavefunctions has been demonstrated in subsection 3.1.3 and in problem 7. The SU(2) subgroups also yield, in straightforward fashion, extremely useful dynamical predictions. Because all the particles of a specified charge within an SU(3) multiplet lie in a single U-spin family, U-spin symmetry has many significant consequences for electromagnetic interactions.

The electromagnetic current is of the form

$$j_\mu = e\left[\tfrac{2}{3}\bar{u}\gamma_\mu u - \tfrac{1}{3}\left(\bar{d}\gamma_\mu d + \bar{s}\gamma_\mu s\right)\right]. \qquad (3.39)$$

The up quark is a U-spin singlet. So too is the combination $\bar{d}d + \bar{s}s$. Thus the photon transforms as a U-spin singlet. Consequently the electromagnetic contribution to a hadron mass must be the same for all members of a U-spin multiplet. This is obviously satisfied by the quark-model predictions given above for mesons.

For baryons, let us write the masses of the $J^P = 1/2^+$ octet as

$$M(\mathrm{p}) = M_\mathrm{N} + \delta m(\mathrm{p}), \tag{3.40}$$

$$M(\mathrm{n}) = M_\mathrm{N} + \delta m(\mathrm{n}), \tag{3.41}$$

etc., where M_N has a nonelectromagentic origin and δm is purely electromagnetic in character. U-spin symmetry immediately implies that

$$\delta m(\mathrm{p}) = \delta m(\Sigma^+), \tag{3.42}$$

$$\delta m(\mathrm{n}) = \delta m(\Xi^0), \tag{3.43}$$

and

$$\delta m(\Sigma^-) = \delta m(\Xi^-). \tag{3.44}$$

Therefore, the hadron mass differences

$$M(\mathrm{p}) - M(\mathrm{n}) = \delta m(\mathrm{p}) - \delta m(\mathrm{n}) = -1.29 \text{ MeV}/c^2, \tag{3.45}$$

$$M(\Sigma^+) - M(\Sigma^-) = \delta m(\mathrm{p}) - \delta m(\Xi^-) = -7.98 \pm 0.08 \text{ MeV}/c^2, \tag{3.46}$$

$$M(\Xi^-) - M(\Xi^0) = \delta m(\Xi^-) - \delta m(\mathrm{n}) = 6.39 \pm 0.62 \text{ MeV}/c^2, \tag{3.47}$$

are simply related:

$$\begin{aligned} M(\Sigma^+) - M(\Sigma^-) + M(\Xi^-) - M(\Xi^0) &= \delta m(\mathrm{p}) - \delta m(\mathrm{n}) \\ &= M(\mathrm{p}) - M(\mathrm{n}). \end{aligned} \tag{3.48}$$

This is the Coleman–Glashow (1961, 1964) relation, which is satisfied within experimental errors, the left-hand side yielding -1.59 ± 0.63 MeV$/c^2$.

U-spin symmetry makes similarly strong and simple predictions for baryon magnetic moments [Coleman and Galshow 1961, Okubo 1962a]. One has immediately that

$$\mu_\mathrm{p} = \mu_{\Sigma^+}, \tag{3.49}$$

$$\mu_\mathrm{n} = \mu_{\Sigma^0_\mathrm{u}} = \mu_{\Xi^0}, \tag{3.50}$$

$$\mu_{\Xi^-} = \mu_{\Sigma^-}, \tag{3.51}$$

where the neutral member of the U-spin triplet has been denoted as

$$|\Sigma^0_\mathrm{u}\rangle = \tfrac{1}{2}\left(|\Sigma^0\rangle + \sqrt{3}\,|\Lambda^0\rangle\right). \tag{3.52}$$

The U-spin singlet in the octet is therefore the orthogonal combination

$$|\Lambda^0_\mathrm{u}\rangle = \tfrac{1}{2}\left(\sqrt{3}\,|\Sigma^0\rangle - |\Lambda^0\rangle\right). \tag{3.53}$$

Using the inverse relations

$$|\Lambda^0\rangle = \tfrac{1}{2}\left(\sqrt{3}\,|\Sigma_u^0\rangle - |\Lambda_u^0\rangle\right)$$ (3.54)

and

$$|\Sigma^0\rangle = \tfrac{1}{2}\left(|\Sigma_u^0\rangle + \sqrt{3}\,|\Lambda_u^0\rangle\right),$$ (3.55)

it is straightforward to compute that

$$\mu_\Lambda = \langle\Lambda|\mu|\Lambda\rangle = \tfrac{3}{4}\mu_{\Sigma_u^0} + \tfrac{1}{4}\mu_{\Lambda_u^0},$$ (3.56)

$$\mu_{\Sigma^0} = \langle\Sigma^0|\mu|\Sigma^0\rangle = \tfrac{3}{4}\mu_{\Lambda_u^0} + \tfrac{1}{4}\mu_{\Sigma_u^0},$$ (3.57)

$$\mu_{\Sigma\Lambda} = \langle\Sigma^0|\mu|\Lambda\rangle = \tfrac{1}{4}\sqrt{3}\left(\mu_{\Sigma_u^0} - \mu_{\Lambda_u^0}\right), = \tfrac{1}{2}\sqrt{3}\left(\mu_\Lambda - \mu_{\Sigma^0}\right).$$ (3.58)

Using the fact that the photon is a combination of isoscalar and isovector, we may define the otherwise unmeasurable Σ^0 moment as [Marshak et al. 1957]:

$$\mu_{\Sigma^0} = \tfrac{1}{2}\left(\mu_{\Sigma^+} + \mu_{\Sigma^-}\right).$$ (3.59)

The comparison with data (sources are listed in table 6 below) is thus (in n.m.):

$$2.79 \overset{?}{=} 2.33 \pm 0.13,$$ (3.49′)

$$-1.91 \overset{?}{=} -1.15 \pm 0.07 \overset{?}{=} -1.25 \pm 0.01,$$ (3.50′)

$$-0.75 \pm 0.06 \overset{?}{=} -1.41 \pm 0.25,$$ (3.51′)

$$-1.82^{+0.25}_{-0.18} \overset{?}{=} -0.93 \pm 0.12.$$ (3.58′)

While there is qualitative agreement, the U-spin argument fails to account for the measured moments in a quantitative fashion. The explicit quark model analysis presented in section 3.4 will reveal, that much, but not all, of the discrepancy is resolved by relaxing the assumption that $\mu_s = \mu_d$.

In contrast to the situation for mesons, enough information exists for baryon mass differences that it is possible to extract and test the reasonableness of all the parameters in the simple quark description of masses. Because particle–antiparticle restrictions do not apply within the baryon octet, there are eight independent masses, to be compared with five for the mesons. It is this added richness that makes for a more incisive confrontation of the model with experiment.

Still approximating $\mu_s = \mu_d$ in the hyperfine term we may write, in the notation of eqs. (2.41–2.43):

$$M(\mathrm{p}) = M_0 + (m_d - m_u) + \tfrac{4}{3}\delta M_m, \tag{2.42}$$

$$M(\mathrm{n}) = M_0 + 2(m_d - m_u) - \tfrac{1}{3}\delta M_C + \delta M_m, \tag{2.43}$$

$$M(\Sigma^+) = M_0 + (m_s - m_u) + \tfrac{4}{3}\delta M_m, \tag{3.60}$$

$$M(\Sigma^0) = M_0 + (m_d - m_u) + (m_s - m_u) - \tfrac{1}{3}\delta M_C, \tag{3.61}$$

$$M(\Sigma^-) = M_0 + 2(m_d - m_u) + (m_s - m_u) + \tfrac{1}{3}\delta M_C - \tfrac{1}{3}\delta M_m, \tag{3.62}$$

$$M(\Lambda) = M_0 + (m_d - m_u) + (m_s - m_u) - \tfrac{1}{3}\delta M_C + \tfrac{2}{3}\delta M_m, \tag{3.63}$$

$$M(\Xi^0) = M_0 + 2(m_s - m_u) - \tfrac{1}{3}\delta M_C + \delta M_m, \tag{3.64}$$

$$M(\Xi^-) = M_0 + (m_d - m_u) + 2(m_s - m_u) + \tfrac{1}{3}\delta M_C - \tfrac{1}{3}\delta M_m, \tag{3.65}$$

$$M(\Sigma^0_u) = M_0 + (m_d - m_u) + (m_s - m_u) - \tfrac{1}{3}\delta M_C + \delta M_m, \tag{3.66}$$

$$M(\Lambda^0_u) = M_0 + (m_d - m_u) + (m_s - m_u) - \tfrac{1}{3}\delta M_C - \tfrac{1}{3}\delta M_m, \tag{3.67}$$

$$\langle \Lambda^0 | M | \Sigma^0 \rangle = (1/\sqrt{3})\delta M_m. \tag{3.68}$$

It is evident at once that the U-spin relations (3.42–3.44) are respected; the electromagnetic terms are equal in (2.42) and (3.60), in (2.43), (3.64), and (3.66), and in (3.62) and (3.65).

A modicum of arithmetic serves to isolate the parameters:

$$m_d - m_u = M(\mathrm{n}) - M(\mathrm{p}) + \tfrac{1}{3}\big[M(\Sigma^+) + M(\Sigma^-) - 2M(\Sigma^0)\big]$$
$$= 1.89 \pm 0.04 \ \mathrm{MeV}/c^2, \tag{3.69}$$

$$\delta M_C = M(\mathrm{p}) - M(\mathrm{n}) + M(\Sigma^-) - M(\Sigma^0)$$
$$= 3.59 \pm 0.06 \ \mathrm{MeV}/c^2, \tag{3.70}$$

$$\delta M_m = M(\mathrm{n}) - M(\mathrm{p}) + M(\Sigma^+) - M(\Sigma^0)$$
$$= -1.81 \pm 0.10 \ \mathrm{MeV}/c^2, \tag{3.71}$$

$$m_s - m_u = \tfrac{1}{2}\big[M(\Xi^0) - M(\mathrm{n})\big] + (m_d - m_u)$$
$$= 189.55 \pm 0.3 \ \mathrm{MeV}/c^2. \tag{3.72}$$

The quark mass differences (3.69) and (3.72) are of the same order of magnitude as the values (3.33) and (3.38) deduced from the mesons, but are not identical with those values. The defining equations (2.40), (2.41) show that the Coulomb term is to be identified as

$$\delta M_C \equiv \langle \alpha/r \rangle, \tag{3.73}$$

from which the effective radius is found to be

$$\langle 1/r \rangle = 1/(0.4 \text{ fm}), \tag{3.74}$$

a reasonable value. Similarly, the hyperfine term is

$$\delta M_m = -2\pi\alpha |\Psi(0)|^2/3m^2, \tag{3.75}$$

which should be, and is in (3.71), a negative energy. If m is chosen as a representative quark mass, say as one third of

$$M_0 = 938.81 \text{ MeV}/c^2, \tag{3.76}$$

bearing in mind our continued approximation of $\mu_s = \mu_d$, one may determine an effective value for

$$|\Psi(0)|^2 = 0.0116 \text{ GeV}^3 = (0.87 \text{ fm})^{-3}, \tag{3.77}$$

which is again a sensible value. Such an estimate for the square of the wavefunction at zero interquark separation is a necessary ingredient in estimates of the proton lifetime [reviewed by Langacker 1981], and of the electric dipole moment of the neutron [Ellis et al. 1981].

The parameters given in eqns. (3.69–3.72) and (3.76) lead to the baryon masses shown in table 5. The mass differences within isospin multiplets are reproduced extremely well, and this success is nontrivial. Three parameters with reasonable values reproduce four mass dif-

Table 5
Baryon masses (MeV/c^2)

Particle	Model		Experiment	
p	938.28		938.28	
n	939.58		939.57	
Σ^+	1125.95	-7.98	1189.36	-7.98 ± 0.08
Σ^-	1133.93		1197.34	
Σ^0	1129.05	4.88	1192.46	4.88 ± 0.06
Λ	1127.84		1115.60	
Ξ^0	1314.90		1314.9	
Ξ^-	1321.60		1321.62	

ferences. There is, however, a fly in the ointment. The $\Lambda-\Sigma^0$ mass difference, here attributed to the hyperfine interaction, is grossly underestimated. Indeed, if the hyperfine term had been estimated from the $\Lambda-\Sigma$ splitting, the result would have been

$$\text{``}\delta M_{\mathrm{m}}\text{''} = \tfrac{3}{2}\left[M(\Lambda)-M(\Sigma^0)\right] \simeq -115 \text{ MeV}/c^2, \tag{3.78}$$

which is decidely not the scale of an electromagnetic mass shift! We shall see in section 5.3 that the N–Δ level splitting and the $\Lambda-\Sigma$ mass difference have a common plausible interpretation as effects of the color hyperfine interaction.

As a closing remark on U-spin invariance, let us note the selection rules

$$Y_1^{*-} \nrightarrow \Sigma^- \gamma, \tag{3.79}$$

and

$$\Xi^{*-} \nrightarrow \Xi^- \gamma, \tag{3.80}$$

and the prediction that the matrix elements for the transitions $\Delta^+ \rightarrow p\gamma$ and $Y_1^{*+} \rightarrow \Sigma^+ \gamma$ should be equal. The high-energy hyperon beams at Fermilab and at the CERN SPS bring tests of these predictions within reach. For further discussion see Kane (1972), Lipkin (1973b) and Quigg and Rosner (1976).

3.3. Strong-interaction mass formulas

3.3.1. The Gell-Mann–Okubo formula

We have just seen that counting strange quarks does not give a perfect description of masses in the baryon octet. For the other multiplets of immediate interest, however, the situation is somewhat more satisfying, as our experience with vector mesons indicated in (3.37), (3.38). In the baryon decimet, for example, strange-quark counting leads to an equal spacing rule

$$M(\Delta) - M(Y_1^*) = M(Y_1^*) - M(\Xi^*) = M(\Xi^*) - M(\Omega^-), \tag{3.81}$$

which agrees reasonably well with the measured charge-averaged intervals of 152, 149, and 139 MeV$/c^2$.

The hypothesis that masses can be determined by counting strange quarks can be interpreted to mean that the departures from exact SU(3) symmetry are governed by the hypercharge operator Y or λ_8. Because

$$Y = U_3 + \tfrac{1}{2}Q, \tag{3.82}$$

and because all members of a U-spin multiplet have the same charge, masses may be expressed as a sum of a U-spin scalar contribution, plus a contribution proportional to U_3, as

$$M = A + BU_3. \tag{3.83}$$

This constitutes an equal spacing rule within U-spin multiplets. In the case of the vector meson octet, relation (3.83) implies that

$$M(K^{*0}) - M(U_8) = M(U_8) - M(\overline{K}^{*0}), \tag{3.84}$$

which, because of the equality of K^{*0} and $\overline{K}^{*0}$ masses, requires that

$$M(U_8) = M(K^{*0}). \tag{3.85}$$

Here I have introduced the notation

$$|U_8\rangle = |U = 1, U_3 = 0\rangle \tag{3.86}$$

for the state defined in (3.23), which according to (3.8) is

$$|U_8\rangle = \tfrac{1}{2}(|\rho^0\rangle + |\omega^0\rangle - \sqrt{2}|\varphi\rangle). \tag{3.87}$$

Thus the implication of (3.85) is that

$$892 \text{ MeV}/c^2 = M(K^*)$$
$$= \tfrac{1}{4}[M(\rho^0) + M(\omega) + 2M(\varphi)] = 898 \text{ MeV}/c^2, \tag{3.88}$$

which is well satisfied.

Application of the equal spacing rule (3.83) to the baryon octet yields the relation

$$M(n) - M(\Sigma_u^0) = M(\Sigma_u^0) - M(\Xi^0), \tag{3.89}$$

where $|\Sigma_u^0\rangle$ is defined in (3.52). When rewritten in the form

$$\tfrac{1}{2}[M(n) + M(\Xi^0)] = \tfrac{1}{4}[M(\Sigma^0) + 3M(\Lambda^0)], \tag{3.90}$$

this connection is known as the Gell-Mann (1961, 1962)–Okubo (1962a, b) mass formula. The measured masses give

$$1127.24 \text{ MeV}/c^2 \overset{?}{=} 1134.82 \text{ MeV}/c^2, \tag{3.91}$$

in excellent agreement. Note, however, that the baryon masses given by the simple quark model in table 5 yield 1128.14 MeV/c^2, for the right-hand side, which is also splendid agreement. Reproducing $M(\Sigma_u^0)$ is thus not the same as successfully describing the $\Lambda - \Sigma$ hyperfine splitting.

Applied to the meson octet, the Gell-Mann–Okubo formula makes a prediction for the mass of the isoscalar member. For the moment, let us assume $\eta(548.8)$ to be the eighth member of the octet. Then the Gell-Mann–Okubo formula predicts

$$M(\eta) = \tfrac{1}{3}\left[4M(\text{K}^0) - M(\pi^0)\right] = 618.6\,\text{MeV}/c^2, \qquad (3.92)$$

which is not a spectacular success. Among many purported explanations for this failure is the suggestion that the Gell-Mann–Okubo formula should be applied to mass-squared, rather than mass, whereupon

$$M^2(\eta) = \tfrac{1}{3}\left[4M^2(\text{K}^0) - M^2(\pi^0)\right] = (569.4\,\text{MeV}/c^2). \qquad (3.93)$$

This proposal has provoked much learned debate and occasional puckish commentary [see footnote 21 of Cahn and Einhorn 1971]. Until we have a true theory of hadron structure, I am willing to concede that ambiguity exists.

3.3.2. Singlet – octet mixing

In our initial discussion of vector meson masses in terms of quarks, we skimmed over the identification of singlet and octet isoscalars. The Gell-Mann–Okubo formula for the vector octet reads

$$M(\text{V}_8) = \tfrac{1}{3}\left[4M(\text{K}^*) - M(\rho)\right] = 930.4\,\text{MeV}/c^2, \qquad (3.94)$$

which describes neither $\omega(782.4)$ nor $\varphi(1019.6)$. It is then natural to conclude that neither physical state is a pure octet member, and that $\omega - \varphi$ (or singlet–octet) mixing must be considered. In the quark basis, the singlet state (3.6) and the octet state (3.8) correspond to

$$|V_1\rangle = (u\bar{u} + d\bar{d} + s\bar{s})/\sqrt{3} \qquad (3.95)$$

and

$$|V_8\rangle = (u\bar{u} + d\bar{d} - 2s\bar{s})/\sqrt{6}, \qquad (3.96)$$

both of which are orthogonal to

$$|\rho^0\rangle = (d\bar{d} - u\bar{u})/\sqrt{2}. \qquad (3.97)$$

The isoscalar mass matrix can be written in the singlet–octet basis as

$$\mathfrak{M} = \begin{pmatrix} M_1 & \Delta \\ \Delta & M_8 \end{pmatrix}, \qquad (3.98)$$

which has eigenvalues

$$\begin{pmatrix} M(\varphi) \\ M(\omega) \end{pmatrix} = \frac{M_1 + M_8}{2} \pm \left(\frac{(M_1 - M_8)^2}{4} + \Delta^2\right)^{1/2}. \qquad (3.99)$$

The value of M_8 has already been determined in (3.94). The eigenvalue expression yields

$$M_1 = M(\varphi) + M(\omega) - M_8 = 871.6 \text{ MeV}/c^2. \tag{3.100}$$

Therefore the mixing parameter Δ is given by

$$\Delta^2 = \tfrac{1}{4}[M(\varphi) - M(\omega)]^2 - \tfrac{1}{4}(M_1 - M_8)^2 = (114.9 \text{ MeV}/c^2)^2. \tag{3.101}$$

It is convenient to parametrize the singlet–octet mixing trigonometrically, as

$$|\varphi\rangle = |V_1\rangle \sin\theta - |V_8\rangle \cos\theta, \tag{3.102a}$$

$$|\omega\rangle = |V_1\rangle \cos\theta + |V_8\rangle \sin\theta. \tag{3.102b}$$

The mixing angle can then be determined from the eigenvalue conditions

$$\mathfrak{M}|\varphi\rangle = M(\varphi)|\varphi\rangle, \tag{3.103a}$$

$$\mathfrak{M}|\omega\rangle = M(\omega)|\omega\rangle, \tag{3.103b}$$

which can be expanded as

$$M(\varphi) = M_1 - \Delta\cot\theta = M_8 - \Delta\tan\theta, \tag{3.104a}$$

$$M(\omega) = M_1 + \Delta\tan\theta = M_8 + \Delta\cot\theta. \tag{3.104b}$$

The solution of these equations is

$$\tan^2\theta = \frac{M(\omega) - M_1}{M(\omega) - M_8} = \frac{M_8 - M(\varphi)}{M_1 - M(\varphi)} = (0.776)^2, \tag{3.105}$$

which is not far from the "ideal mixing" value

$$\tan\theta_{\text{ideal}} = 1/\sqrt{2}. \tag{3.106}$$

The implied angles are $\theta = 37.8°$, corresponding to (3.105), and $\theta_{\text{ideal}} = 35.3°$, corresponding to (3.106). For ideal mixing, the physical particle wavefunctions are those given in (3.5):

$$|\omega\rangle = (u\bar{u} + d\bar{d})/\sqrt{2} \tag{3.107a}$$

and

$$|\varphi\rangle = s\bar{s}. \tag{3.107b}$$

The quark composition of these states provides immediate insight into the pattern of masses. The ω^0 and ρ^0 contain the same light quarks in equal weights, and are nearly degenerate in mass. The hidden strangeness state φ is more massive because it contains the heavier strange quark, as was made quantitative in (3.37).

The $\eta'(957.57)$ suggests itself as a ninth pseudoscalar meson to be considered within a nonet containing $\eta(548.8)$. The singlet–octet mixing analysis can be transcribed directly from the vector meson case, with the definitions

$$|\eta'\rangle = |\eta_1\rangle \sin\theta - |\eta_8\rangle \cos\theta, \tag{3.108a}$$

$$|\eta\rangle = |\eta_1\rangle \cos\theta + |\eta_8\rangle \sin\theta. \tag{3.108b}$$

For a linear mass formula we have, upon reinterpreting (3.92) as M_8,

$$M_1 = 887.8 \text{ MeV}/c^2, \tag{3.109}$$

$$\Delta^2 = \left(153.8 \text{ MeV}/c^2\right)^2, \tag{3.110}$$

$$\theta = 65.6° \tag{3.111}$$

and the convenient approximate forms

$$|\eta'\rangle = (1/8)^{1/2}(u\bar{u} + d\bar{d}) + (3/4)^{1/2} s\bar{s}, \tag{3.112a}$$

$$|\eta\rangle = (3/8)^{1/2}(u\bar{u} + d\bar{d}) - \tfrac{1}{2} s\bar{s}. \tag{3.112b}$$

An analysis using quadratic mass relations, with (3.93) interpreted as M_8^2, yields:

$$M_1^2 = \left(945 \text{ MeV}/c^2\right)^2, \tag{3.113}$$

$$\Delta^2 = \left(343 \text{ MeV}/c^2\right), \tag{3.114}$$

$$\theta = 78.85°, \tag{3.115}$$

and the approximate flavor wavefunctions

$$|\eta'\rangle = \tfrac{1}{2}(u\bar{u} + d\bar{d}) + s\bar{s}/\sqrt{2}, \tag{3.116a}$$

$$|\eta\rangle = \tfrac{1}{2}(u\bar{u} + d\bar{d}) - s\bar{s}/\sqrt{2}. \tag{3.116b}$$

In neither case is a particularly transparent interpretation of hadron masses in terms of quark masses possible. For the linear mass formula the implied mass difference $m_s - m_u$ is more than twice as large as the values encountered before in (3.37), (3.38) and (3.72). Quark counting with a quadratic mass formula would lead to the expectation of degenerate η and η' states. This may be taken as a hint that there is something yet to be understood about the pseudoscalar mesons. Whether the clue lies in the special role of the almost massless pion, or in a deeper understanding of the $\eta-\eta'$ mixing mechanism, or both is a question to which we shall have to return.

3.4. Electromagnetic properties redux

In this section we shall discuss the electromagnetic properties of hadrons in considerable quantitative detail. Such detail is justified by the impressive quality of recent measurements of baryon magnetic moments, transition moments in mesons, and meson charge radii. The first two subjects have been reviewed by Rosner (1980b, 1981a). What emerges from the comparison of the quark model and experiment is a level of agreement which is at least to my eye miraculous, but at the same time imperfect. The challenge to deeper theoretical approaches is to explain the successes while repairing the failures. To a limited extent this will be achieved in subsequent sections.

3.4.1. Baryon magnetic moments

Problems 8 and 9 provide the opportunity to compute the baryon magnetic moments using the quark-model wavefunctions derived in problem 7. The results are given in table 6, together with the experimental measurements. Three sets of quark model predictions are shown. The first, designated "exact SU(3)", is based on the assumption that the quark magnetic moments are proportional only to the quark charges, so that

$$\mu_u = -2\mu_d = -2\mu_s = 2\mu_p/3. \tag{3.117}$$

This simple description reproduces the trends of the data, insofar as signs and relative sizes are concerned, but it is not adequate quantitatively.

To attempt to improve the degree of agreement, we may break SU(3) symmetry by allowing μ_s to differ from μ_d. It is then natural to fix μ_s by fitting to the well-measured Λ^0 magnetic moment as

$$\mu_s = \mu_\Lambda = -0.614 \text{ n.m.} \tag{3.118}$$

This is smaller in magnitude than $\mu_d = -0.931$ n.m., consistent with the evidence from the spectrum that $m_s > m_d$. The assumption (3.118) makes a noticable improvement in the predictions, but leaves us short of a perfect description. Finally if we indulge in the fine tuning of μ_u and μ_d discussed in section 2.3, eqs. (2.65), (2.66), the overall situation is not markedly changed.

Although the general pattern of baryon moments is extremely well reproduced, there are noticeable quantitative failures. Particularly bad is

Table 6
Magnetic moments of the baryon octet. Numerical values are given in nuclear magnetons.
Underlined quantities are inputs

Baryon	Quark model	Exact SU(3) $\mu_u = -2\mu_d$ $= -2\mu_s = 2\mu_p/3$	$\mu_u = -2\mu_d$ $\mu_s \equiv \mu_\Lambda$	$\mu_u \neq -2\mu_d$ $\mu_s \equiv \mu_\Lambda$	Measured values
p	$(4\mu_u - \mu_d)/3$	$\mu_p = \underline{2.793}$	2.793	2.793	2.793
n	$(4\mu_d - \mu_u)/3$	$-2\mu_p/3 = -1.862$	-1.862	-1.913	-1.913
Λ	μ_s	$-\mu_p/3 = -0.931$	-0.614	-0.6138	-0.6138 ± 0.0047[a] -0.6129 ± 0.0045[b]
$\Lambda - \Sigma^0$	$(\mu_d - \mu_u)/\sqrt{3}$	$-\mu_p/\sqrt{3} = -1.612$	-1.612	-1.633	$-1.82^{+0.18}_{-0.25}$[c]
Σ^+	$(4\mu_u - \mu_s)/3$	$\mu_p = 2.793$	2.687	2.673	2.33 ± 0.13[d]
Σ^0	$(2\mu_u + 2\mu_d - \mu_s)/3$	$\mu_p/3 = 0.931$	0.825	0.791	0.46 ± 0.28[e]
Σ^-	$(4\mu_d - \mu_s)/3$	$-\mu_p/3 = -0.931$	-1.037	-1.091	-1.41 ± 0.25[f]
Ξ^0	$(4\mu_s - \mu_u)/3$	$-2\mu_p/3 = -1.862$	-1.439	-1.436	-1.250 ± 0.014[b]
Ξ^-	$(4\mu_s - \mu_d)/3$	$-\mu_p/3 = -0.931$	-0.508	-0.494	-0.75 ± 0.06[g]

[a] Schachinger et al. (1978).
[b] Cox et al. (1981).
[c] Dydak et al. (1977).
[d] Settles et al. (1979) and Particle Data Group (1980).
[e] Defined by eq. (3.59).
[f] Roberts et al. (1975) and Particle Data Group (1980).
[g] Handler et al. (1980).

the combination

$$3(\mu_{\Xi^-} - \mu_{\Xi^0}) = \mu_u - \mu_d = \tfrac{3}{5}(\mu_p - \mu_n), \qquad (3.119)$$

for which the measurements yield

$$1.50 \pm 0.18 \text{ n.m.} \neq 2.82 \text{ n.m.} \qquad (3.120)$$

However, if the effective moments of the up and down quarks are altered in strange hadrons, it is not in a systematic way, at least at the level of present measurements. The quark-model relation

$$\tfrac{3}{4}(\mu_{\Sigma^+} - \mu_{\Sigma^-}) = \mu_u - \mu_d = \tfrac{3}{5}(\mu_p - \mu_n) \qquad (3.121)$$

is well satisfied by experiment:

$$2.81 \pm 0.21 \text{ n.m.} = 2.82 \text{ n.m.} \qquad (3.122)$$

Forthcoming measurements of the Σ moments will provide a more incisive test of (3.121).

We have already found in section 2.3, eqs. (2.63), (2.64), that if the quarks are regarded as Dirac particles, the inferred masses of the up and down quarks are reasonable—about one third of a proton mass. From (3.118) and (2.66) we conclude that

$$m_{\mathrm{s}} \approx 1.58\, m_{\mathrm{d}} \approx 510 \; \mathrm{MeV}/c^2. \tag{3.123}$$

This both is a sensible value on its own (half the φ mass) and implies that

$$m_{\mathrm{s}} - m_{\mathrm{u}} \approx 172 \; \mathrm{MeV}/c^2, \tag{3.124}$$

consistent with the determination (3.72) from the baryon spectrum of approximately 190 MeV/c^2.

The availability of precise data on the hyperon moments and the suggestion that the elementary quark model is close to, but not exactly the truth has stimulated many attempts at improving the model. The suggestions include relativistic corrections, configuration mixing, various dilution effects, and anomalous quarks moments. Many are interesting, but none is yet compelling. A representative sampling may be gleaned from the mini-review by Rosner (1980b), the papers by Bohm and Teese (1981), Cohen and Lipkin (1980), Dothan (1981), Geffen and Wilson (1980), Isgur and Karl (1980), Lichtenberg (1981), Lipkin (1981a), Teese (1981), and references therein.

The now-standard method for measuring hyperon magnetic moments by exploiting the polarization of inclusively produced hyperons is described in Schachinger et al. (1978), and in the Les Houches seminar by Cox (1981). These spin-rotation measurements rely upon the self-analyzing weak decays of hyperons, which are the subject of problems 10 and 11.

3.4.2. When fermions are confined

The foregoing analysis has demonstrated that if quarks are regarded as Dirac particles with masses that seem reasonable for constituents of hadrons a systematic understanding of the baryon magnetic moments emerges. On the one hand, this state of affairs provides motivation to elaborate or "improve" the simple quark model. On the other, we are impelled to ask whether it is indeed reasonable that the simple picture should work so well.

Because quarks appear [Jones 1977, Lyons 1981] to be securely bound within hadrons, it is important to ask whether a bound fermion behaves as a Dirac particle. That there is room for discussion is shown by an elementary example due to Lipkin and Tavkhelidze (1965). The Dirac equation for a particle of mass m in an electromagnetic field may be written in the form

$$\gamma^{\mu}(p_{\mu} - eA_{\mu})\Psi = m\Psi. \tag{3.125}$$

It is obvious from common experience that the magnetic moment of an electron does not depend upon the strength of the magnetic field in which it is measured. In the same way, if the particle is subjected to an additional four-vector interaction V_{μ}, the ensuing Dirac equation,

$$\gamma^{\mu}(p_{\mu} - eA_{\mu} + V_{\mu})\Psi = m\Psi, \tag{3.126}$$

is still that of a particle with mass m, but with a redefined momentum. In the specific case of a static potential

$$V_0 = \mathcal{V}, \qquad V = 0, \tag{3.127}$$

the change amounts merely to a shift in the energy scale,

$$E \rightarrow E^* = E + \mathcal{V}. \tag{3.128}$$

Consider instead the case of a fermion interacting with a four-scalar potential $\mathcal{V}$. The Dirac equation is then

$$\left[\gamma^{\mu}(p_{\mu} - eA_{\mu}) + \mathcal{V}\right]\Psi = m\Psi, \tag{3.129}$$

which has the look,

$$\gamma^{\mu}(p_{\mu} - eA_{\mu})\Psi = (m - \mathcal{V})\Psi, \tag{3.130}$$

and the effect of a Dirac equation for a particle with mass $m^* = m - \mathcal{V}$, which is to say that the fermion has acquired an apparent anomalous moment. Thus it is proper to be concerned about the effect of confinement upon the apparent magnetic moment of a quark.

More insight into the effective properties of a quark within a hadron may be gained by considering a free fermion confined within a rigid sphere. This is a textbook problem in relativistic quantum mechanics [cf. Akhiezer and Berestetskii 1965] and has been applied in one guise [Bogoliubov 1967] or another [Chodos et al. 1974b, DeGrand et al. 1975] to the quark model by many authors. (A brief summary of the problem is also to be found in ch. 18 of Close 1979, but watch out for misprints!)

Suppose that within a rigid static sphere of radius R the fermion behaves as a free particle, and satisfies

$$\not{p}\Psi = m\Psi. \tag{3.131}$$

The confinement hypothesis requires that no probability current flow across the boundary of the sphere, which is characterized covariantly by the outward normal n_μ. The boundary condition is thus

$$n^\mu \bar\Psi \gamma_\mu \Psi = 0|_{r=R},\qquad(3.132)$$

or simply $\bar\Psi \hat{n} \Psi = 0$ at $r = R$. This lowest mode solution of the Dirac equation is:

$$\Psi(r,t) = \mathfrak{N}(x)\left(\begin{array}{c} \left(\dfrac{\omega+m}{\omega}\right)^{1/2} i\, j_0(xr/R)\chi \\[3mm] -\left(\dfrac{\omega-m}{\omega}\right)^{1/2} j_1(xr/R)\boldsymbol\sigma\cdot\hat{r}\chi \end{array}\right) e^{-i\omega t},\quad(3.133)$$

where ω is the particle energy, x/R its momentum, χ is a two-component spinor, the j_u are spherical Bessel functions, and $\mathfrak{N}(x)$ is a normalization factor.

The boundary condition (3.132) will be satisfied if at $r = R$

$$\hat{n}\Psi = i\Psi,\qquad(3.134)$$

for then:

$$\bar\Psi \hat{n} = -i\bar\Psi,\qquad(3.135)$$

and thus:

$$\bar\Psi \hat{n} \Psi = i\bar\Psi\Psi = -i\bar\Psi\Psi = 0.\qquad(3.136)$$

Since the outward normal is

$$n_\mu = (0, \hat{r}),\qquad(3.137)$$

the boundary condition (3.134) is

$$i\boldsymbol\gamma\cdot\hat{r}\Psi = \Psi|_{r=R}\qquad(3.138)$$

or explicitly,

$$\left(\begin{array}{cc} 0 & -i\boldsymbol\sigma\cdot\hat{r} \\ i\boldsymbol\sigma\cdot\hat{r} & 0 \end{array}\right)\left(\begin{array}{c} i(\omega+m)^{1/2} j_0(x) \\ -(\omega-m)^{1/2}\boldsymbol\sigma\cdot\hat{r} j_1(x) \end{array}\right)$$

$$= \left(\begin{array}{c} i(\omega+m)^{1/2} j_0(x) \\ -(\omega-m)^{1/2}\boldsymbol\sigma\cdot\hat{r} j_1(x) \end{array}\right),\quad(3.139)$$

which implies, because $\boldsymbol\sigma\cdot\boldsymbol{a}\,\boldsymbol\sigma\cdot\boldsymbol{b} = \boldsymbol{a}\cdot\boldsymbol{b} + i\boldsymbol\sigma\cdot\boldsymbol{a}\times\boldsymbol{b}$, that:

$$j_1(x) = [(\omega+m)/(\omega-m)]^{1/2} j_0(x).\qquad(3.140)$$

Enforcing the boundary condition (3.139) thus leads to the eigenvalue condition

$$\tan x = \frac{x}{1 - mR - \left[x^2 + (mR)^2\right]^{1/2}}.$$ (3.141)

In passing from (3.140) to (3.141) we have used the explicit forms of the spherical Bessel functions

$$j_0(x) = \sin x / x,$$ (3.142a)

$$j_1(x) = (\sin x - x \cos x)/x^2,$$ (3.142b)

and the connection

$$\omega = \left(m^2 + x^2/R^2\right)^{1/2}.$$ (3.143)

The first eigenvalue of (3.141), corresponding to the ground-state momentum (in units of $1/R$) of the confined fermion, is plotted in fig. 12a. It ranges between the values

$$x(mR = 0) = 2.043,$$ (3.144a)

$$x(mR \to \infty) = \pi.$$ (3.144b)

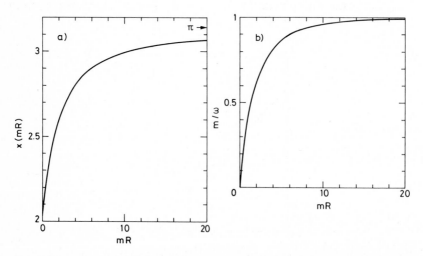

Fig. 12. Properties of the lowest mode of a fermion confined within a rigid sphere. (a) Fermion momentum as a function of its mass m and the sphere radius R. (b) Ratio of the fermion mass to the energy of its lowest confined mode.

For any value of mR, the ratio of the free-particle mass to the confined-particle energy depends only upon mR. This is shown in fig. 12b. In the extreme-relativistic limit $mR \to 0$, all the energy is a consequence of the confinement, and

$$\omega = 2.043/R, \quad mR = 0. \tag{3.145}$$

In the nonrelativistic, or free particle limit $mR \to \infty$, the confined-particle energy approaches the free-particle mass:

$$\lim_{mR \to \infty} \omega = m. \tag{3.146}$$

It is of interest to study the behavior of the confined-particle energy in two special cases. In fig. 13 we see the energy of a massless particle confined within a sphere of radius R. Note some typical values: a massless particle confined within a radius of 1 fm acquires an energy, or effective mass, of 400 MeV/c^2. A system confined to a fixed radius of 4/3 fm is portrayed in fig. 14. The two limits described by (3.145) and (3.146) are readily apparent.

What implications does confinement have for the magnetic moment? With the Dirac wavefunction (3.133) for the ground state in hand, it is

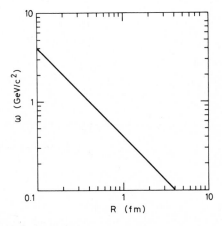

Fig. 13. Lowest-mode energy of a massless fermion confined to a rigid, static sphere of radius R [see eq. (3.145)].

C. Quigg

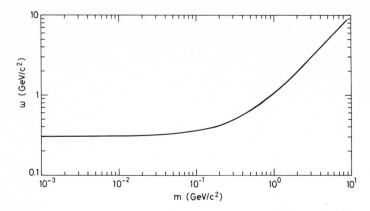

Fig. 14. Lowest-mode energy of a fermion of mass m, confined within a rigid, static sphere of radius $4/3$ fm.

straightforward to compute the magnetic moment from the definition

$$\boldsymbol{\mu} \equiv \tfrac{1}{2} \int_{|r| < R} d^3r \, \boldsymbol{r} \times \boldsymbol{J} = \tfrac{1}{2} e \int_{|r| < R} d^3r \, \boldsymbol{r} \times \bar{\Psi} \boldsymbol{\gamma} \Psi. \tag{3.147}$$

I note in passing an elegant discussion of the nature of intrinsic dipole moments [Jackson, 1977a], from which everyone can learn something. The result [Chodos et al. 1974 for the massless case, Allen 1975, DeGrand et al. 1975, Golowich 1975] is

$$\mu = \frac{e}{2m} \left\{ \frac{mR}{3} \cdot \frac{4\omega R + 2mR - 3}{2(\omega R)^2 - 2\omega R + mR} \right\}, \tag{3.148}$$

which can lead to a large reduction of the magnetic moment, compared with that of the free particle. In particular, a confined massless fermion acquires a finite magnetic moment.

This result has been cited by the developers of the MIT Bag [e.g., DeGrand et al. 1975] as an example of the unreliability of the nonrelativistic quark model. While I do not disagree, I would emphasize a different interpretation. We may recast (3.148) as

$$\mu = \frac{e}{2\omega} \left\{ \frac{\omega R}{3} \cdot \frac{4\omega R + 2mR - 3}{2(\omega R)^2 - 2\omega R + mR} \right\}, \tag{3.149}$$

recognizing that the energy ω plays the role of an effective mass for the confined fermion. Comparing the magnetic moment of the confined fermion with the Dirac moment of a free fermion with mass ω, we find (see the plot of the ratio in fig. 15) that the two differ by less than 20%, even for the extreme case of a confined massless fermion. I take this very stylized calculation to indicate that it is not nonsensical for confined quarks to display Dirac moments characteristic of their constituent masses. Quark model phenomenology is therefore likely to make sense, although it may well be the case that a quark manifests slightly different moments in different hadrons, and that the model succeeds for complicated reasons. DeGrand (1980) is in grudging agreement with this view. This issue recurs for models of composite quarks and fermions. See a recent comment by Bander et al. (1981).

3.4.3. Axial charges

The simileptonic decays of hadrons provide information about both the weak current and the properties of the hadrons themselves. For exhaustive reviews, see Willis and Thompson (1968), and Chounet et al. (1972). In the limit of zero momentum transfer, the matrix element for the baryon semileptonic decay $B \to B' \ell \nu$ is of the form

$$\bar{u}(B')\gamma_\mu(g_V + g_A\gamma_5)u(B). \tag{3.150}$$

The vector couplings are determined by the Cabibbo (1963) hypothesis and by the assumption that the vector charges are generators of SU(3). They are protected from symmetry breaking effects [Ademollo and Gatto 1964, Bouchiat and Meyer 1964] and thus yield little information about baryon structure. The axial charges may have a greater sensitivity to hadron physics.

In the nonrelativistic quark model [e.g. Kokkedee 1969], the ratio g_A/g_V is given by

$$g_A/g_V = \frac{\langle B'|I_+\sigma_z|B\rangle}{\langle B'|I_+|B\rangle} \tag{3.151}$$

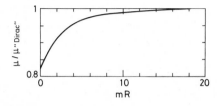

Fig. 15. Magnetic moment of the confined fermion in units of the Dirac moment for a free fermion with mass equal to the energy ω of the confined fermion.

C. Quigg

for $\Delta S = 0$ decays, and by

$$g_A/g_V = \frac{\langle B'|V_+\sigma_z|B\rangle}{\langle B'|V_+|B\rangle} \tag{3.152}$$

for $\Delta S = 1$ decays. The predictions for various semileptonic decays are gathered in table 7. The quark model predictions correspond to those of the SU(3) algebra, also given in table 7, with the specific choice of symmetric and antisymmetric octet couplings $D \equiv 1$, $F/D = 2/3$. Measurements of individual decay rates determine only linear combinations of g_A^2 and g_V^2, so can only constrain g_A within broad limits. All existing measurements are compatible with the Cabibbo parametrization, with

$$F/D = 0.54 \pm 0.02, \tag{3.153}$$

[Shrock and Wang 1978], not terribly far from the quark model value.

For the cases in which g_A/g_V or $|g_A/g_V|$ has been determined directly by correlation or polarization measurements, the pattern of the data is systematically that of the quark model. However, the quark model overestimates the axial charges by approximately one third. (Note that the sign convention adopted by Particle Data Group 1980, is opposite to mine). Existing measurements are shown in table 7. Here again, experiments using high-energy hyperon beams hold considerable promise for dramatically extending our knowledge.

The effects of confinement, which were treated schematically in the preceding section, are significant for quark matrix elements $\langle q|\sigma_z|q\rangle$ as well because the lower components of the Dirac spinor may have the "wrong" spin projection. Confinement tends to increase the effective

Table 7
Axial charges g_A/g_V in semileptonic decays of baryons

Decay	Cabibbo/SU(3)	Quark model	Experiment
$n \to pe\nu$	$D + F$	$5/3$	1.254 ± 0.007[a]
$\Lambda \to pe\nu$	$F + D/3$	1	0.62 ± 0.05[a]
			$\pm (0.734 \pm 0.03)$[b]
$\Sigma^- \to \Sigma^0 e\nu$	F	$2/3$	
$\Sigma^\pm \to \Lambda e\nu$	pure axial	pure axial	$g_{\bar{V}}/g_{\bar{A}} = -0.10 \pm 0.22$[a]
$\Sigma^- \to n e\nu$	$F - D$	$-1/3$	$\pm (0.385 \pm 0.070)$[a]
$\Xi^- \to \Xi^0 e\nu$	$F - D$	$-1/3$	
$\Xi^- \to \Lambda e\nu$	$F - D/3$	$1/3$	

[a] Particle Data Group (1980).
[b] Jensen et al. (1980).

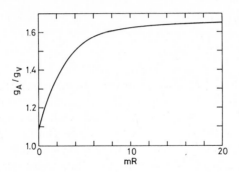

Fig. 16. Axial charge of a nucleon composed of equal-mass quarks confined within a rigid spherical cavity, as a function of the dimensionless parameter mR.

mass of a fermion, and thus to increase the importance of the lower components. The static spherical cavity was studied by Golowich (1975) who found for the nucleon ground state

$$g_A/g_V = \frac{5}{3}\left\{\frac{2(\omega R)^2 + 4mR\omega R - 3mR}{6(\omega R)^2 - 6\omega R + 3mR}\right\}, \tag{3.154}$$

which is plotted in fig. 16 as a function of the dimensionless parameter mR. The values range from $g_A/g_V = 1.09$ for massless quarks to the free-fermion value of $5/3$ as $mR \to \infty$. The extension to other systems was made by Donoghue et al. (1975) and by DeGrand et al. (1975). See also the treatment by Le Yaouanc, et al. (1977).

3.4.4. M1 transitions in mesons

In section 2.4 we discussed the interpretation of radiative (M1) decays of vector mesons as single-quark spinflip transitions, and compared the relative rates for $(\rho, \omega) \to \pi\gamma$. In the course of extending the quark model predictions to the full vector meson nonet, we may also consider the expectations of the quark model for absolute decay rates. The treatment given in section 2.4 was complete so far as the flavor and spin aspects of the simple quark model are concerned, but did not consider the degree of overlap between initial and final spatial wavefunctions. In the static limit, which corresponds to initial and final hadrons of equal mass, this overlap is likely to be complete. However, the vector and pseudoscalar mesons are decidedly different in mass, so recoil effects and possible differences in the radial wavefunctions due to the distortion of the strong hyperfine splitting can be expected to make the overlap incomplete. A true dynamical theory of hadrons should permit the computation of the overlap integral. Lacking that, we shall merely parametrize the anticipated effect by writing

$$\langle P|M1|V\rangle = \langle P|M1|V\rangle_{\text{flavor-spin}} \cdot \mathcal{O}, \tag{3.155}$$

where P and V are generic labels for pseudoscalar and vector mesons, $M1$ is the transition operator of eq. (2.98), and $\mathcal{O}$ will be called the overlap factor. It will be considered encouraging if, with the values of quark magnetic moments determined from the baryon magnetic moments, the overlap factor is less than unity, but of order unity. See also Rosner (1980b, 1981a), and O'Donnell (1981).

The quark model predictions and the comparison with experiment are given in table 8, which requires a good deal of explanation and comment. After tabulating the energy ω of the emitted photon and the measured decay rate Γ, I have chosen to characterize the experimental matrix-element-squared by the dimensionless quantity

$$\frac{3\pi\Gamma(V \to P\gamma)}{\omega^3\mu_N}, \tag{3.156}$$

where the nuclear magneton μ_N is defined for a Dirac proton, so that

$$\mu_N^2 = \pi\alpha/m_p^2. \tag{3.157}$$

This is free of the trivial kinematic dependence, and can readily be compared with the square of the flavor-spin matrix element, $3\pi\Gamma/\omega^3\mu_N^2\mathcal{O}$, evaluated in the quark model. Comparison of the experimental and theoretical numbers leads to a determination of the overlap factor $\mathcal{O}$. In table 8, the quark model prediction has been evaluated for two sets of assumptions about the quark magnetic moments: the SU(3)-symmetric case

$$\mu_u = -2\mu_d = -2\mu_s = 2\mu_p/3, \tag{3.117}$$

and the broken-symmetry case with the strange quark moment given by

$$\mu_s = \mu_\Lambda. \tag{3.118}$$

This parallels our discussion of baryon magnetic moments.

The isoscalar pseudoscalar mesons were left in some disarray in subsection 3.3.2. As a result the flavor wavefunctions of η and η' are not well specified. I have therefore presented calculations for two cases, labelled Q (for quadratic mass formula) and L (for linear mass formula)

Table 8
M1 Transitions in mesons

Process	Photon energy ω (GeV)	Decay rate Γ (keV)	$3\pi\Gamma/\omega^3\mu_N^2$	$|\mathcal{M}_{\text{flavor–spin}}|^2$ Quark model	SU(3) $3\pi\Gamma/\omega^3\mu_N^2$	Θ	$\mu_s \neq \mu_d$ $3\pi\Gamma/\omega^3\mu_N^2$	Θ
$\omega \to \pi^0\gamma$	0.380	889 ± 50[a] 789 ± 92	5.87 ± 0.35 5.21 ± 0.61	$(\mu_u - \mu_d)^2$	7.80	0.75 ± 0.04 0.67 ± 0.08		
$\rho \to \pi\gamma$	0.375	67 ± 7[b]	0.461 ± 0.038	$(\mu_u + \mu_d)^2$	0.87	0.53 ± 0.04		
$K^{*+} \to K^+\gamma$	0.309	62 ± 14[c]	0.753 ± 0.170	$(\mu_u + \mu_s)^2$	0.87	0.87 ± 0.20	1.56	0.48 ± 0.11
$K^{*0} \to K^0\gamma$	0.309	75 ± 35[a]	0.921 ± 0.422	$(\mu_d + \mu_s)^2$	3.47	0.27 ± 0.12	2.39	0.39 ± 0.18
$\varphi \to \pi^0\gamma$	0.501	5.7 ± 2[a]	0.017 ± 0.006	0	0	—		
$\omega \to \eta\gamma$	0.199	$3^{+2.5}_{-1.8}$[a]	$0.138^{+0.115}_{-0.084}$	Q: $(\mu_u + \mu_d)^2/2$ L: $3(\mu_u + \mu_d)^2/4$	0.43 0.65	$0.32^{+0.28}_{-0.20}$ $0.21^{+0.18}_{-0.13}$		
$\rho^0 \to \eta\gamma$	0.194	50 ± 13[a]	2.48 ± 0.64	Q: $(\mu_u - \mu_d)^2/2$ L: $3(\mu_u - \mu_d)^2/4$	3.90 5.85	0.63 ± 0.17 0.42 ± 0.11		
$\varphi \to \eta\gamma$	0.362	62 ± 9[a]	0.47 ± 0.07	Q: $2\mu_s^2$ L: μ_s^2	1.73 0.87	0.27 ± 0.04 0.54 ± 0.08	0.75 0.38	0.63 ± 0.09 1.24 ± 0.18
$\varphi \to \eta'\gamma$	0.060			Q: $2\mu_s^2$ L: $3\mu_s^2$	1.73 2.60		0.75 1.13	
$\eta' \to \rho^0\gamma$	0.164	83 ± 30[a,d]	6.81 ± 2.46	Q: $3(\mu_u - \mu_d)^2/2$ L: $3(\mu_u - \mu_d)^2/4$	11.70 5.85	0.58 ± 0.21 0.29 ± 0.11		
$\eta' \to \omega\gamma$	0.159	7.6 ± 3[a,d]	0.68 ± 0.27	Q: $3(\mu_u + \mu_d)^2/2$ L: $3(\mu_u + \mu_d)^2/4$	1.30 0.65	0.53 ± 0.21 1.05 ± 0.41		

[a] Particle Data Group (1980).
[b] Berg et al. (1980a).
[c] Berg et al. (1980b, 1981).
[d] Adjusted by Rosner (1980) using total η' width measured by Binnie et al. (1979) and by Abrams et al. (1979).

in table 8. The approximate flavor wavefunctions (3.116) and (3.112) have been used in the calculations.

A reading of table 8 shows that the quark model is rather successful in predicting the ratios of $M1$ transition rates, and that the reduced strange-quark moment of (3.118) improves the agreement between theory and experiment. The quark-model-forbidden transition $\varphi \to \pi^0 \gamma$ is enormously suppressed. The predicted absolute rates are also sensible; an overlap factor of approximately $1/2$ brings predictions and observations into reasonable agreement. This is shown in graphic form in fig. 17, which compares the overlap factors deduced from individual decay rates, for the broken symmetry case. An average value of

$$\mathbb{O} = 0.56 \tag{3.158}$$

accommodates most of the measurements. Ohshima's (1980) reanalysis of the $\omega \to \pi^0 \gamma$ rate, quoted above in eq. (2.102), leads to a reduced overlap factor for that decay of

$$\mathbb{O}(\omega \to \pi^0 \gamma)_{\text{Ohshima}} = 0.67 \pm 0.07, \tag{3.159}$$

which is more in line with the other inferred values. This underscores the already obvious remark that there is much room for improved experiments. The advantages of high-energy beams, already apparent from recent data, are stressed in the Les Houches seminar by T. Jensen (1981); see also Jensen (1980). An application of quark model techniques to $M1$ transitions among charmed mesons is posed as problem 12.

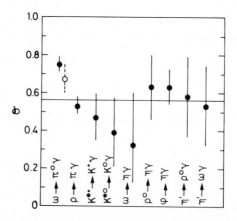

Fig. 17. Overlap factor $\mathbb{O}$ defined in eq. (3.155) as measured in various $M1$ decays of mesons. The dashed entry for $\omega \to \pi^0 \gamma$ is for the reanalysis by Ohshima (1980).

3.4.5. *Electromagnetic form factors*

Hadron form factors constitute an important piece of evidence that hadrons have a finite size and are composite in nature. A brief but lucid introduction to the subject appears in ch. 19 of the book by Perl (1974), and a thorough review with special emphasis on the vector dominance interpretation has been given by Gourdin (1974).

For spinless hadrons, the cross section for electron–hadron elastic scattering can be written as

$$\left.\frac{d\sigma}{dq^2}\right|_{\text{elastic}} = \left.\frac{d\sigma}{dq^2}\right|_{\text{point}} \times |F_h(q^2)|^2, \tag{3.160}$$

where the form factor $F_h(q^2)$ is the Fourier transform of the charge distribution within the hadron. For a spherically symmetric charge distribution,

$$
\begin{aligned}
F_h(q^2) &= \int d^3r e^{iq\cdot r}\rho(r)\\
&= \int_0^\infty r^2 dr \int_{-1}^1 dz \int_0^{2\pi} d\varphi\, e^{iqrz}\rho(r)\\
&= 4\pi \int_0^\infty r^2 dr \rho(r)\sin(qr)/qr, \tag{3.161}
\end{aligned}
$$

so that

$$F_h(0) = \int d^3r \rho(r) = Q_h. \tag{3.162}$$

At small values of the momentum transfer q^2, it is appropriate to approximate the integral by Taylor-expanding $\sin(qr)$:

$$
\begin{aligned}
F_h(q^2) &\approx 4\pi \int_0^\infty r^2 dr \rho(r)\left[1 - q^2r^2/3! + q^4r^4/5! - \cdots\right]\\
&\approx Q_h - q^2\langle r_{\text{EM}}^2\rangle_h/3!. \tag{3.163}
\end{aligned}
$$

The mean-squared charge radius of the hadron has been defined in a natural way as

$$\langle r_{\text{EM}}^2\rangle_h = \int d^3r \rho(r) r^2. \tag{3.164}$$

(Sometimes a factor of Q_h is divided out of F_h and $\langle r_{\text{EM}}^2\rangle_h$. This is an unappealing convention for neutral particles.)

It is frequently assumed [e.g., Chou and Yang 1968] that the charge distribution within a hadron is representative of the matter distribution.

This is a simple, but not unavoidable assumption. It appears [Amaldi et al. 1976] not to be grossly misleading.

The definitions (3.163), (3.164) show that a measurement of the slope of the form factor as a function of q^2 yields a determination of the charge radius as

$$\langle r_{EM}^2 \rangle_h = -6 \left. \frac{\partial F_h(q^2)}{\partial q^2} \right|_{q^2=0}.$$
(3.165)

Experiments to measure the electromagnetic form factors are of three basic types.

The form factors of stable targets, which is to say nucleons and nuclei, are measured directly by the scattering of high energy electron beams. The classic experiments have been reviewed by Hofstadter (1963). For spin-1/2 particles, there are of course two independent form factors, which may be conveniently defined as electric and magnetic. Useful summaries are given by Weber (1967) and Rutherglen (1969). We have already failed in section 2.3 to explain the charge distribution within the neutron, and again promise enlightenment only in section 5.3.

To determine the form factors of unstable particles it has until recently been necessary to rely upon indirect means, analogous to the Goebel (1958)–Chew and Low (1959) extrapolation method. To study the interactions of hadrons with the pion, it is useful to regard the nucleon as surrounded by a pion cloud, and to treat the virtual pions in the cloud as target particles. This is illustrated in fig. 18a for the case of $\pi\pi$ scattering, which is studied in the reaction

$$\pi N \rightarrow \pi\pi N'.$$
(3.166)

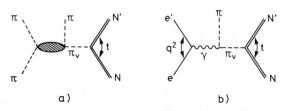

a)   b)

Fig. 18. Method of pole extrapolation for studying the interactions of unstable target particles. (a) Pion scattering from a virtual pion. (b) Measurement of the electromagnetic form factor of a virtual pion. Additional diagrams are required to contribute to ensure gauge invariance.

The momentum transfer $t \equiv (p_N - p_{N'})^2$ between initial and final nucleons gives a measure of the mass of the virtual pion. The proposal of Goebel, Chew and Low is to measure the properties of $\pi\pi$ scattering as a function of the virtual pion mass and to determine by extrapolating to the pion pole the properties of physical $\pi\pi$ scattering. Despite numerous technical complications, this suggestion has been enormously fruitful. It played an important part in the development of the peripheral exchange picture [e.g., Jackson 1966, Fox and Quigg 1973) and created an industry devoted to the study of meson-meson scattering [e.g., Williams and Hagopian 1973, Estabrooks 1977]. In similar fashion, one may study single pion electroproduction ·

$$eN \rightarrow e\pi N', \tag{3.167}$$

for which the pion pole contribution is depicted in fig. 18(b). Typical of attempts to measure the pion form factor using this technique is the work of Mistretta et al. (1969).

Until the advent of high-energy pion beams, electroproduction was the only method available for the measurement of the pion form factor in the spacelike region. Why high energies make a difference can be seen from an elementary kinematical argument. In pion scattering from a stationary (i.e. atomic) electron,

$$s \approx 2m_e p_\pi \tag{3.168}$$

is kept small by the smallness of the electron mass. The maximum momentum transfer is characterized by

$$q^2_{max} \approx s. \tag{3.169}$$

If the charge radius is characterized by a typical hadronic dimension, perhaps on the order of 1 GeV, demanding that $q^2_{max}\langle r^2_{EM} \rangle \sim 0.1$ imposes the requirement that $p_\pi \sim 100$ GeV/c. With the availability of such beams, the direct study of reactions such as

$$\pi e \rightarrow \pi e \tag{3.170}$$

has become possible. The experimental results are summarized in table 9.

Two interpretations of these data are instructive. First, let us consider the consequences of SU(3) symmetry and vector meson dominance. We assume that the photon couples to hadrons through the ideally mixed vector mesons ρ^0, ω, and φ, and that the photon transforms as a U-spin

Table 9
Electromagnetic charge radii of mesons

Particle	Beam momentum (GeV/c)	$\langle r_{EM}^2 \rangle$ (fm^2)	Reference
π^-	100	0.31 ± 0.04	Dally et al. (1977)
	250	0.43 ± 0.03	Dally et al. (1980a)
	Combined fit	0.39 ± 0.04	Dally et al. (1980a)
K^-	250	0.28 ± 0.05	Tsyganov (1979), Dally et al. (1980c)
	π/K Comparison	0.25 ± 0.05	Dally et al. (1980b)
K^0	30–100	-0.054 ± 0.026	Molzon et al. (1978)

singlet, so that

$$|\gamma\rangle \sim \frac{3|\rho^0\rangle + |\omega\rangle - \sqrt{2}|\varphi\rangle}{\sqrt{12}} \tag{3.171}$$

Thus the pion form factor will be controlled only by the ρ^0, by G-parity, but the kaon will be influenced by ρ^0, ω, and φ. An elementary SU(3) calculation gives

$$F_{\pi^+}(q^2) = m_\rho^2 / (q^2 + m_\rho^2), \tag{3.172}$$

$$F_{K^+}(q^2) = \frac{1}{2} \frac{m_\rho^2}{q^2 + m_\rho^2} + \frac{1}{6} \frac{m_\omega^2}{q^2 + m_\omega^2} + \frac{1}{3} \frac{m_\varphi^2}{q^2 + m_\varphi^2}, \tag{3.173}$$

$$F_{K^0}(q^2) = -\frac{1}{2} \frac{m_\rho^2}{q^2 + m_\rho^2} + \frac{1}{6} \frac{m_\omega^2}{q^2 + m_\omega^2} + \frac{1}{3} \frac{m_\varphi^2}{q^2 + m_\varphi^2}. \tag{3.174}$$

The definition (3.165) of the mean-squared charge radius then yields

$$\langle r_{EM}^2 \rangle_{\pi^+} = 6/m_\rho^2 = 0.388 \text{ fm}^2, \tag{3.175}$$

$$\langle r_{EM}^2 \rangle_{K^+} = 3/m_\rho^2 + 1/m_\omega^2 + 2/m_\varphi^2 = 0.332 \text{ fm}^2, \tag{3.176}$$

$$\langle r_{EM}^2 \rangle_{K^0} = -3/m_\rho^2 + 1/m_\omega^2 + 2/m_\varphi^2 = -0.055 \text{ fm}^2. \tag{3.177}$$

Comparing with the experimental results in table 9, we find quite good agreement, though it must be said that the data are perhaps still not definitive.

A second kind of analysis involves the explicit use of constituents. This has been a popular approach since the early days of the quark model [see Gerasimov 1966]. Recent work [Greenberg et al. 1977, Isgur 1978] has been concentrated on the neutral kaon. In a nonrelativistic description of mesons, we have

$$\langle r_{EM}^2 \rangle = \langle \sum_{i=q,\bar{q}} e_i (r_i - R)^2 \rangle, \tag{3.178}$$

where r_i is the coordinate of the ith constituent and

$$R = (m_q r_q + m_{\bar{q}} r_{\bar{q}})/(m_q + m_{\bar{q}}) \tag{3.179}$$

is the CM coordinate. We may also define the relative coordinate

$$\rho = r_q - r_{\bar{q}}. \tag{3.180}$$

With these definitions, the mean-squared charge radius is

$$\langle r_{EM}^2 \rangle_h = \frac{1}{(m_q + m_{\bar{q}})^2} \left(e_q m_{\bar{q}}^2 + e_{\bar{q}} m_q^2 \right) \langle \rho^2 \rangle_h. \tag{3.181}$$

For the cases of interest, we then have

$$\langle r_{EM}^2 \rangle_{\pi^+} = \frac{(2m_d^2 + m_u^2)}{3(m_u + m_d)^2} \langle \rho^2 \rangle_\pi, \tag{3.182}$$

$$\langle r_{EM}^2 \rangle_{K^+} = \frac{(2m_s^2 + m_u^2)}{3(m_u + m_s)^2} \langle \rho^2 \rangle_K, \tag{3.183}$$

$$\langle r_{EM}^2 \rangle_{K^0} = \frac{(m_d^2 - m_s^2)}{3(m_d + m_s)^2} \langle \rho^2 \rangle_K. \tag{3.184}$$

We see at once that the K^0 charge radius is negative because the planetary d quark, negatively charged, orbits the more massive s quark. The numerically successful vector dominance calculation did not give such a transparent explanation. The ratio of charged and neutral kaon charge radii is predicted to be

$$\langle r_{EM}^2 \rangle_{K^0}/\langle r_{EM}^2 \rangle_{K^+} = \frac{m_d^2 - m_s^2}{2m_s^2 + m_u^2}. \tag{3.185}$$

In the approximation that $m_\rho = m_\omega$, the vector dominance result is

$$\langle r_{EM}^2 \rangle_{K^0}/\langle r_{EM}^2 \rangle_{K^+} = \frac{m_\rho^2 - m_\varphi^2}{2m_\varphi^2 + m_\rho^2} = -0.16. \tag{3.186}$$

If we set $m_s = m_\varphi/2$ and $m_u = m_\rho/2$, which is consistent with our earlier estimates, the quark model result coincides exactly. Both models thus agree with the experimental result.

The quark model in elementary form is less definite about the comparison of π and K form factors:

$$\langle r_{EM}^2\rangle_{K^+}/\langle r_{EM}^2\rangle_{\pi^+} = \frac{2m_s^2 + m_u^2}{(m_u + m_s)^2} \cdot \frac{(m_u + m_d)^2}{(2m_d^2 + m_u^2)} \frac{\langle\rho^2\rangle_K}{\langle\rho^2\rangle_\pi} \qquad (3.187)$$

$$\approx \frac{2m_s^2 + m_u^2}{(m_u + m_s)^2} \frac{4}{3} \cdot \frac{\langle\rho^2\rangle_K}{\langle\rho^2\rangle_\pi}. \qquad (3.188)$$

With $m_s = m_\varphi/2$ and $m_u = m_\rho/2$, this implies

$$\langle r_{EM}^2\rangle_{K^+}/\langle r_{EM}^2\rangle_{\pi^+} \approx 1.11\langle\rho^2\rangle_K/\langle\rho^2\rangle_\pi. \qquad (3.189)$$

It is natural to expect that $\langle\rho^2\rangle_K < \langle\rho^2\rangle_\pi$, but we require a more specific model to predict the difference. The data imply

$$\langle\rho^2\rangle_{K^+}/\langle\rho^2\rangle_{\pi^+} \approx 2/3, \qquad (3.190)$$

which is not unreasonable.

Thus, we see that the quark model gives an immediate understanding of the sign of the neutral kaon charge radius, and properly predicts the ratio of charged and neutral kaon form factors. It is poor for observables that rely on some dynamics, such as absolute sizes and π/K ratios. We may nevertheless ask what hadron sizes, as represented by $\langle\rho^2\rangle$, are required to describe the data. The results are

$$\langle\rho^2\rangle_\pi^{1/2} \approx 2\langle r_{EM}^2\rangle_\pi^{1/2} = 1.25 \pm 0.06 \text{ fm}, \qquad (3.191)$$

$$\langle\rho^2\rangle_{K^+}^{1/2} \approx 1.9\langle r_{EM}^2\rangle_{K^+}^{1/2} = 1.00 \pm 0.09 \text{ fm}, \qquad (3.192)$$

$$\langle\rho^2\rangle_{K^0}^{1/2} \approx 4.7\langle r_{EM}^2\rangle_{K^0}^{1/2} = 1.09 \pm 0.26 \text{ fm}. \qquad (3.193)$$

which are again reasonable hadronic dimensions. Thus the confined fermion approach of the MIT Bag Model can be expected to yield sensible absolute sizes for the charge radii. The results presented by DeGrand et al. (1975) are somewhat smaller than current measurements require.

4. Orbitally excited hadrons

In the preceding long yet incomplete section, the quark model has been seen to give a creditable account of the mesons and baryons in the SU(6) ground state: $L = 0$ configurations of (q$\bar{\text{q}}$) for mesons and (qqq) for baryons. In this section, two things are done. First, the spectra of excited mesons and baryons will be shown to match the expectations of the quark model with orbital excitations. This will by necessity not be carried out in infinite detail. For the rich microstructure, I refer the reader to the review articles by Protopopescu and Samios (1979), Rosner (1974a), and Samios et al. (1974); to the recent conference talks by Close (1981), Hey (1979), and Montanet (1980); to the proceedings of the latest spectroscopy conferences [Chung 1980, Isgur 1980b], and to ch. 5 of Close (1979). The consonance between quark-model predictions and experimental observations strongly motivates a serious consideration of the quark model as a basis for hadron spectroscopy.

If the quark model is to be taken seriously, it must be made free from internal inconsistencies. Thus the problem of the exclusion principle (i.e. of symmetric fermion wavefunctions), which has been held in abeyance since section 2.2, must be faced. It is now seen to exist not only for the ground state, but for the excited states as well. This calls for action; the action taken will be the introduction of *color*, a degree of freedom not directly observed. A formulation of the color hypothesis and a brief résumé of the evidence for color will make up the second principal topic of this section.

4.1. Mesons

With respect to the flavor–spin symmetry SU(6), the (q$\bar{\text{q}}$) meson states lie in the

$$6 \otimes 6^* = 1 \oplus 35 \tag{4.1}$$

dimensional representations. In the useful notation of Young Tableaux, the arithmetic is

$$\square \otimes \begin{array}{c}\square\\\square\\\square\\\square\\\square\end{array} = \begin{array}{c}\square\\\square\\\square\\\square\\\square\end{array} \oplus \begin{array}{cc}\square&\square\\\square&\\\square&\\\square&\\\square&\end{array} \; . \tag{4.2}$$

Decomposed with respect to $SU(3)_{\text{flavor}}$ and $SU(2)_{\text{spin}}$, the quark and antiquark states are

$$6 = \{[3], (2)\}, \tag{4.3a}$$

$$6^* = \{[3^*], (2)\}, \tag{4.3b}$$

where we follow our earlier practice (section 3.1) of denoting SU(3) representations by square brackets and SU(2) representations by parentheses. The quark–antiquark product (4.1) is therefore

$$\mathbf{6} \otimes \mathbf{6}^* = \{[\mathbf{1}],(\mathbf{1})\} \oplus \{[\mathbf{1}],(\mathbf{3})\} \oplus \{[\mathbf{8}],(\mathbf{1})\} \oplus \{[\mathbf{8}],(\mathbf{3})\}. \tag{4.4}$$

Consequently the SU(6) **35** consists of a spin-triplet nonet (which for the $L = 0$ ground state is simply the vector mesons $\rho, \omega, K^*, \varphi$) and a spin-singlet octet (π, η_8, K). The SU(6) singlet ground state is to be identified with the η_1.

For orbital excitations of the $(q\bar{q})$ systems, it suffices to observe (compare problem 1) that the discrete quantum numbers are

$$P(q\bar{q})_L = (-1)^{L+1}, \tag{4.5}$$

$$C(q\bar{q})_L = (-1)^{L+s}, \tag{4.6}$$

which are of course consistent with the identification of pseudoscalar and vector mesons with the $(q\bar{q})$ ground state. The excited states through $L = 3$ are listed in table 10, together with the observed meson resonances with which they are identified. Except as otherwise noted, the experimental results are taken from Particle Data Group (1980). For the isoscalar mesons, the first column contains the dominantly $u\bar{u} + d\bar{d}$ states and the second contains the dominantly $s\bar{s}$ states. In several cases $(^1S_0, {}^3P_0, {}^3D_1)$ the amount of mixing is uncertain, and so too are the $I = 0$ assignments. Although numerous openings remain unfilled, the general multiplet structure is quite nicely confirmed by experiment.

Several comments are in order:

(i) The sequence $J^P = 0^+, 1^-, 2^+ \ldots$ for which $P = (-1)^J$ is called natural parity. The $P = -(-1)^J$ sequence of $J^P = 0^-, 1^+, 2^- \ldots$ is known as unnatural parity.

(ii) Some combinations of J^{PC} cannot occur in the $(q\bar{q})$ picture. The state with $J^{PC} = 0^{--}$ would require $L = s$, to arrive at $J = 0$, but that implies by (4.6) that $C = +1$. In addition, the sequence $J^{PC} = 0^{+-}, 1^{-+}, 2^{+-}, \ldots$ must have $s = 0$, in order that $CP = -1$. This means that J must be equal to L, but then by (4.5) the parity must be unnatural. Hence the sequence cannot be reached in $(q\bar{q})$. These "C-exotic states" would be good signatures for mesons that cannot be accomodated in the $(q\bar{q})$ scheme. None has yet been observed.

(iii) Mixing among the strange particle states with the same J^P (e.g. $^1P_1, {}^3P_1$ or $^1D_2, {}^3D_2$) is not forbidden by the discrete symmetries C or G. Only SU(3) symmetry breaking is required for the mixing to occur. This

Table 10
Light mesons as quark–antiquark bound states

State	Mixing?	J^{PC}	$I=1$	$I=0$		$I=1/2$
$^1S_0$		0^{-+}	$\pi(140)$	$\eta(549)$	$\eta'(958)$	K(496)
$^3S_1$	$^3D_1$	1^{--}	$\rho(776)$	$\omega(784)$	$\varphi(1019)$	K*(892)
$^1P_1$		1^{+-}	B(1231)	H(1190)[a]		$Q_B(1355)$[b]
$^3P_0$		0^{++}	$\delta(981)$	$\varepsilon(1300)$?[h]	S*(980)	$\kappa(1500)$?
$^3P_1$		1^{++}	$A_1(1240)$[a]	D(1285)	E(1418)	$Q_A(1340)$[b]
$^3P_2$	$^3F_2$	2^{++}	$A_2(1317)$	$f^0(1273)$	f*(1516)	K**(1430)
$^1D_2$		2^{-+}	$A_3(1660)$			L(1765)?
$^3D_1$	$^3S_1$	1^{--}	$\rho'(1600)$		$\varphi'(1634)$	K*(1650)?
$^3D_2$		2^{--}				
$^3D_3$	$^3G_3$	3^{--}	g(1700)	$\omega(1670)$	$\varphi_3(1870)$[c]	K*(1753)[d]
$^1F_3$		3^{+-}				
$^3F_2$	$^3P_2$	2^{++}	$f\pi(1700)$[e]		$\theta(1640)$[g]?	
$^3F_3$		3^{++}				
$^3F_4$		4^{++}	$K^+K_s(2060)$[f]	h(2040)		K*(2070)[d]

[a] Dankowych et al. (1981).
[b] Leith (1977).
[c] Armstrong et al. (1982).
[d] Aston et al. (1981), Cleland et al. (1980b), Dorsaz (1981).
[e] Cashmore (1980), see also Montanet (1980).
[f] Cleland et al. (1980a), see also Montanet (1980).
[g] Seen in $\psi \to \gamma\eta\eta$; Scharre (1981).
[h] According to Wicklund et al. (1980), the pole lies at 1425 MeV/c^2.

phenomenon is observed in the axial strange particles, in which the physical resonances $Q_1(1280)$ and $Q_2(1400)$ are mixtures of the quark-model states Q_A and Q_B [Leith 1977].

(iv) Radial excitations are also possible, and many are observed in the heavy meson systems known as quarkonium (cf. section 6). It is likely that $\rho'(1600)$ is, or is mixed with, a 2^3S_1 radial excitation of $\rho(776)$, because it is seen prominently in $e^+e^- \to$ hadrons, which suggests a nonvanishing wavefunction at the origin. A $^3D_1$ state would be coupled only weakly to e^+e^-. A similar statement can be made for $\varphi(1634)$ as a radially excited $\varphi(1019)$. A candidate for a radial excitation of the pion, $\pi'(1342) \to \varepsilon\pi$ has recently been reported by Bonesini et al. (1981).

(v) The hyperfine splitting so prominent between the pseudoscalar and vector states is less apparent for the L-excitations. This seems consistent with an elementary picture of hyperfine structure.

(vi) The natural-parity states, including $I = 1$ candidates with $J^P = 5^-$ at 2300 MeV/c^2 [Cashmore 1980] and $J^P = 6^+$ at 2515 MeV/c^2 [Cleland et al. 1980a], lie on linear Regge trajectories of the form

$$J(M^2) = \alpha_0 + \alpha' M^2, \tag{4.7}$$

as shown in fig. 19. The data suggest [e.g., Field and Quigg 1975] the possibility that Regge slopes are not universal, but depend systematically

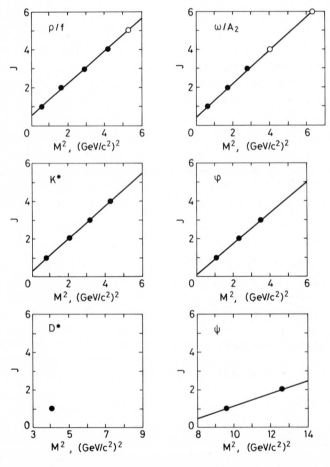

Fig. 19. Regge trajectories of the natural-parity mesons. Uncertain states are indicated by open circles.

upon the constituent quark masses, as

$$\alpha'_\rho \rangle \alpha'_{K*} \rangle \alpha'_\varphi \rangle \alpha'_{D*} \rangle \alpha'_\psi. \tag{4.8}$$

If this trend is not an illusion, it is not obvious whether it is transitory [Igi 1977] or deeply connected with the flavor-independence of inter-quark forces [Close 1981]. For a summary of Regge Pole phenomenology, see Collins (1977).

(vii) The scalar mesons do not present an unambiguous picture. While their quantum numbers are those of the $(q\bar{q})$ scheme it is problematic to correlate their masses and decays. Mixing with the vacuum may be pronounced, and may thus be a complicating factor. Jaffe (1977a) has argued that the scalars are in fact $(qq\bar{q}\bar{q})$ states, but at the time of his proposal the s-wave isoscalar state was generally believed to occur at around 700 MeV$/c^2$. Present fashion favors 1300 MeV$/c^2$ and does not as neatly fit the original Jaffe scheme. For additional developments, see Jaffe and Low (1979) and Jaffe (1979).

4.2. Baryons

We now turn to a similarly sketchy review of the baryon resonances. To discuss the SU(6) classification it is necessary to expand the product $\mathbf{6} \otimes \mathbf{6} \otimes \mathbf{6}$. As in problem 5 for flavorspin SU(4), this will be done explicitly, and in stages. From two quarks, the representations

$$\mathbf{6} \otimes \mathbf{6} = \mathbf{15} \oplus \mathbf{21} \tag{4.9}$$

may be formed. In Young tableaux, this is

$$\square \otimes \square = \begin{array}{c}\square\\\square\end{array} \oplus \square\square . \tag{4.10}$$

By decomposing $\mathbf{6} \otimes \mathbf{6}$ under SU(3)$_{\text{flavor}}$ and SU(2)$_{\text{spin}}$,

$$\{[3],(2)\} \otimes \{[3],(2)\}$$
$$= \{[6],(3)\} \oplus \{[6],(1)\} \oplus \{[3^*],(3)\} \oplus \{[3^*],(1)\}, \tag{4.11}$$

we may identify

$$\mathbf{15} = \{[6],(1)\} \oplus \{[3^*],(3)\}, \tag{4.12}$$
$$\mathbf{21} = \{[6],(3)\} \oplus \{[3^*],(1)\}, \tag{4.13}$$

which reflect the symmetry of the Young tableaux.

The product

$$6 \otimes 15 = 20 \oplus 70, \qquad (4.14)$$

represented as

$$\square \otimes \begin{array}{c}\boxminus\end{array} = \begin{array}{c}\square\end{array} \oplus \begin{array}{c}\square\end{array} , \qquad (4.15)$$

is expanded as

$$\{[3],(2)\} \otimes \{[6],(1)\} = \{[10],(2)\} \oplus \{[8],(2)\} \qquad (4.16)$$

and

$$\{[3],(2)\} \otimes \{[3^*],(3)\}$$
$$= \{[8],(4)\} \oplus \{[8],(2)\} \oplus \{[1],(4)\} \oplus \{[1],(2)\}. \qquad (4.17)$$

This permits the identifications

$$20 = \{[8],(2)\} \oplus \{[1],(4)\} \qquad (4.18)$$

and

$$70 = \{[10],(2)\} \oplus \{[8],(4)\} \oplus \{[8],(2)\} \oplus \{[1],(2)\}. \qquad (4.19)$$

Similarly the product

$$6 \otimes 21 = 70 \oplus 56 \qquad (4.20)$$

or

$$\square \otimes \square\square = \begin{array}{c}\square\end{array} \oplus \begin{array}{c}\square\square\square\end{array} \qquad (4.21)$$

may be expanded as

$$\{[3],(2)\} \otimes \{[6],(3)\}$$
$$= \{[10],(4)\} \oplus \{[10],(2)\} \oplus \{[8],(4)\} \oplus \{[8],(2)\}, \qquad (4.22)$$

and

$$\{[3],(2)\} \otimes \{[3^*],(1)\} = \{[8],(2)\} \oplus \{[1],(2)\}, \qquad (4.23)$$

from which we conclude that

$$56 = \{[10],(4)\} \oplus \{[8],(2)\}. \qquad (4.24)$$

Three quarks thus lead to the SU(6) representations

$$\mathbf{6} \otimes \mathbf{6} \otimes \mathbf{6} = \mathbf{20} \oplus \mathbf{70} \oplus \mathbf{70} \oplus \mathbf{56}, \qquad (4.25)$$

which is written symbolically as

$$\square \otimes \square \otimes \square = \begin{smallmatrix}\square\\\square\\\square\end{smallmatrix} \oplus \begin{smallmatrix}\square\square\\\square\end{smallmatrix} \oplus \begin{smallmatrix}\square\square\\\square\end{smallmatrix} \oplus \square\square\square \,. \qquad (4.26)$$

It was found in section 3 that the symmetric state **56** describes the ground state baryons. The decomposition (4.24) shows immediately that in the absence of any orbital angular momentum, the **56** contains a spin-1/2 octet and a spin-3/2 quartet, as the explicit construction of section 3.1.2 illustrated. Consequently, it is natural to suppose [Greenberg 1964, Dalitz 1965, Greenberg and Resnikoff 1967], that excited states are given by orbital (or radial) excitations which lead to wavefunctions symmetric under the interchange of any two particles. The flavor–spin [or SU(6)] part of the wavefunction and the spatial part of the wavefunction must therefore have matching symmetry properties.

The permutation group for the three objects has been investigated in problem 4. For three distinct objects (such as quarks numbered 1, 2, 3), three representations can be constructed. One of these (S) is symmetric under the interchange of any pair. Another (A) is antisymmetric under the interchange of any pair. The last (M) is of mixed symmetry, like the three-quark isospinor wavefunctions of problem 4. Three-quark wavefunctions with definite permutation symmetry and orbital angular momentum were constructed by Karl and Obryk (1968) for equal-mass quarks. A radial wavefunction is characterized by a polynomial of degree N with specified permutation and rotation (i.e. angular momentum) symmetries times a smooth, symmetric, scalar function of the quark coordinates.

The Karl–Obryk classification of radial wavefunctions has been put to use in constructing fig. 20, which displays the SU(6) baryon multiplets corresponding to particular values of the degree N, which governs the parity of the state, and angular momentum L. An analogy with atomic physics suggests that baryon masses should be increasing functions of N and L.

Many baryon resonances have been identified as members of these multiplets on the grounds of their masses and their production and decay characteristics. The populations of the first few multiplets are indicated in table 11, which is based on a similar compilation in Rosner (1981a), supplemented by recent information [Gopal 1980, Kelly 1980, Kinson

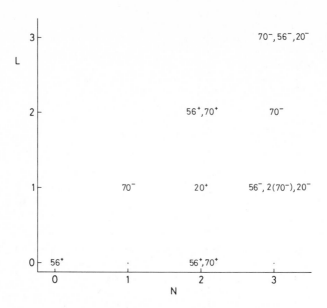

Fig. 20. Expected SU(6) multiplets of baryons.

1980, Montanet 1980]. The ground-state $\mathbf{56}_{0^+}$ is of course completely filled. All of the nucleons and many of the $S = -1$ hyperons are known for the $\mathbf{70}_{1^-}$. The number of known resonances diminishes rapidly with increasing strangeness. This is undoubtely a consequence of the difficulties in producing hyperons, which must couple to the $\overline{\mathrm{K}}\mathrm{N}$ entrance channel in formation experiments, and to the general shortcomings of production experiments. There is good reason to hope [Quigg and Rosner, 1976] that high-energy hyperon beams may fill in the gaps, particularly for Ξ-spectroscopy, which seems perpetually in its infancy. [See Biagi et al. 1981.]

Absent from table 11, because no candidates are known, is the $\mathbf{20}_{1^+}$. There is a specific excuse, in this case, because the $\mathbf{20}$ cannot be reached in meson–baryon scattering. The product

$$\mathbf{35} \otimes \mathbf{56} = \mathbf{56} \oplus \mathbf{70} \oplus \mathbf{700} \oplus \mathbf{1134} \tag{4.27}$$

does not contain $\mathbf{20}$.

There is some evidence for the other $N = 2$ families. Most of the nucleon resonances are known for the $\mathbf{56}_{0^+}$ and $\mathbf{56}_{2^+}$ states, and a few

Table 11
SU(6) classification of the baryon resonances

N	$SU(6)_{L^P}$	$^{(2J+1)}SU(3)$	Members
0	$\mathbf{56}_{0^+}$	$^2[8]$	$N(939), \Lambda(1115), \Sigma(1193), \Xi(1318)$
		$^4[10]$	$\Delta(1232), \Sigma(1385), \Xi(1533), \Omega(1672)$
1	$\mathbf{70}_{1^-}$	$^2[1]$	$\Lambda(1405)$
		$^4[1]$	$\Lambda(1520)$
		$^2[8]$	$N(1535), \Lambda(1670), \Sigma(1750), \Xi(1684)?$
		$^2[8]$	$N(1700), \Lambda(1870)$
		$^4[8]$	$N(1520), \Lambda(1690), \Sigma(1670), \Xi(1820)?$
		$^4[8]$	$N(1700), \qquad \Sigma(1940)?$
		$^6[8]$	$N(1670), \Lambda(1830), \Sigma(1765)$
		$^2[10]$	$\Delta(1650)$
		$^4[10]$	$\Delta(1670)$
2	$\mathbf{56}_{2^+}$	$^4[8]$	$N(1810), \Lambda(1860)$
		$^6[8]$	$N(1688), \Lambda(1815), \Sigma(1915), \Xi(2030)?$
		$^2[10]$	$\Delta(1910)$
		$^4[10]$	
		$^6[10]$	$\Delta(1890)$
		$^8[10]$	$\Delta(1950), \Sigma(2030)$
	$\mathbf{56}_{0^+}$	$^2[8]$	$N(1470), \qquad \Sigma(1660)$
		$^4[10]$	$\Delta(1690)$

candidates can be attributed to the $\mathbf{70}_{0^+}$ and $\mathbf{70}_{2^+}$ multiplets [Hey, 1980a]. Like the natural parity mesons, the $I = 1/2$ and $I = 3/2$ nucleon resonances and the Λ states appear to lie on linear Regge trajectories, shown in fig. 21.

4.3. Color at last

All spectroscopic evidence thus points to the utility of the symmetric quark model as a classification scheme and as a model for the static and dynamical properties of hadrons. Faced with the inconsistency between the model as formulated and the Pauli principle, one has three possible courses of action:

(i) Dismiss the successes of the quark model as coincidence or illusion.

(ii) Deny that the spin and statistics connection applies to confined particles.

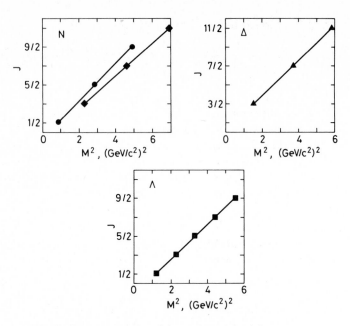

Fig. 21. Regge trajectories of the nucleon, Δ, and Λ resonances.

(iii) Envisage a hitherto unnoticed degree of freedom, in terms of which the wavefunction can be antisymmetrized. The third course was advocated by Greenberg (1964). Although it seems at first sight arbitrary and extravagant, it has gained experimental support and now forms the foundation of our understanding of the strong interactions.

To reconcile the quark model with the Pauli principle it is necessary that each quark flavor exist in no less than three distinct varieties, so that three-quark wavefunctions can be antisymmetrized. The number of varieties must in fact be precisely three. If there were more, several distinguishable varieties of proton could exist, in violation of common experience and the experimental evidence on specific heats. Other evidence for three species will be produced below. We shall refer to the new degree of freedom as *color* and label the three colors as

$$C_1 = \text{red} = R,$$

$$C_2 = \text{blue} = B,$$

$$C_3 = \text{green} = G. \tag{4.28}$$

If each of the explicit baryon wavefunctions is multiplied by

$$\frac{1}{\sqrt{6}} \varepsilon_{ijk} C_i C_j C_k = \frac{1}{\sqrt{6}} [RBG - RGB + GRB - GBR + BGR - BRG],$$

$$(4.29)$$

where ε_{ijk} is the antisymmetric three-index symbol (the elements of the antisymmetric representation of the permutation group), the resulting wavefunctions will be antisymmetric with respect to the interchange of any pair of quarks. As an example, consider the Δ^{++} with maximal spin projection, which was in the colorless formulation the most symmetric of states. Its wavefunction is now

$$|\Delta^{++}; s_z = 3/2\rangle = (uuu)(\uparrow \uparrow \uparrow) \varepsilon_{ijk} C_i C_j C_k / \sqrt{6} . \quad (4.30)$$

It is useful to consider the idea of a color symmetry group (represented by 3×3 matrices $\mathfrak{M}_{ij}$) which generates the color transformations

$$C_i \rightarrow C_i' \equiv \mathfrak{M}_{ij} C_j. \quad (4.31)$$

Under such a transformation, the antisymmetric combination (4.29) becomes:

$$\varepsilon_{ijk} C_i' C_j' C_k' = \varepsilon_{ijk} \mathfrak{M}_{im} \mathfrak{M}_{jn} \mathfrak{M}_{kp} C_m C_n C_p. \quad (4.32)$$

If the transformation has unit determinant,

$$|\mathfrak{M}| = 1, \quad (4.33)$$

so that

$$\varepsilon_{ijk} \mathfrak{M}_{im} \mathfrak{M}_{jn} \mathfrak{M}_{kp} = \varepsilon_{mnp}, \quad (4.34)$$

then

$$\varepsilon_{ijk} C_i' C_j' C_k' = \varepsilon_{mnp} C_m C_n C_p. \quad (4.35)$$

In other words, the color wavefunction is invariant under color transformations, which is to say that the wavefunction is a *color singlet*. This would not be the case, had there been more than three colors allowed.

For meson wavefunctions, we assign anticolors to the antiquarks, and construct the color singlet wavefunctions

$$\frac{1}{\sqrt{3}} C_i \bar{C}_i = \frac{1}{\sqrt{3}} [R\bar{R} + B\bar{B} + G\bar{G}]. \quad (4.36)$$

Depending upon the choice of the color symmetry group, the colors and anticolors may or may not be equivalent. The different implications of the two choices will be commented upon below.

The conclusion of this construction is that a formally consistent picture of mesons and baryons can be obtained by requiring that hadrons be color singlets. What then is the evidence for three quark colors? A thorough review has been given by Greenberg and Nelson (1977); see also Bardeen et al. (1973). I will therefore be brief.

Apart from the baryon spin and statistics connection, there are two basic observables that rule in favor of the three-color hypothesis. The first category includes the ratio

$$R \equiv \sigma(e^+e^- \to \text{hadrons})/\sigma(e^+e^- \to \mu^+\mu^-) \tag{4.37}$$

and related quantities. In the quark–parton model [Feynman 1972], this ratio is given by

$$R = \sum_{\substack{\text{quark} \\ \text{species}}} e_i^2 = N_c \sum_i e_i^2, \tag{4.38}$$

where N_c is the number of colors, and the sum runs over the energetically accessible flavors. Thus we predict

$$R(\text{no color}) = 2/3, 10/9, 11/9, \tag{4.39a}$$

and

$$R(\text{color}) = 2, 10/3, 11/3 \tag{4.39b}$$

for c.m. energies below the charm, bottom, and top quark thresholds, respectively. The experimental results are shown in fig. 22 for energies below (a) and above (b) charm threshold. The three-color predictions are decisively favored. Notice that within the framework of quantum chromodynamics, the ratio R receives strong-interaction corrections which are small, positive, and decreasing with c.m. energy.

A related observable is the branching ratio for semileptonic τ decays, which is governed by the branching ratios of the virtual W in the transition $\tau \to \nu_\tau W$. An elementary calculation then leads to

$$B_\ell \equiv \frac{\Gamma(\tau \to \ell \bar{\nu}_\ell \nu_\tau)}{\Gamma(\tau \to \text{all})} = \frac{1}{(1+1+N_c)} = \begin{cases} 1/3, & N_c = 1, \\ 1/5, & N_c = 3. \end{cases} \tag{4.40}$$

The experimental result [Particle Data Group 1980],

$$B_\ell = 17.44 \pm 0.85\%, \tag{4.41}$$

is in accord with the color hypothesis. A refined theoretical estimate

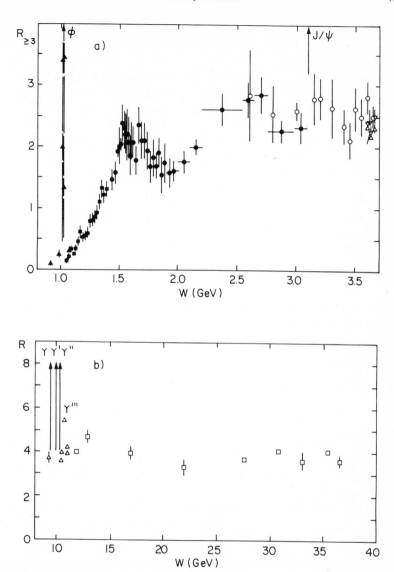

Fig. 22. The ratio $R = \sigma(e^+e^- \to \text{hadrons})/\sigma(e^+e^- \to \mu^+\mu^-)$ compared with the predictions of the quark–parton model. (a) $W < 4$ GeV [after Spinetti 1979]; (b) 8 GeV $< W <$ 40 GeV [after Schamberger 1981].

[Gilman and Miller 1978, Kawamoto and Sanda 1978] of the branching ratio, $B_\ell = 17.75\%$, is in excellent agreement with experiment. It is hoped that direct measurements of the total widths and branching ratios of the intermediate bosons $W^\pm$ and Z^0 will soon be available from high-energy colliders.

The other important measure of the number of quark colors is the $\pi^0 \to \gamma\gamma$ decay rate. A good discussion of the theoretical details is given by Llewellyn Smith (1980). A calculation for π^0 decay via a quark-anti-quark loop is straightforward to carry out, although the justification is subtle. The result is

$$\Gamma(\pi^0 \to \gamma\gamma) = \left(\frac{\alpha}{2\pi}\right)^2 \left[N_c\left(e_u^2 - e_d^2\right)\right]^2 \frac{M_\pi^3}{8\pi f_\pi^2} \tag{4.42}$$

where the pion decay constant is normalized so that

$$f_\pi = 130 \text{ MeV}. \tag{4.43}$$

The rate (4.42) is then

$$\Gamma(\pi^0 \to \gamma\gamma) = \begin{cases} 0.86 \text{ eV}, & N_c = 1, \\ 7.75 \text{ eV}, & N_c = 3, \end{cases} \tag{4.44}$$

to be compared with the measured decay rate of (7.86 ± 0.54) eV. We therefore conclude that the hidden color degree of freedom is indeed present. It is tempting then to suppose that color is what distinguishes quarks from leptons, and might play the role of a strong interaction charge.

The mention of charge calls to mind tests of the (average) electric charges of the quarks. In addition to the baryon spectrum itself, there are two checks that are easily explained. The production of massive dileptons in hadron–hadron collisions is viewed [Drell and Yan 1970, 1971] as the elementary process

$$q\bar{q} \to \gamma \to e^+e^-. \tag{4.45}$$

Consequently the reactions

$$\pi^\pm C \to \mu^+\mu^- + \text{anything}, \tag{4.46}$$

which entail dilepton production off an isoscalar target correspond to the elementary transitions

$$\pi^+: d\bar{d} \to \mu^+\mu^-, \tag{4.47a}$$

$$\pi^-: u\bar{u} \to \mu^+\mu^-. \tag{4.47b}$$

The rates for reactions (4.47) are simply proportional to the charge-squared of the annihilating quarks: 1/9 for $\bar{d}d \rightarrow \gamma$ and 4/9 for $\bar{u}u \rightarrow \gamma$. Thus it is expected that

$$\frac{\sigma(\pi^+ C \rightarrow \mu^+\mu^- + \text{anything})}{\sigma(\pi^- C \rightarrow \mu^+\mu^- + \text{anything})} = \frac{1}{4}. \tag{4.48}$$

Numerous experiments have by now confirmed this expectation [Pilcher 1980].

A second, and also model-dependent, test of the quark charges is supplied by the leptonic decay rates of the light vector mesons. As you will show in problem 13, in a nonrelativistic picture the rate for the decay $V^0 \rightarrow e^+e^-$ is given by

$$\Gamma(V^0 \rightarrow e^+e^-) = \frac{\text{const.}}{M_V^2} \cdot e_q^2 |\Psi(0)|^2, \tag{4.49}$$

where $\Psi(0)$ is the quark wavefunction at zero separation, and e_q^2 is the mean-squared charge of the quarks in V^0. For ρ^0, ω, and φ, this is

$$\langle e_q^2 \rangle_\rho = \left[\frac{1}{\sqrt{2}} \left(\frac{2}{3} + \frac{1}{3} \right) \right]^2 = 1/2, \tag{4.50a}$$

$$\langle e_q^2 \rangle_\omega = \left[\frac{1}{\sqrt{2}} \left(\frac{2}{3} - \frac{1}{3} \right) \right]^2 = 1/18, \tag{4.50b}$$

$$\langle e_q^2 \rangle_\varphi = (1/3)^2 = 1/9, \tag{4.50c}$$

where the ideally mixed wavefunctions (3.5) have been used. If $|\Psi(0)|^2$ is the same for all three of these states, the reduced leptonic widths

$$\tilde{\Gamma}_V \equiv M_V^2 \Gamma(V^0 \rightarrow e^+e^-) \tag{4.51}$$

will be in the ratio

$$\tilde{\Gamma}_\rho : \tilde{\Gamma}_\omega : \tilde{\Gamma}_\varphi :: 9 : 1 : 2. \tag{4.52}$$

This expectation compares favorably with the experimental result

$$(8.7 \pm 2.9) : 1 : (2.8 \pm 0.8). \tag{4.53}$$

Experimental results are therefore consistent with the canonical charge assignments.

Showing the uniqueness of the fractionally charged quark model is another matter, however. An alternative scheme and some possible experimental distinctions are the subject of problem 14. The most decisive evidence for fractionally charged quarks comes from an analysis of the decay rate for $\eta' \to \gamma\gamma$ [Chanowitz, 1980].

5. Some implications of QCD

In this section we begin to explore the ways in which color can be more than a mere label which serves as a convenient cure for the theoretical ills of the symmetric quark model. Color has been seen in section 4.3 to be a hidden quantum number which is manifested, albeit indirectly, in experimental observables. If hadrons must be color singlets, then the nonoccurrence of stable diquarks and other exotic configurations can be understood. In order to understand why hadrons must be color singlets, it is necessary to give color a dynamical standing. This is easily done, at least at a conceptual level, by regarding color as a local gauge symmetry. To do so, it is necessary to choose a color group and so we shall review some arguments in favor of SU(3) color.

Once the theory of colored quarks interacting by means of colored gluons, Quantum ChromoDynamics, has been formulated, it is desirable to derive its consequences. This is more easily said than done. The lectures here at Les Houches by Brower (1981) take stock of efforts to solve pure or sourceless QCD on the lattice. All such efforts have until now been restricted to configurations of pure glue, or static configurations of infinitely massive quarks. A direct computation of the hadron spectrum thus lies in the future, although considerable progress appears to have been made recently. In the absence of a complete solution to QCD it is necessary to proceed by means of schematic partial calculations and pictures abstracted from general principles. We shall first verify in the framework of a straightforward "maximally attractive channel" analysis the plausibility of the idea that color singlets are preferred configurations. Then we shall indicate how a string picture might emerge from QCD, and why it provides an appealing scheme for the spectrum of light hadrons. With those two results in hand, we shall assume that color singlets are confined and investigate QCD-inspired descriptions of the fine and hyperfine structure of hadron spectra.

Discussion of the light-hadron spectrum will be carried out at what seems to be an essentially nonrelativistic level. This is justified in the first instance by faith and subsequently by the results. Many consequences appear likely to survive the transition to a relativistic theory. Indeed, much of the arithmetic to be carried out is common to the nonrelativistic and MIT bag approaches, so I will not meticulously observe the distinctions. A simple implementation of the nonrelativistic reduction of one-gluon exchange gives a generally good account of the ground-state hadrons. It resolves some difficulties of the elementary model of section 3, notably the issue of $\Lambda - \Sigma$ splitting and the problem of the neutron charge radius. Isoscalar pseudoscalars remain a source of puzzlement, but a systematic understanding of hyperfine splittings is achieved. This success stimulates the extension of the evolving model to multiquark hadrons, lightly bound by the hyperfine interaction alone.

Extensive work has been devoted to the development of QCD-inspired pictures that may give comprehensive descriptions of the complete spectrum of light hadrons. These pictures contain much that is arbitrary, and the implications for QCD of their successes and failures are at best indirect. To the extent that they bring order and understanding to a rich spectrum of hadrons, they command our attention. The history of physics in general and of the quark model in particular contains many reminders that the simplest picture often teaches the most.

Many excellent reviews deal with aspects of QCD that are relevant to the topics in this section. In addition to the specific articles cited below, the following works contain material of general interest: Appelquist (1978), Bjorken (1980), Close (1980), Feynman (1977), Fritzsch (1979), Marciano and Pagels (1978, 1979), Morgan (1981), Quigg (1981), Rosner (1981a), and Wess (1981).

5.1. Toward QCD

Having noticed the possibility that color may function as the charge of strong interactions, it is natural to seek to formulate a dynamical theory based on color symmetry. In the present climate it is obvious that what is required is a color gauge theory. Early steps in this direction were taken by Nambu (1966). It was recognized by Greenberg and Zwanziger (1966), and later emphasized by Lipkin (1969, 1973c, 1979b), and by Fritzsch et al. (1973), that such a theory might provide a basis for understanding the simple rules that mesons are ($q\bar{q}$) states and baryons are (qqq) states. Let us see why SU(3) is a promising choice for the color symmetry group.

5.1.1. Choosing the gauge group
When we entertain the possibility that the color quantum number reflects a continuous symmetry of the strong-interaction Lagrangian, three candidates for the symmetry group come immediately to mind: SO(3),

SU(3), and U(3). Simple arguments discourage the use of SO(3) and U(3), as we shall now see.

Color SO(3) was considered long ago by Tati (1966). In SO(3) there is no difference between color and anticolor, so in the computation of forces there will be no distinction between quarks and antiquarks. The existence of (q$\bar{\text{q}}$) mesons thus implies the existence of (qq) diquarks, which will be fractionally charged. Because fractionally charged matter appears to be less commonplace than ordinary mesons, SO(3) does not seem an apt choice for the color symmetry group. One may also be concerned that the asymptotic freedom of SO(3) gauge theory is less secure that that of SU(3). Recently Slansky et al. (1981) have proposed that SU(3)$_{\text{color}}$ QCD is spontaneously broken down to an SO(3) subgroup to which they refer as "glow," liberating fractionally charged diquarks. Whether this can be accomplished in a manner compatible with experimental limits on the production of fractionally charged objects without upsetting the possibility of grand unification below the Planck mass is left as an exercise for the student.

In U(3) color gauge theory, the color singlet gauge boson which occurs in the product

$$[3] \otimes [3^*] = [1] \oplus [8], \tag{5.1}$$

in SU(3) notation, cannot be dispensed with. It would mediate long-range strong interactions between color singlet hadrons, and is thus ruled out by experiment. Thus U(3) is excluded, and we are left with SU(3) as a candidate gauge group.

5.1.2. The QCD Lagrangian

The possibility of constructing a theory based on a local gauge symmetry more complicated than the U(1) phase symmetry of electromagnetism was demonstrated by Yang and Mills (1954), [see also Shaw 1955] for the flavor-SU(2) symmetry isospin. The problem of extending the Yang–Mills construction to a general gauge symmetry was dealt with by Gell-Mann and Glashow (1961). It is, as the literature attests, one thing to write down a mathematically consistent gauge theory and quite another to choose a gauge symmetry that leads to experimentally acceptable consequences. Having sounded that note of humility, let us formulate the gauge theory of color-triplet quarks that interact by means of vector gluons which belong to the octet representation of color SU(3). The

Lagrangian will have the standard Yang–Mills form,

$$\mathcal{L} = -\tfrac{1}{4}G^a_{\mu\nu}G^{a\mu\nu} + \bar{\psi}_\alpha\left(i\gamma^\mu\mathcal{D}^{\alpha\beta}_\mu - m\delta^{\alpha\beta}\right)\psi_\beta, \tag{5.2}$$

where α and β, the color indices for the quark fields ψ, run over the values 1, 2, 3 or R, B, G and $a = 1, 2, \ldots, 8$ is the gluon color label. The field-strength tensor is given by

$$G_{\mu\nu} = (ig)^{-1}\left[\mathcal{D}_\mu, \mathcal{D}_\nu\right] = \tfrac{1}{2}\lambda^a G^a_{\mu\nu}, \tag{5.3}$$

and the gauge-covariant derivative is

$$\mathcal{D}_\mu = 1\partial_\mu - \tfrac{1}{2}ig\lambda^a B^a_\mu, \tag{5.4}$$

where B^a_μ is the color gauge field (the gluon field) and the λ^a are the eight 3×3 matrix representations of the $SU(3)_c$ octet:

$$\lambda_1 = \begin{pmatrix} 1 & 0 & 0 \\ 0 & -1 & 0 \\ 0 & 0 & 0 \end{pmatrix}\begin{matrix}\bar{R} \\ \bar{B} \\ \bar{G}\end{matrix}, \qquad \lambda_2 = \begin{pmatrix} 0 & -i & 0 \\ i & 0 & 0 \\ 0 & 0 & 0 \end{pmatrix},$$

$$\quad\;\; R \quad B \quad G$$

$$\lambda_3 = \begin{pmatrix} 1 & 0 & 0 \\ 0 & -1 & 0 \\ 0 & 0 & 0 \end{pmatrix}, \qquad \lambda_4 = \begin{pmatrix} 0 & 0 & 1 \\ 0 & 0 & 0 \\ 1 & 0 & 0 \end{pmatrix},$$

$$\lambda_5 = \begin{pmatrix} 0 & 0 & -i \\ 0 & 0 & 0 \\ i & 0 & 0 \end{pmatrix}, \qquad \lambda_6 = \begin{pmatrix} 0 & 0 & 0 \\ 0 & 0 & 1 \\ 0 & 1 & 0 \end{pmatrix},$$

$$\lambda_7 = \begin{pmatrix} 0 & 0 & 0 \\ 0 & 0 & -i \\ 0 & i & 0 \end{pmatrix}, \qquad \lambda_8 = \frac{1}{\sqrt{3}}\begin{pmatrix} 1 & 0 & 0 \\ 0 & 1 & 0 \\ 0 & 0 & -2 \end{pmatrix}. \tag{5.5}$$

These are of course the matrices familiar from flavor SU(3), for which the indices 1, 2, 3, correspond to quark flavors u, d, s. In that case the matrices $\lambda_1, \lambda_2, \lambda_3$ are proportional to the generators of $SU(2)_{\text{isospin}}$.

With the normalization adopted in eq. (5.5), the λ-matrices have a number of simple properties, including

$$\text{Tr}(\lambda^a) = 0, \tag{5.6}$$

$$\text{Tr}(\lambda^a\lambda^b) = 2\delta^{ab}, \tag{5.7}$$

and

$$[\lambda^a, \lambda^b] = 2if^{abc}\lambda^c. \tag{5.8}$$

Using (5.7) and (5.8), it is easy to compute the structure constants

$$f^{abc} = (4\mathrm{i})^{-1}\mathrm{Tr}\left(\lambda^c[\lambda^a, \lambda^b]\right).$$ (5.9)

The field-strength tensor is then

$$G_{\mu\nu}^a = \partial_\nu B_\mu^a - \partial_\mu B_\nu^a - gf^{abc}B_\nu^b B_\mu^c.$$ (5.10)

Although the normalization adopted in (5.5) is a canonical choice, it is not universally employed. For example, Buras (1980) follows a different and also widespread convention in his review of QCD corrections beyond leading order. Forewarned is forearmed!

The form of the QCD Lagrangian (5.2) may compared with that of the QED Lagrangian,

$$\mathcal{L}_{\mathrm{QED}} = -\tfrac{1}{4}F_{\mu\nu}F^{\mu\nu} + \bar{\psi}\left(\mathrm{i}\gamma^\mu\mathcal{D}_\mu - m\right)\psi,$$ (5.11)

for which

$$\mathcal{D}_\mu = \partial_\mu - \mathrm{i}qA_\mu,$$ (5.12)

for fermions of electric charge q, and the field-strength tensor has the familiar form

$$F_{\mu\nu} = \partial_\nu A_\mu - \partial_\mu A_\nu,$$ (5.13)

characteristic of an Abelian gauge theory. The essential difference between the theories is the existence of gluon–gluon interactions in QCD, which contrasts with the absence of photon–photon interactions in QED.

5.2. Consequences for the hadron spectrum

5.2.1. Stability of color singlets

Knowing the QCD Lagrangian, we may now study the properties of the interactions among quarks, at least in a very simplified—perhaps over-simplified—fashion. The point of the exercise is to verify that color singlets enjoy a preferred status. This encourages the hope that the spectrum of QCD, when it is computed, will display the systematics that inspired the invention of the theory.

The quark–gluon interaction term in the QCD Lagrangian is

$$\mathcal{L}_{\mathrm{inter}} = \tfrac{1}{2}g\bar{\psi}\lambda^a\gamma^\mu B_\mu^a\psi,$$ (5.14)

in matrix notation with

$$\psi \equiv \begin{pmatrix} \psi_R \\ \psi_B \\ \psi_G \end{pmatrix}.$$ (5.15)

The Feynman rule for the quark–quark–gluon vertex is given in fig. 23. Thus, the one-gluon-exchange force between quarks is proportional to

$$\mathcal{E} \equiv \tfrac{1}{4} \sum_a \lambda^a_{\alpha\beta} \lambda^a_{\gamma\delta}$$ (5.16)

for the transition $\alpha + \gamma \rightarrow \beta + \delta$. We shall take the quantity (5.16) as representative of the interaction energy between quarks, and proceed to deduce the consequences of QCD for the hadron spectrum, according to that measure. More heuristic, but completely equivalent treatments may be found in the lectures by Feynman (1977) and Quigg (1981).

To compute the interaction it is necessary to evaluate the expectation value of products such as $\tfrac{1}{4}\lambda^{(1)} \cdot \lambda^{(2)}$, where the superscripts label the interacting quarks, and the $\lambda^{(i)}$ are eight-vectors in color space. The $SU(N)$ techniques are quite standard. An explicit and accessible reference is Rosner (1981a), to whose notation I will adhere. Jaffe (1977b) uses a different normalization convention. It will save writing to define the $SU(3)$ generators

$$T \equiv \tfrac{1}{2}\lambda$$ (5.17)

and to evaluate the expectation value $\langle T^2 \rangle$ in various representations of interest.

In $SU(N)$, it is equivalent to average the square of any single generator over the representation, or to perform the sum over the all generators. The former tactic is simpler, and it is particularly convenient to choose I_z, the third component of isospin in the flavor analogy, as the designated generator. Consequently the expectation value $\langle T^2 \rangle$ in a representation

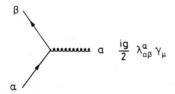

Fig. 23. The quark–quark–gluon interaction in QCD.

Table 12
Value of the color Casimir operator in small representations of SU(3)

Representation	$\langle T^2 \rangle$
[1]	0
[3] or [3*]	4/3
[6] or [6*]	10/3
[8]	3
[10] or [10*]	6
[27]	8

of dimension d is

$$\langle T^2 \rangle_d = (N^2 - 1) \sum_{\substack{\text{members} \\ \text{of rep.}}} I_z^2/d \qquad (5.18)$$

where $N^2 - 1$ is the number of generators of SU(N). Results for the low-dimensioned representations of SU(3) are given in table 12. To evaluate $\langle T^{(1)} \cdot T^{(2)} \rangle$, we use the familiar identity

$$\langle T^{(1)} \cdot T^{(2)} \rangle = \tfrac{1}{2} (\langle T^2 \rangle - \langle T^{(1)2} \rangle - \langle T^{(2)2} \rangle) \qquad (5.19)$$

The "interaction energies" for two-body systems composed of quark–quark and quark–antiquark are given in table 13. For the (q$\bar{\text{q}}$)

Table 13
"Interaction energies" for few-quark systems

Configuration	$\left\langle \sum_{i<j} T^{(i)} \cdot T^{(j)} \right\rangle$
(q$\bar{\text{q}}$)$_{[1]}$	$-4/3$
(q$\bar{\text{q}}$)$_{[8]}$	$+1/6$
(qq)$_{[3*]}$	$-2/3$
(qq)$_{[6]}$	$1/3$
(qqq)$_{[1]}$	-2
(qqq)$_{[8]}$	$-1/2$
(qqq)$_{[10]}$	$+1$
(qqqq)$_{[3]}$	-2

Fig. 24. A baryon configuration which is not considered in the sum over two-body forces.

systems, the one-gluon-exchange contribution is attractive for the color singlet but repulsive for the color octet. Similarly for diquark systems the color triplet is attracted but the color sextet is repelled. Of all the two-body channels, the color singlet ($q\bar{q}$) is the most attractive. On the basis of this analysis, one may choose to believe that colored mesons should not exist, whereas color singlets should be found.

To analyze three (or more)-body systems, let us assume that the interaction is merely the sum of two-body forces, so that

$$\mathcal{E} = \sum_{i<j} \langle T^{(i)} \cdot T^{(j)} \rangle, \tag{5.20}$$

which is easily computed as

$$2 \sum_{i<j} \langle T^{(i)} \cdot T^{(j)} \rangle = \langle T^2 \rangle - \sum_i \langle T^{(i)2} \rangle. \tag{5.21}$$

For three-quark systems, the results in table 13 show that the color singlet is again the most attractive channel. This is as desired.

Several potentially important effects have been neglected in these calculations:

(i) Multiple gluon exchanges between quarks have been ignored, and we have put forward no arguments for faith in lowest-order perturbation theory.

(ii) Configurations involving the three-gluon vertex, such as the (qqq) color-singlet shown in fig. 24, have not been taken into account. This is related to the incompleteness of the calculation noted in (i).

(iii) What may be the most serious shortcoming of the toy calculation is its neglect of the energetics associated with the creation of an isolated color non-singlet state. In section 7 we shall review a plausibility argument that implies an infinite cost in energy to isolate a colored system. If that is so, the attraction provided by one-gluon exchange will be insufficient to bind colored states.

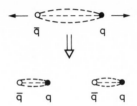

Fig. 25. Attempting to separate a quark and antiquark results in the creation of a quark–antiquark pair from the vacuum, so that color is always neutralized locally.

In spite of these shortcomings, the elementary calculation we have just completed does make it plausible that color singlets are energetically favored states. In addition, it is easy to see that there is no long-range interaction with color singlets. As an example, consider whether a quark is bound to a baryon. The final entry in table 13 shows that the interaction energy of the quark plus baryon system is precisely that which binds the baryon, with no additional attraction.

It is quite generally believed, but not proved, that QCD is in fact a confining theory, and that colored objects cannot be liberated. Seeking proofs, loopholes, and interpretations of the experimental indications [LaRue et al. 1981] for fractionally-charged matter is an occupation of key importance.

5.2.2. The string picture of hadrons

Suppose that the interaction among quarks is so strong at large distances that a $(q\bar{q})$ pair is always created when the quarks are widely separated, as depicted in fig. 25. By analogy with the hadronic clusters typically inferred from experiments on multiple production [Dremin and Quigg 1978], it is reasonable to expect that a quark is accompanied by an antiquark in a typical hadron of mass ~ 1 GeV/c^2 at a separation of ~ 1 fm. That would imply that between every quark and antiquark there is a linear energy density of order

$$k = \Delta E/\Delta r \approx 1 \text{ GeV/fm}$$

$$\approx 0.2 \text{ GeV}^2 \approx 5/\text{fm}^2. \tag{5.22}$$

This picture is supported by the evidence for linear Regge trajectories of the light hadrons, which have already been displayed in figs. 19 and 21. For the families of hadrons composed entirely of light quarks, the

Regge trajectories are given by

$$J(M^2) = \alpha_0 + \alpha' M^2, \qquad (5.23)$$

with

$$\alpha' \approx 0.8\text{--}0.9 \left(\text{GeV}/c^2\right)^{-2}. \qquad (5.24)$$

The connection between linear energy density and the linear Regge trajectories is provided by the string model formulated by Nambu (1974).

Consider a massless quark and antiquark connected by a string of length r_0, which is characterized by an energy density per unit length k. The situation is sketched in fig. 26. For a given value of the length r_0, the largest achievable angular momentum L occurs when the ends of the string move with the velocity of light. In this circumstance, the speed at any point along the string will be

$$\beta(r) = 2r/r_0. \qquad (5.25)$$

The total mass of the system is then

$$M = 2\int_0^{r_0/2} dr\, k\left[1 - \beta(r)^2\right]^{-1/2} = kr_0\pi/2, \qquad (5.26)$$

while the orbital angular momentum of the string is

$$L = 2\int_0^{r_0/2} dr\, kr\beta(r)c\left[1 - \beta(r)^2\right]^{-1/2} = kcr_0^2\pi/8. \qquad (5.27)$$

Using the fact (5.26) that $r_0^2 = 4M^2/k^2\pi^2$, we find that

$$L = M^2/2\pi k, \qquad (5.28)$$

which corresponds to a linear Regge trajectory, with

$$\alpha' = 1/2\pi k. \qquad (5.29)$$

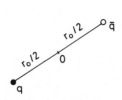

Fig. 26. A massless quark and antiquark connected by a linear string.

This connection yields

$$k = \begin{cases} 0.18 \text{ GeV}^2 \\ 0.20 \text{ GeV}^2 \end{cases} \text{ for } \alpha' = \begin{cases} 0.9 \text{ GeV}^{-2} \\ 0.8 \text{ GeV}^{-2}, \end{cases} \qquad (5.30)$$

consistent with our heuristic estimate of the energy density. Thus we see that a linear energy density implies linearly rising Regge trajectories, and that the connection makes quantitative sense.

How does a linear energy density arise in Quantum Chromo-Dynamics? In fact, I am not certain that it does, but the tendency is indicated in perturbation theory, and there is some similarity between hadron strings and pinned magnetic flux lines in superconductors [Nielsen and Olesen 1973]. (It now seems apparent [Mandelstam 1979] that the analogy is more properly between the chromoelectric field and the magnetic field. See also the discussion in section 7.1.) In ordinary quantum electrodynamics, the pattern of equipotentials and electric field lines is similar at large and small separations between charges. QCD at short distances exhibits a nearly identical flux pattern. At larger separations, a collimated flux tube begins to emerge in QCD. The mechanism for attraction among the chromoelectric flux lines resembles the Biot–Savart effect in classical electrodynamics, the attraction between like currents. Whether this suggestion of an approach to the string picture persists to very large separations is an interesting question.

5.3. A picture of hadron masses

In their paper with the perhaps too ambitious title "Hadron masses in a gauge theory," De Rújula et al. (1975) proposed to take seriously the proposition that QCD has something to say about hadron masses. Although it would be exaggeration to claim that the mass of any hadron has yet been computed in QCD, it is plain that QCD-inspired models have provided many important insights into the pattern of hadron masses. The picture put forward by De Rújula, Georgi, and Glashow was an early and influential example of this genre. Their proposal [which is thoroughly discussed in the SLAC Summer School lectures by Jackson, (1976)] is that all spin effects arise from the short-distance one-gluon-exchange interaction expected in QCD. This may seem an extreme position and indeed it can readily be relaxed. Let us, however, see where it leads.

The Fermi–Breit Hamiltonian (toward which problems 15–17 lead) can be transcribed from the familiar QED case. For s-wave states, the structure will be the same as that of eq. (2.40) which was used to describe hadron electromagnetic mass differences in section 2.3. The "interaction energies" which appeared in the discussion of the stability of color singlets in section 5.2.1 can be recast as Coulomb potentials. With the introduction of a strong coupling constant α_s, we write

$$V(r) = \frac{\alpha_s}{4r} \langle \lambda^{(1)} \cdot \lambda^{(2)} \rangle. \tag{5.31}$$

In the specific cases of interest for two-body interactions, this yields

$$V_{q\bar{q} \in [1]} = -\tfrac{4}{3}\alpha_s/r, \tag{5.32a}$$

$$V_{q\bar{q} \in [8]} = \tfrac{1}{6}\alpha_s/r, \tag{5.32b}$$

$$V_{qq \in [3^*]} = -\tfrac{2}{3}\alpha_s/r, \tag{5.32c}$$

$$V_{qq \in [6]} = \tfrac{1}{3}\alpha_s/r. \tag{5.32d}$$

We shall assume that the explicit effect of this Coulomb interaction, which is the same for all quark flavors, can be subsumed into the consequences of confinement. Of more immediate concern is the color hyperfine interaction, which takes the form

$$\Delta E_{\text{HFS}} = \frac{-\pi\alpha_s}{6m_1 m_2} |\Psi_{12}(0)|^2 \langle \lambda^{(1)} \cdot \lambda^{(2)} \sigma_1 \cdot \sigma_2 \rangle \tag{5.33}$$

between two quarks in a relative s-wave. Thus, for $q\bar{q}$ in a color singlet, relevant for mesons, we have

$$\Delta E_{\text{HFS}}(q\bar{q} \in [1]) = \frac{8\pi\alpha_s}{9m_1 m_2} |\Psi_{12}(0)|^2 \langle \sigma_1 \cdot \sigma_2 \rangle, \tag{5.34}$$

while for qq in a color antitriplet, relevant for baryons, we have

$$\Delta E_{\text{HFS}}(qq \in [3^*]) = \frac{4\pi\alpha_s}{9m_1 m_2} |\Psi_{12}(0)|^2 \langle \sigma_1 \cdot \sigma_2 \rangle. \tag{5.35}$$

The intrinsic strength of the color hyperfine interaction is thus half as large in baryons as in mesons.

5.3.1. The light hadrons
We recall that in mesons

$$\langle \sigma_1 \cdot \sigma_2 \rangle = \begin{cases} -3, & s = 0, \\ 1, & s = 1, \end{cases} \tag{2.26}$$

so that the color hyperfine interaction is attractive for pseudoscalars and repulsive for vector mesons. The inverse factors of mass in (5.34) mean that the hyperfine splitting will be smaller between K and K* than between π and ρ.

In the case of baryons, we have already computed

$$\left\langle \sum_{i<j} \sigma_i \cdot \sigma_j \right\rangle = \begin{cases} -3, & s = 1/2, \\ 3, & s = 3/2. \end{cases} \tag{2.29}$$

We therefore expect states in the decimet to be more massive than those in the octet. This expectation and the corresponding one for the mesons are in accord with observation. Two other qualitative facts are also correlated with these by the Ansatz (5.33).

First, consider the problem of the $\Lambda - \Sigma$ splitting, which electromagnetic considerations decisively failed to explain in section 3.2. In the $J^P = 1/2^+$ Σ-hyperon, the two nonstrange quarks are in a configuration with $I = 1$, color [3*], and angular momentum = 0. Under interchange of the two quarks, this state is symmetric $\times$ antisymmetric $\times$ symmetric = antisymmetric. The spin part of the wavefunction must therefore be symmetric, or in other words $s = 1$. Thus by (2.26) we conclude that

$$\langle \sigma_n \cdot \sigma_n \rangle_\Sigma = 1, \tag{5.36}$$

where the subscript n stands for non-strange. Because according to (2.29) the quantity

$$\left\langle \sum_{i<j} \sigma_i \cdot \sigma_j \right\rangle_\Sigma = \langle \sigma_n \cdot \sigma_n \rangle_\Sigma + 2\langle \sigma_n \cdot \sigma_s \rangle_\Sigma = -3, \tag{5.37}$$

evidently

$$\langle \sigma_n \cdot \sigma_s \rangle_\Sigma = -2. \tag{5.38}$$

The hyperfine shift in the Σ mass is therefore

$$\Delta E_{\text{HFS}}(\Sigma) = \frac{4\pi\alpha_s|\Psi(0)|^2}{9}\left[\frac{1}{m_n^2} - \frac{4}{m_n m_s}\right], \tag{5.39}$$

where we have assumed that $|\Psi(0)|^2$ is the same for any pair of quarks. This cannot be quite right, but we may hope it is not misleading.

In contrast, for the Λ, the nonstrange quarks are in an isoscalar, color-antitriplet, s-wave state, which is antisymmetric $\times$ antisymmetric $\times$ symmetric = symmetric under particle interchange. The spin state must therefore be antisymmetric, $s = 0$, for which

$$\langle \sigma_n \cdot \sigma_n \rangle_\Lambda = -3, \tag{5.40}$$

by eq. (2.26). Again according to (2.29) we know that

$$\left\langle \sum_{i<j} \sigma_i \cdot \sigma_j \right\rangle_\Lambda = \langle \sigma_n \cdot \sigma_n \rangle_\Lambda + 2\langle \sigma_n \cdot \sigma_s \rangle_\Lambda = -3; \qquad (5.41)$$

we deduce that

$$\langle \sigma_n \cdot \sigma_s \rangle_\Lambda = 0. \qquad (5.42)$$

As a consequence, the hyperfine shift in the Λ mass is

$$\Delta E_{\mathrm{HFS}}(\Lambda) = \frac{4\pi\alpha_s |\Psi(0)|^2}{9} \left[\frac{-3}{m_n^2} \right]. \qquad (5.43)$$

This is more negative than the Σ hyperfine shift, so long as $m_s > m_n$. This is another qualitative success.

The second conspicuous failure of the quark model described in sections 2 and 3 was the prediction (2.39) of the vanishing of the neutron's charge radius. In this case as well the color hyperfine interaction provides some enlightenment [Carlitz et al. 1977]. The two down quarks in the neutron must be in an $I = 1$, color [3*], angular momentum zero state, which is symmetric $\times$ antisymmetric $\times$ symmetric $=$ antisymmetric under particle interchange. They must therefore be in a symmetric spin-triplet state, for which the hyperfine interaction is repulsive. Since the overall hyperfine interaction is attractive, the up-down pairs must attract. Hence the up quark will be drawn to the center of the neutron, while the down quarks are pushed toward the periphery. As a result the neutron's mean-squared charge radius will be negative, in agreement with experiment [Krohn and Ringo 1973, Berard et al. 1973, Borkowski et al. 1974, Koester et al. 1976]. One may, at the price of additional assumptions, attempt to estimate the ratio $\langle r_{\mathrm{EM}}^2 \rangle_n / \langle r_{\mathrm{EM}}^2 \rangle_p$. This has been done with reasonable success by Carlitz et al. (1977), and by Isgur et al. (1978). Isgur et al. (1981) have gone yet further and produced a fit to $G_E^n(q^2)$ within the framework of a harmonic oscillator model for the confining interaction.

Notice that the mean-squared magnetic radius, defined with respect to the magnetic form factor as

$$\langle r_{\mathrm{mag}}^2 \rangle = - \partial G_M(q^2) / \partial q^2 \big|_{q^2=0}, \qquad (5.44)$$

need not be identical to the charge radius, because the contributions of the quarks are weighted differently. An estimate of the ratio $\langle r_{\mathrm{mag}}^2 \rangle_p / \langle r_{\mathrm{charge}}^2 \rangle_p$ has been made by Carlitz et al. (1977).

The idea of a color hyperfine interaction has been shown to yield a qualitative understanding of the pseudoscalar–vector splitting, the octet–decimet splitting, the $\Lambda - \Sigma$ splitting, and the neutron charge radius. Can it also give a quantitative description of the masses of the ground-state hadrons? To examine this question, we ignore electromagnetic mass differences which are easily restored according to the procedure followed in sections 2.3, 2.4, 3.1 and 3.2.

We first consider baryons. Let us write

$$M = M_0 + N_s(m_s - m_u) + m_u^2 \left\langle \sum_{i<j} \frac{\sigma_i \cdot \sigma_j}{m_i m_j} \right\rangle \delta M_{\text{c.m.}}, \tag{5.45}$$

where the chromomagnetic hyperfine shift is given by

$$\delta M_{\text{c.m.}} = 4\pi\alpha_s |\Psi(0)|^2 / 9m_u^2. \tag{5.46}$$

M_0 is the common unperturbed mass of the octet and decimet baryons, and N_s is the number of strange quarks in a baryon. The hyperfine and strange-quark mass shifts are given in table 14.

Instead of making a global fit, we determine the parameters as follows:

$$M_0 = (M_N + M_\Delta)/2 = 1085.5 \text{ MeV}/c^2, \tag{5.47}$$

$$\delta M_{\text{c.m.}} = (M_\Delta - M_N)/6 = 48.83 \text{ MeV}/c^2, \tag{5.48}$$

$$(m_s - m_u) = (M_\Sigma + M_\Lambda + M_{\Sigma^*})/4 - M_0 = 183.75 \text{ MeV}/c^2. \tag{5.49}$$

Table 14
Baryon masses including the color hyperfine interaction.
For definitions see (5.44) and (5.45)

Baryon	$\Delta E_{\text{HFS}}/\delta M_{\text{c.m.}}$	N_s	Fitted mass (MeV/c^2)
N(939)	-3	0	939[a]
Λ(1116)	-3	1	1123
Σ(1193)	$1 - 4m_u/m_s$	1	1189
Ξ(1318)	$-4m_u/m_s + m_u^2/m_s^2$	2	1345
Δ(1232)	$+3$	0	1232[a]
Σ^*(1384)	$1 + 2m_u/m_s$	1	1383
Ξ^*(1533)	$2m_u/m_s + m_u^2/m_s^2$	2	1539
Ω(1672)	$3m_u^2/m_s^2$	3	1701

[a] Input value.

If we now interpret M_0 as three times the up-quark mass, we have

$$m_u = M_0/3 = 361.83 \text{ MeV}/c^2, \tag{5.50}$$

and

$$m_s = 545.6 \text{ MeV}/c^2. \tag{5.51}$$

These numbers are reasonably consistent with the values of 338 and 510 MeV/c^2 deduced from the fit to baryon magnetic moments in section 3.4.1. They will enter our calculation of baryon masses only in the ratio and difference, and the results would be effectively unchanged if we adopted the smaller values.

Overall the agreement between the model and experiment is good, and would be improved slightly if we were to determine parameters by making a global fit. The strange particle hyperfine splittings are just slightly too small, and the $m_s - m_u$ mass difference is slightly too large. These defects are of course correlated. What has been achieved is a unified understanding of the $\Lambda - \Sigma$ splitting and the splitting between the $1/2^+$ and $3/2^+$ baryon multiplets. This has been done in spite of the fact that we have somewhat cavalierly ignored the possibility of variations in kinetic energies, binding energies, and the wavefunction at the origin. These effects probably tend to reduce the discrepancies we have noted. See, for example, Cohen and Lipkin (1981).

Using the definition (5.46) for the chromomagnetic level shift and the somewhat casual inference from electromagnetic mass splittings in the baryon octet that

$$|\Psi(0)|^2 \simeq 0.0116 \text{ GeV}^3, \tag{3.77}$$

it is straightforward to compute that

$$\alpha_s \simeq 0.4. \tag{5.52}$$

This value is large enough to be regarded as a strong interaction coupling constant, but not so large as to make the one-gluon-exchange picture seem entirely ridiculous.

A similar analysis can be carried out for the mesons, for which we write

$$M = M_0 + N_s(m_s - m_u) + m_u^2 \left\langle \frac{\sigma_q \cdot \sigma_{\bar{q}}}{m_q m_{\bar{q}}} \right\rangle \delta M_{\text{c.m.}}, \tag{5.53}$$

Table 15
Meson masses including the color hyperfine interaction.
For definitions see (5.52) and (5.53)

Meson	$\Delta E_{\text{HFS}}/\delta M_{\text{c.m.}}$	N_s	Fitted mass (MeV/c^2)
$\pi(138)$	-3	0	138[a]
K(496)	$-3m_u/m_s$	1	489
$\eta(549)$	$-9/4 - \frac{3}{4}(m_u/m_s)^2$	1/2	297
$\eta'(958)$	$-3/4 - \frac{9}{4}(m_u/m_s)^2$	3/2	616
$\rho(776)$	1	0	776[a]
$\omega(784)$	1	0	776
K*(892)	m_u/m_s	1	894
$\varphi(1020)$	m_u^2/m_s^2	2	1034

[a] Input value.

where the color–magnetic hyperfine shift is

$$\delta M_{\text{c.m.}} = 8\pi\alpha_s |\Psi(0)|^2/9m_u^2. \tag{5.54}$$

The hyperfine and strange-quark mass shifts are given in table 15. The strange-quark content of the η and η' are determined from the wavefunctions (3.112) deduced from the linear mass formula in subsection 3.3.2.

To fit parameters, we proceed as for the baryons:

$$M_0 = (M_\pi + 3M_\rho)/4 = 617 \text{ MeV}/c^2, \tag{5.55}$$

$$\delta M_{\text{c.m.}} = (M_\rho - M_\pi)/4 = 160 \text{ MeV}/c^2, \tag{5.56}$$

$$(m_s - m_u) = (M_K + 3M_{K^*})/4 - M_0 = 176 \text{ MeV}/c^2. \tag{5.57}$$

If M_0 is interpreted as twice the up-quark mass, we have

$$m_u = M_0/2 = 308.5 \text{ MeV}/c^2, \tag{5.58}$$

and

$$m_s = 484 \text{ MeV}/c^2. \tag{5.59}$$

These numbers, like those deduced for the baryons in (5.50) and (5.51), lie within 10% of the values inferred from baryon magnetic moments. Given that we are playing fast and loose with the idea of a quark mass, the agreement seems quite satisfying.

The fitted masses agree rather well with experiment except for the η and η' masses for which the predictions are disastrous. These states are too heavy for the interpretation of masses that we have given. This problem persists whether or not the pion is regarded as exceptionally light. It is unreasonable in the present framework for the η to outweigh the kaon. A possible description of this phenomenon in terms of communication with (pure gluon) quarkless channels will be examined in section 8.3.

Meson electromagnetic mass differences permit a rather crude determination of the mesonic wavefunction at the origin as

$$|\Psi(0)|^2 \approx 0.017 \text{ GeV}^3, \tag{5.60}$$

One may therefore estimate

$$\frac{\delta M_{\text{c.m.}}(\text{meson})}{\delta M_{\text{c.m.}}(\text{baryon})} \approx \frac{2|\Psi_{\text{meson}}(0)|^2}{|\Psi_{\text{baryon}}(0)|^2} \approx 2.93, \tag{5.61}$$

which compares favorably with the ratio of the values (5.56) and (5.48), which is 3.28. The similarity of these numbers is very tantalizing. It underscores the desirability of better experimental determinations of electromagnetic mass differences, which would justify a less casual analysis than I have made here.

5.3.2. Extension to charm and beauty

Mesons and baryons may also be formed from the heavy quarks c (charm), b (beauty), and t (truth)–if it exists–either alone or in combination with the light quarks. Characteristics of the heavy quarks are listed in table 16. The resulting particles may be classified according to an enlarged, and badly broken, flavor symmetry group $SU(6)_{\text{flavor}}$. To display the weight diagrams, it is convenient to decompose

$$SU(6)_{\text{flavor}} \rightarrow SU(4)_{\text{udsc}} \otimes U(1)_b \otimes U(1)_t. \tag{5.62}$$

The 36 species of mesons are exhibited in fig. 27. For ground-state baryons, there are 70 spin-1/2 states, illustrated in fig. 28, and 56 spin-3/2 states, shown in fig. 29. The classic introduction to charm

Table 16
Some properties of the heavy quarks

Quark	I	Q	Charm	Beauty	Truth
c	0	2/3	1	0	0
b	0	$-1/3$	0	1	0
t	0	2/3	0	0	1

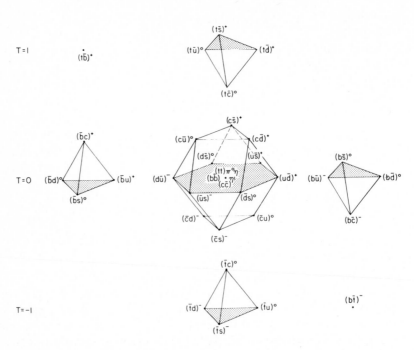

Fig. 27. Meson states in flavor SU(6), decomposed into $SU(4)_{udsc} \otimes U(1)_b \otimes U(1)_t$. The additive quantum numbers are denoted by B(beauty) and T(truth).

spectroscopy is the preview by Gaillard et al. (1975). Some details of the flavor-SU(4) symmetry may be found in the lectures by Einhorn (1975), and in my lectures at the Gomel summer school [Quigg, 1979]. See also Rosner (1981a).

For charmed mesons, eq. (5.53) is immediately generalized to

$$M = M_0 + N_s(m_s - m_u) + N_c(m_c - m_u) + m_u^2 \left\langle \frac{\sigma_q \cdot \sigma_{\bar{q}}}{m_q m_{\bar{q}}} \right\rangle \delta M_{c.m.},$$

(5.63)

where N_c counts the number of charmed quarks. It is then straightfor-

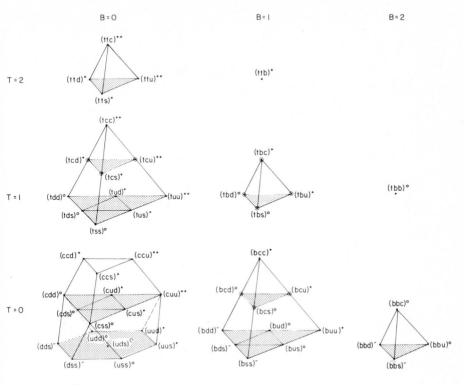

Fig. 28. $J^P = 1/2^+$ baryon states in flavor SU(6). The circled states occur twice, as do those that lie in both [6] and [3*] of SU(3)$_{uds}$. There are 70 states in all.

ward to evaluate the charmed-quark mass as

$$(m_c - m_u) = (3M_{D^*} + M_D)/4 - (3M_\rho + M_\pi)/4$$
$$= 1354.5 \text{ MeV}/c^2,$$ (5.64)

whence

$$m_c \approx 1663 \text{ MeV}/c^2.$$ (5.65)

We therefore expect

$$(M_{D^*} - M_D) = (M_{K^*} - M_K)m_s/m_c$$
$$= (M_\rho - M_\pi)m_u/m_c \approx 119 \text{ MeV}/c^2.$$ (5.66)

which is roughly in agreement with the observed splitting of approxi-

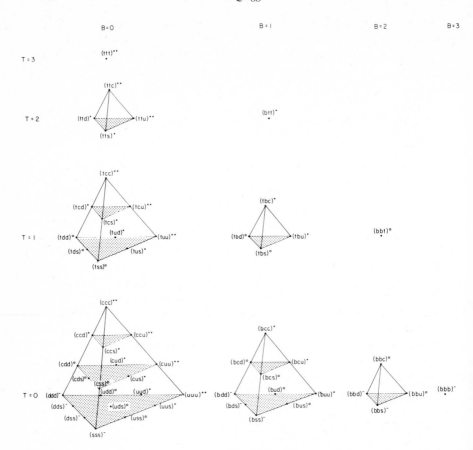

Fig. 29. $J^P = 3/2^+$ baryon states in flavor SU(6). There are 56 states in all.

mately 142 MeV/c^2. A similar computation leads to the expectation that

$$(M_{F^*} - M_F) = (M_{D^*} - M_D) m_u / m_s = 90 \text{ MeV}/c^2, \qquad (5.67)$$

which is to be compared with the experimental hint of approximately 110 MeV/c^2. For the $\psi - \eta_c$ interval, the same line of argument leads to

$$M_\psi - M_{\eta_c} \approx (M_{D^*} - M_D)^2 / (M_\rho - M_\pi) \approx 32 \text{ MeV}/c^2, \qquad (5.68)$$

which is far from the observed spacing [Schamberger, 1981] of 115 MeV/c^2. This is a sign that we cannot with impunity ignore the variation of $|\Psi(0)|^2$ with quark mass.

For the charmed baryons, we proceed in analogy with eq. (5.45). It is convenient to fix the charmed quark mass by comparing the Λ and the charmed Λ (better known as C_0^+), for which the color hyperfine shift is identical up to variations in the wavefunction. This yields

$$(m_c - m_s) = M_{C_0^+} - M_\Lambda \simeq 1169 \text{ MeV}/c^2, \tag{5.69}$$

from which we conclude using (5.51) that

$$m_c \approx 1715 \text{ MeV}/c^2, \tag{5.70}$$

which is not far from (5.65). The hyperfine splitting between the spin-1/2 charmed baryons C_1 and C_0 (or Σ_c and Λ_c) is then analogous to the $\Sigma - \Lambda$ splitting. The appropriate arithmetic yields

$$(M_{C_1} - M_{C_0}) \approx 154 \text{ MeV}/c^2, \tag{5.71}$$

in good agreement with the observed splitting of approximately 155 MeV/c^2. We have as well the simple relation

$$(M_{C_1^*} - M_{C_1}) = (M_{\Sigma^*} - M_\Sigma) m_s / m_c \approx 62 \text{ MeV}/c^2, \tag{5.72}$$

which cannot yet be given a meaningful test. More extensive discussions of charmed masses appear in the papers by De Rújula et al. (1975), Sakharov (1975); Lee et al. (1977); and Copley et al. (1979).

The procedure to be followed for hadrons with beauty or truth is now obvious. Of most immediate interest is the expectation that

$$(M_{B^*} - M_B) \approx (M_{D^*} - M_D) m_c / m_b \approx 50 \text{ MeV}/c^2, \tag{5.73}$$

which suggests the decay

$$B^* \to B\gamma \tag{5.74}$$

as a possible experimental tag.

5.4. Further applications of chromomagnetism

It has been made plausible that color-singlet states correspond to stable hadrons and we have also seen, to a very limited extent, that the dynamical consequences of quark–gluon interactions correspond to reality. Although color-singlet configurations exist that are more complicated than $(q\bar{q})$ or (qqq), the one-gluon-exchange arguments of subsection 5.2.1 give no reason to expect that these will be appreciably bound by the color force. In ancient times [Rosner, 1968], duality diagrams suggested the necessity of $(qq\bar{q}\bar{q})$ states coupled to the baryon–antibaryon channel.

Is it possible that specific configurations of this kind, or (4q$\bar{\mathrm{q}}$) or (6q) configurations, might benefit from a large hyperfine interaction and thus be bound?

The parameter of interest, the expectation value $\langle \sum_{i<j} \lambda^{(i)} \cdot \lambda^{(j)} \sigma_i \cdot \sigma_j \rangle$ in a multiquark state, is conveniently computed using the color–spin technique introduced by Jaffe (1977a, b, 1979). A color $\otimes$ spin SU(6) algebra can be constructed out of the 35 operators $\lambda_a \otimes \sigma_b$ (24 operators), $\mathbf{1} \otimes \sigma_b$ (3 operators), and $\lambda_a \otimes \mathbf{1}$ (8 operators). It is convenient to define the generators of SU(6)$_{\mathrm{color-spin}}$ as

$$G^{(6)} \equiv \begin{cases} \lambda_a \otimes \sigma_b / 2\sqrt{2} , \\ \lambda_a \otimes 1_{2 \times 2} / 2\sqrt{2} , \\ 1_{3 \times 3} \otimes \sigma_b / 2\sqrt{3} , \end{cases} \tag{5.75}$$

which are normalized so that

$$\mathrm{Tr}\left(G_i^{(6)} G_j^{(6)} \right) = \tfrac{1}{2} \delta_{ij} \tag{5.76}$$

Similarly, we continue to define the normalized generators of SU(2) and SU(3) as

$$G^{(2)} = \sigma/2, \tag{5.77}$$

and

$$G^{(3)} = \lambda/2. \tag{5.78}$$

Again I caution that these normalizations are widely [e.g., Rosner 1981a], but not universally [Buras 1980, Jaffe 1977a, b, 1979] used in the literature.

With the conventions (5.75–5.78), it is easy to compute that

$$G_1^{(6)} \cdot G_2^{(6)} = \tfrac{1}{8}\lambda^{(1)} \cdot \lambda^{(2)} \sigma_1 \cdot \sigma_2 + \tfrac{1}{8}\lambda^{(1)} \cdot \lambda^{(2)} + \tfrac{1}{12}\sigma_1 \cdot \sigma_2, \tag{5.79}$$

so that

$$\lambda^{(1)} \cdot \lambda^{(2)} \sigma_1 \cdot \sigma_2 = 8G_1^{(6)} \cdot G_2^{(6)} - 4G_1^{(3)} \cdot G_2^{(3)} - \tfrac{8}{3}G_1^{(2)} \cdot G_2^{(2)}. \tag{5.80}$$

Each dot-product on the right-hand side can be evaluated as usual as

$$G_1^{(N)} \cdot G_2^{(N)} = \tfrac{1}{2}\left[\left(G_1^{(N)} + G_2^{(N)} \right)^2 - G_1^{(N)2} - G_2^{(N)2} \right]. \tag{5.81}$$

Consequently the quantity of interest is

$$
\sum_{i<j} \boldsymbol{\lambda}^{(i)} \cdot \boldsymbol{\lambda}^{(j)} \boldsymbol{\sigma}_i \cdot \boldsymbol{\sigma}_j = 4\left[G_{\text{tot}}^{(6)2} - \sum_i G_i^{(6)2} \right]
$$

$$
- 2\left[G_{\text{tot}}^{(3)2} - \sum_i G_i^{(3)2} \right] - \tfrac{4}{3}\left[G_{\text{tot}}^{(2)2} - \sum_i G_i^{(2)2} \right]
$$

$$
= 4G_{\text{tot}}^{(6)2} - 2G_{\text{tot}}^{(3)2} - \tfrac{4}{3}s(s+1) - 4\sum_i G_i^{(6)2}
$$

$$
+ 2\sum_i G_i^{(3)2} + \tfrac{4}{3}\sum_i s_i(s_i+1). \tag{5.82}
$$

For color singlet states, $\langle G_{\text{tot}}^{(3)2} \rangle = 0$, while for a color-triplet, spin-1/2 quark $\langle G_i^{(3)2} \rangle = 4/3$, $s_i(s_i+1) = 3/4$, and $\langle G_i^{(6)2} \rangle = 35/12$. In passing, I note that a convenient source for properties of group representations is the volume by Patera and Sankoff (1973). The quantity ℓ listed in their table II is related to the SU(N) Casimir operators required in (5.82) by

$$
G^{(N)2} = (N^2 - 1)\ell/2d, \tag{5.83}
$$

for a representation of dimension d. See also McKay and Patera (1981). For a state of n quarks in a color-singlet, (5.82) becomes

$$
\mathcal{C} \equiv \left\langle \sum_{i<j} \boldsymbol{\lambda}^{(i)} \cdot \boldsymbol{\lambda}^{(j)} \boldsymbol{\sigma}_i \cdot \boldsymbol{\sigma}_j \right\rangle = 4G_{\text{tot}}^{(6)2} - \tfrac{4}{3}s(s+1) - 8n. \tag{5.84}
$$

It is easy to recover the result of subsection 5.3.1 for the octet–decimet splitting in this formalism. In s-wave configurations, the SU(3)$_{\text{flavor}} \otimes$ SU(6)$_{\text{color-spin}}$ wavefunctions must be antisymmetric. The possible anti-symmetric combinations for three quarks are shown in table 17. Note first that there is no appropriate color singlet for the flavor singlet state. It will therefore not exist in the ground state, as we have already seen in section 4.2. For either the flavor octet or the flavor decimet, only a single color–spin representation has the requisite symmetry properties. This is rather generally the case for states of interest. Together with the rarity of color-singlet configurations, it explains much of the practical value of color–spin.

In the remaining cases, the color–spin parameter that controls the strength of the color-hyperfine interaction is

$$
\mathcal{C}_{20} = -8, \quad \text{for the flavor } [\mathbf{10}], \tag{5.85}
$$

Table 17
Symmetry properties of flavor and color–spin wavefunctions for three quarks

Flavor SU(3)	Color–spin SU(6)	$G^{(6)2}$	$SU(3)_{color} \otimes SU(2)_{spin}$
[10] ▭▭▭ S	20 ▯ A	21/4	[1],(4)
			[8],(2)
[8] ▭▯ M	70 ▭▯ M	33/4	[1],(2)
			[8],(4)⊕[8],(2)
			[10],(2)
[1] ▯ A	56 ▭▭▭ S	45/4	[8],(2)
			[10],(4)

and

$$\mathcal{C}_{70} = +8, \quad \text{for the flavor } [8]. \tag{5.86}$$

The color hyperfine interaction is thus equal and opposite in the baryon octet and decimet.

Two color-singlet, octet baryons will have color–spin $2\mathcal{C}_{70} = +16$. We have seen before that no residual color force is likely to bind them. Is there a more attractive six-quark state? The answer is yes; the color–spin

$$\textbf{490} \quad \boxplus \quad \supset [1],(1) \tag{5.87}$$

has color-spin

$$\mathcal{C}_{490} = +24. \tag{5.88}$$

A flavor-singlet $(uds)^2$ state therefore has the possibility of being bound by the excess hyperfine attraction. A $\Lambda\Lambda$ bound state was conjectured by Jaffe (1977c) on the basis of these arguments. Its properties were estimated by him in the MIT bag model. A first experimental search [Carroll et al. 1978] has produced no evidence for such a state.

Further applications of color–spin are made in problems 18–20. For specific discussions of $(qq\bar{q}\bar{q})$ states, see Jaffe and Johnson (1976), Jaffe (1977a), and Jaffe and Low (1979). A brief review of this approach is given by Hey (1980b). Many other analyses have led to expectations of $(qq\bar{q}\bar{q})$ baryonium states for which there is, at the moment, no experimental evidence. A comprehensive review of both experiment and theory has been made by Montanet et al. (1980).

At the current stage of understanding of hadronic structure, theory is unable to deny that multiquark states should exist. In many model calculations, such states do emerge. To my knowledge, the only active experimental candidate for a multiquark state is the observation by Amirzadeh et al. (1979) of a narrow ($\Gamma < 20$ MeV) hyperon structure with a mass of 3.17 GeV/c^2 in the reaction

$$K^- p \to \pi^- Y^{*+}, \tag{5.89}$$

at 6.5 and 8.25 GeV/c^2. The narrowness of the state and its propensity for decays into strange particle channels such as $(\Lambda, \Sigma)K\overline{K}$ + pions and ΞK + pions are taken as hints that this is a (4q$\overline{q}$) state.

5.5. Excited mesons and baryons

Some expectations for excited hadron states have been given in section 4. In this section we shall review the status of radially-excited mesons very briefly, and then turn to the question of mass formulas for the orbitally excited states. The last topic deserves a more complete treatment than it will receive here, because it is a principal focal point for the interplay between QCD-based inspiration and experimental results. To maintain the finiteness of these lecture notes, I shall merely summarize the main ideas and provide an entrée to the considerable recent literature.

5.5.1. Radially excited mesons

I have commented in section 4.1 on the likelihood that the vector mesons $\rho'(1600)$ and $\varphi'(1634)$ are, or are considerably mixed with, 2^3S_1 radial excitations of the familiar vector mesons. Essentially pure radial excitations are commonplace in the ψ and Υ families. These will be treated in some detail in section 6. Increasing attention is being devoted to the study of pseudoscalar states beyond the familiar nonet. If these are indeed (q$\overline{q}$) states, they are necessarily radial excitations, because the quantum numbers $J^{PC} = 0^{-+}$ do not occur in orbitally excited states. The other likely interpretations of new pseudoscalar levels as glueballs or multiquark states are only tenable if the (q$\overline{q}$) interpretation can be ruled out. This is therefore an issue of more than passing importance.

Cohen and Lipkin (1979) have presented a comprehensive analysis of the pseudoscalars within the framework of two simple models, one of which is in essence that of section 5.3. They argue that the masses of η and η' can be understood if η is identified almost entirely as a member of the ground-state octet, but η' is appreciably mixed with the 2^1S_0 levels.

Table 18
Candidates for radially excited pseudoscalars

State	I	Seen in	References
$\pi'(1342)$	1	$\varepsilon\pi$	Bonesini et al. (1981)
$\eta(1275)$	0	$\eta\pi\pi$	Stanton et al. (1979)
$\eta'(1400)$	0	$\eta\varepsilon$	Stanton et al. (1979) via Close (1981)
$\iota(1440)$	0	$\psi \to \gamma + (K\bar{K}\pi)$	Scharre (1981)
$K'(1400)$	1/2	$K\pi\pi(K\varepsilon)$	Brandenburg et al. (1976), Aston et al. (1981)

This interpretation also makes less acute the failure of SU(3) sum rules for the peripheral production of η and η'. The remaining isoscalar levels, corresponding roughly to 2^1S_0 levels, are expected in the neighborhood of 1280 and 1500 MeV/c^2.

Experimental sightings of unfamiliar pseudoscalars are summarized in table 18. It is worth emphasizing that not one of these states yet appears in the Particle Data Tables. However, I do not believe the experimental claims to be entirely frivolous, either. In addition to verifying the existence and quantum numbers of these states, it is of some importance to understand whether $\eta'(1400)$ and $\iota(1440)$ are distinct objects. The pseudoscalar masses coincide approximately with the projections by Cohen and Lipkin (1979). Whether they behave as ordinary ($q\bar{q}$) states remains to be understood, particularly in view of the enthusiasm (see section 8.4) for $\iota(1440)$ as a glueball candidate. Watch this space!

5.5.2. Orbitally excited mesons

The hypothesis of De Rújula et al. (1975) that spin effects are governed entirely by the short-range Coulomb-like interaction cannot be complete. After the discovery of the charmonium p-states, it was immediately recognized that the spin splittings were not those of a Coulomb potential [Schnitzer 1976]. Therefore, it was reasoned, a study of the ^3P_J intervals might provide insights into the nature of the quarkonium potential [Schnitzer 1975]. It soon became apparent that although the ^3P_J intervals cannot be predicted as such, they can be given a sensible interpretation in terms of a short-range interaction arising from Lorentz vector exchange and a confining potential with a scalar Lorentz structure [Henriques et al. 1976, see the brief summary in Quigg 1980].

Although a nonrelativistic description of the light mesons cannot so easily be justified, an extensive program to determine the detailed properties of the spin-dependent interactions has been carried out. Accessible reviews, with complete lists of references, have recently been given by Schnitzer (1981a, b). Briefly stated, the idea of this line of investigation is to examine the most general form of the spin-dependent Hamiltonian, to determine the various contributions, and to understand the implications for a theory of strong interactions. The inverse procedure of deriving the interaction is obviously desirable, but considerably less advanced. [See Eichten and Feinberg 1979, 1981, Buchmüller et al. 1981].

To order $(v/c)^2$, the most general two-body interaction is of the form

$$V(r) = V_{\text{central}}(r) + V_{\text{spin}}(r) \frac{\sigma_1 \cdot \sigma_2}{m_1 m_2} + V_{\text{tensor}}(r) \frac{S_{12}}{m_1 m_2}$$

$$+ \left[\frac{V_1(r)}{m_1 m_2} + V_2(r) \left(\frac{1}{m_1^2} + \frac{1}{m_2^2} \right) \right] \mathbf{L} \cdot \mathbf{s}$$

$$+ V_2(r) \left(\frac{1}{m_1^2} - \frac{1}{m_2^2} \right) \mathbf{L} \cdot (\mathbf{s}_1 - \mathbf{s}_2), \tag{5.90}$$

where m_1 and m_2 are the quark masses, and the tensor operator is

$$S_{12} \equiv 3\sigma_1 \cdot \hat{\mathbf{r}} \sigma_2 \cdot \hat{\mathbf{r}} - \sigma_1 \cdot \sigma_2. \tag{5.91}$$

It is instructive to compare (5.90) with the standard Eisenbud and Wigner (1941) expression for nuclear forces [see alternatively Blatt and Weisskopf 1952, Bohr and Mottelson 1969, or Preston 1962]. Given a central potential and its Lorentz structure, one may make a nonrelativistic reduction (at least in the absence of non-Abelian complications) and determine V_{spin}, V_{tensor}, and the spin–orbit potentials V_1 and V_2. This has been done most completely by Gromes (1977).

While the meson spectrum portrayed in table 10 is not as well known as the spectrum of nucleon resonances, the available information on well-established s-wave and p-wave $(q\bar{q})$ states is neatly correlated by the expression (5.90). Schnitzer (1978) has emphasized a characteristic consequence of this nonrelativistic form: the prediction of "inverted multiplets" in light-quark–heavy-quark systems. If $m_2 \gg m_1$, the interaction (5.90) becomes

$$V(r) \rightarrow V_{\text{central}}(r) + \frac{2V_2(r)}{m_1^2} \mathbf{L} \cdot \mathbf{s}_1, \tag{5.92}$$

in which the spin–orbit interaction is determined by the Thomas term V_2. If a scalar confining interaction dominates at medium to large distances, then $V_2(r)$ will be repulsive, and one concludes that, for example,

$$M\left({}^3P_0\right) > M\left({}^3P_1\right) > M\left({}^3P_2\right). \tag{5.93}$$

This would be a dramatic reversal of what is expected on the basis of nonspecific intuition, and well worth searching for in the charmed mesons. The general picture of meson masses that inspires this expectation is made more appealing by the fact that a similar scheme leads to many insights in baryon spectroscopy, to which we now turn.

5.5.3. Orbitally excited baryons

Because of the availability of direct-channel formation experiments and extensive phase-shift analyses, the baryon spectrum has traditionally been more thoroughly studied than the meson spectrum. In addition to a much larger number of states, the baryons provide a richer variety of information on decay modes and decay amplitudes. That situation continues at the present time, in spite of steady progress in meson spectroscopy. There are, as a comparison of tables 10 and 11 (see sections 4.1, 4.2) reveals, more pedigreed p-wave baryons than mesons, and the baryons have been known and studied for a longer time on the average than their mesonic counterparts.

It is thus both natural and desirable that parametrizations inspired by Quantum ChromoDynamics should be applied to the excited baryon masses and decay amplitudes, where a pre-existing fabric of flavor–spin SU(6) folklore awaits derivation, extension, and occasional modification [Rosner 1974a, b, Close 1979]. An extensive program of using the Fermi–Breit interaction in combination with basis states arising from a harmonic oscillator confining interaction has been undertaken by Isgur and Karl and collaborators. These efforts typify much of the current work on the subject. For reviews, see Karl (1979), Hey (1979, 1980a), Isgur (1980a, c), and Close (1981).

For nonstrange baryons [Isgur and Karl 1977], the Fermi–Breit program amounts to a straightforward variation on a classical theme. In order to treat strange baryons systematically, it is necessary to adopt an unperturbed basis in which the difference in mass between the strange and nonstrange quarks is acknowledged explicitly [Isgur and Karl 1978b]. This breaks the exact permutation symmetry of the spatial wavefunctions described in section 4.2. Once this has been done, one may hope to derive connections among spin splittings theoretically, or to deduce them phenomenologically and draw inferences about the underlying interaction.

We have described far too little of the regularities and conundrums of the baryon spectrum to make a detailed analysis meaningful. The following conclusions however seem sound to me. First, the splitting of resonances in the $N = 1$ **70** is characteristic of the spin–spin splitting of an s-wave (qq) pair produced by a short-range Coulomb potential due to the exchange of a vector gluon. Secondly [Isgur and Karl 1978a, b], the inversion of the $J^P = 5/2^-$ states $\Lambda(1830)$ and $\Sigma(1765)$ relative to the ground-state $J^P = 1/2^+$ states $\Lambda(1115)$ and $\Sigma(1193)$ is readily interpreted as a consequence of the nondegeneracy of the appropriate orbital states. This overcomes the color-hyperfine interaction. Third, a successful overall fit can be made to the baryon spectrum [see also Isgur and Karl 1979, Chao et al. 1981]. The fit requires that the spin–orbit interaction be negligible. As emphasized by Close (1981), it is by no means clear how this circumstance might arise in QCD. However, it seems worthwhile to try to understand the remarkable numerical success of these fits, which subsume a great many earlier results of broken-SU(6) phenomenology.

5.6. Hadron – hadron interactions?

If QCD is indeed the complete theory of the strong interactions, it should be possible to derive the interactions among hadrons as collective effects of the interactions among quarks and gluons. How to achieve this worthy goal is not obvious, nor is it apparent that the resulting description should be economical. It is pleasant to realize that the Bardeen–Cooper–Schrieffer (1957) picture of superconductivity must emerge from Quantum ElectroDynamics, but it would be less pleasant to deduce the properties of a specific superconductor from QED. Similarly, the hadron exchange picture of nuclear forces may be the neatest parametrization one may hope to find, even after the hadron spectrum has been derived from QCD.

Anticipating the millenium, many people have attempted to apply QCD-inspired pictures of the hadrons to the problem of interactions. These efforts fall into two principal categories: attempts to extend [DeTar 1978, 1979, 1981a, b, Fairley and Squires 1975] or modify [Brown and Rho 1979, Brown et al. 1979] the MIT bag model; and applications of quark-potential models to hadronic interactions [Fishbane and Grisaru 1978, Willey 1978, Matsuyama and Miyazawa 1979, Fujii and Mima 1978, Gavela et al. 1979, Stanley and Robson 1980, 1981]. I think it is fair to say that this line of research is still at the groping stage. In particular for the potential-model discussions, insufficient attention to the consequences of t-channel analyticity seems a serious shortcoming. A critical discussion from a different orientation has been given by Greenberg and Lipkin (1981).

6. Quarkonium

We have seen in the preceding sections that many aspects of the spectroscopy of light mesons and baryons have elements in common with nonrelativistic potential descriptions. This is so in spite of the fact that the motion of light quarks cannot be argued to be nonrelativistic, and our understanding of the success of nonrelativistic descriptions is only partial. For systems composed exclusively of heavy quarks, the situation may be quite different, in that a potential model may be both adequate and justifiable.

As Appelquist and Politzer (1975) were first to note, the asymptotic freedom of QCD implies that the strong interaction between quarks becomes feeble at very short distances. For bound states of extremely massive quarks, it is possible to imagine that the natural scale of the system is so small that the quarks are bound by a Coulomb potential characteristic of one gluon exchange. The so-called quarkonium system would then be a nonrelativistic hadronic analog of positronium, the well known electron–positron bound state. This vision has not been fulfilled, at least not in the sense of meson states that fit a Coulomb spectrum. However, the ψ and Υ families of heavy mesons are systems in which manifestly nonrelativistic techniques of bound-state quantum mechanics are warranted by the kinematics. These methods are also successful, not only in correlating experimental information, but also in bringing to hadron spectroscopy an element of predictive power.

Since the discovery of the ψ/J [Aubert et al. 1974, Augustin et al. 1974] and of the ψ' [Abrams et al. 1974], the study of quarkonium physics has flourished. Further stimulus was provided by the discovery [Herb et al. 1977] of the upsilon resonances. These states have subsequently been the subject of many conference reports [Jackson 1977b, Gottfried 1977, Eichten 1978, Jackson et al. 1979, Quigg 1980, Rosner 1980a, Gottfried 1980, Berkelman 1980, Schamberger 1981], summer school lectures [Jackson 1976, Krammer and Krasemann 1979a, b, Rosner 1981a] and review articles [Novikov et al. 1978, Appelquist et al. 1978, Quigg and Rosner 1979, Grosse and Martin 1980]. As is the case for many of the topics treated in these lecture notes, quarkonium physics is rich enough that it could fill an entire course of lectures. To be both brief and intelligible, however, I will restrict my attention to some of the most elementary points and to a quick review of the most recent experimental results.

The study of quarkonium levels may be divided into two broad areas in which different methods of analysis are profitably applied. It is to be hoped, of course, that lessons learned in one area are transferable to the other. The first topic, which I shall not discuss in detail, is the application of perturbative QCD to the strong and electromagnetic decays of quarkonium states. This approach is closely connected with the original motivation of Appelquist and Politzer (1975). The evidence for three-jet events interpreted as

$$\Upsilon \rightarrow ggg, \tag{6.1}$$

has been reviewed by Wiik (1981). We shall return to this interpretation briefly in section 8.3 in connection with the Zweig rule. Another aspect of the perturbative approach is the determination, in either relative or absolute terms, of the rates for various quarkonium decays. Although the analogy with positronium is a powerful tool, quantitative analysis is highly nontrivial because of the importance of QCD radiative corrections and the difficulty of separating them unambiguously from wavefunction effects. For a summary of recent progress, I refer to the rapporteur talk by Buras (1981). The second strategy, to which I shall give an introduction, is the application of nonrelativistic quantum mechanics to the spectra and nonstrong decays of quarkonium states.

The assumption that a nonrelativistic analysis is admissible is an extremely strong one, which must be examined critically. If it is justifiable, it confers a great simplification of the problem as well as the important advantage that rigorous statements can be proved within the framework of potential theory. Work along this line has been divided between efforts to make statements that hold for wide classes of potentials and attempts to determine the nature of the interaction between quarks. Two examples of results which are largely independent of details of the potential will be given below. Within the attempts to determine the potential explicitly, there is yet another division between efforts to derive or deduce the interaction from theory and efforts to infer the interaction from experiment.

The earliest work in this field [e.g., Eichten et al. 1975] drew upon theoretical inspiration, or theoretical prejudice, in assuming a potential of the form

$$V(r) = -\tfrac{4}{3}\alpha_s/r + ar. \tag{6.2}$$

The Coulomb term is inspired by the one-gluon-exchange picture that led to (5.32a). The linear term is contrived to ensure quark confinement, and is consistent with the relativistic string interpretation of the light mesons developed in subsection 5.2.2. In applications, the coefficients α_s and a and the heavy-quark mass are regarded as parameters. It is important to realize that although the limiting forms of the Coulomb-plus-linear potential (6.2) at short and long distances may be very well motivated, there is no corresponding reason to believe that (6.2) can be relied upon at intermediate distances. The same caution can be raised for subsequent work aimed at using perturbative QCD plus some notions of elegance or simplicity to fix the potential over all values of r [Celmaster and Henyey 1978, Carlitz and Creamer 1979, Levine and Tomozawa 1979, 1980, Richardson 1979, Krasemann and Ono 1979, Fogleman et al. 1979, Buchmüller et al. 1980]. These caveats stated, it must be said that early work using the Coulomb-plus-linear form was of great importance in demonstrating the viability of the nonrelativistic approach. In addition, the later work has yielded quite satisfactory descriptions of the ψ and Υ spectra. We shall, however, not discuss this approach further.

I shall stress instead another facet of the work based upon definite potentials, which may be called a model-independent approach. Emphasis is placed on this style of analysis because it permits easy insight into the nature of the interaction between quarks and a test of the self-consistency of the potential model approach. In the end, I believe it is necessary to blend the lessons learned from many techniques. For present purposes, "model-independent" will be taken to mean rather elementary manipulations of the Schrödinger equation. A loftier program based upon an inverse scattering algorithm [Thacker et al. 1978a, b, Grosse and Martin 1979, Quigg, et al. 1980, Schonfeld et al. 1980, Quigg and Rosner 1981] leads to similar results. Although I regard this method as especially powerful, it requires an extended introduction which time does not permit here. I shall also not discuss the well-known problem of El transition rates in charmonium. Let us await the upsilons!

6.1. Scaling the Schrödinger equation

For simple potentials, including power laws and other monotonic wells, rather far-reaching results can be derived using quite elementary techniques. This mode of analysis has been reviewed by Quigg and Rosner (1979), and exploited by many authors. I shall summarize here a few of the results with direct applications to experiment.

6.1.1. Dependence on constituent mass and coupling constant

The reduced radial Schrödinger equation for a particle with mass μ and angular momentum ℓ moving in a central potential V may be written in the form

$$\frac{\hbar^2}{2\mu} u''(r) + \left[E - V(r) - \frac{\ell(\ell+1)\hbar^2}{2\mu r^2} \right] u(r) = 0, \tag{6.3}$$

subject to the boundary conditions

$$u(0) = 0, \qquad u'(0) = R(0). \tag{6.4}$$

A prime is used to denote derivatives with respect to the argument, and the reduced radial wavefunction $u(r)$ is related to the three-dimensional wavefunction,

$$\Psi(r) = R(r) Y_{\ell m}(\theta, \varphi), \tag{6.5}$$

by

$$u(r) = rR(r). \tag{6.6}$$

The familiar substitution (6.6) places the radial equation in three dimensions in formal correspondence with the one-dimensional Schrödinger equation.

For the special case of a power-law potential,

$$V(r) = \lambda r^\nu, \tag{6.7}$$

equation (6.3) can be divested of all its dimensionful parameters. To see this, we first introduce a scaled measure of length

$$\rho \equiv \left(\hbar^2/2\mu|\lambda| \right)^p r, \tag{6.8}$$

where the exponent p is to be chosen to eliminate dimensions from (6.3). The choice

$$p = -1/(2+\nu), \tag{6.9}$$

when accompanied by the substitutions

$$E \equiv \left(\frac{\hbar^2}{2\mu} \right) \left(\frac{\hbar^2}{2\mu|\lambda|} \right)^{2p} \varepsilon, \tag{6.10}$$

where ε is dimensionless, and

$$w(\rho) \equiv u(r) \tag{6.11}$$

accomplishes precisely this. The ensuing equation is

$$w''(\rho) + \left[\varepsilon - \mathrm{sgn}(\lambda)\rho^\nu - \ell(\ell+1)/\rho^2\right]w(\rho) = 0, \tag{6.12}$$

which depends only on pure numbers.

Several consequences follow immediately from this legerdemain. Lengths and quantities with the dimensions of lengths depend upon the constituent mass and coupling strength as

$$L \propto \left(\mu|\lambda|\right)^{-1/(2+\nu)}. \tag{6.13}$$

As a result, the particle density at the origin of coordinates behaves as

$$|\Psi(0)|^2 \sim L^{-3} \propto \left(\mu|\lambda|\right)^{3/(2+\nu)}. \tag{6.14}$$

Level spacings have a similarly definite behavior, according to (6.10):

$$\Delta E \propto \mu^{-\nu/(2+\nu)}|\lambda|^{2/(2+\nu)}. \tag{6.15}$$

The limiting behavior of the scaled Schrödinger equation as $\nu \to 0$ is the subject of problem 21. It is easy to see that the scaling laws (6.13–6.15) contain many well known results. Recall, for example, that in the Coulomb potential, for which $\nu = -1$,

$$\Delta E(\nu = -1) \propto \mu\alpha^2 = \mu|\lambda|^2. \tag{6.16}$$

Likewise, the conclusion that in a linear potential

$$|\Psi(0)|^2|_{\nu=1} \propto \mu|\lambda| \tag{6.17}$$

can be derived at once, using the identity (see problem 22)

$$|\Psi(0)|^2 = \frac{\mu}{2\pi\hbar^2}\left\langle \frac{dV}{dr}\right\rangle. \tag{6.18}$$

The scaling laws (6.13–6.15) have many applications in quarkonium physics. For the moment let us merely note that electric multipole matrix elements vary as

$$\langle n'|Ej|n\rangle \sim L^j \propto \left(\mu|\lambda|\right)^{-j/(2+\nu)}, \tag{6.19}$$

so that transition rates behave as

$$\Gamma(Ej) \sim k^{2j+1}|\langle n'|Ej|n\rangle|^2, \tag{6.20}$$

where k is the energy of the radiated photon, which is just a level spacing

ΔE. Using (6.13) and (6.15) we then deduce that

$$\Gamma(Ej) \propto \mu^{-[2j(1+\nu)+\nu]/(2+\nu)} |\lambda|^{2(j+1)/(2+\nu)}. \tag{6.21}$$

This has the interesting consequence that for fixed potential strength $|\lambda|$, $\Gamma(Ej)$ is a decreasing function of j as $\mu \to \infty$ for potentials less singular than the Coulomb potential.

Using the expression

$$\Gamma(V^0 \to e^+ e^-) = \frac{16\pi\alpha^2}{M_V^2} |\Psi(0)|^2 \langle e_q^2 \rangle \tag{6.22}$$

derived in problem 13, one may easily show that for $\nu > -1$ (for which binding energies are asymptotically negligible)

$$\Gamma(Ej)/\Gamma(V^0 \to e^+ e^-) \propto \mu^{-(2j-1)(\nu+1)/(2+\nu)} |\lambda|^{2(j-1)/(2+\nu)}, \tag{6.23}$$

which implies the dominance of leptonic over radiative decays as $\mu \to \infty$ for fixed potential strength $|\lambda|$.

6.1.2. Dependence on principal quantum number

To investigate how observables depend upon the principal quantum number with some degree of generality it is convenient to adopt the semiclassical, or JWKB approximation. This turns out to be rather less of a compromise than one might at first surmise. Judiciously applied, the semiclassical approximation is in fact highly accurate for the sort of nonpathological potentials one hopes to encounter for quarkonium. This accuracy is documented in Quigg and Rosner (1979), where additional references may be found.

The semiclassical results all follow from the quantization condition

$$\int_0^{r_c} dr \{2\mu[E - V(r)]\}^{1/2} = (n - 1/4)\pi\hbar, \tag{6.24}$$

where n is the principal quantum number and the classical turning point r_c is defined through $V(r_c) = E$. Although it is both possible and useful to be more general, it is appropriate to retain the spirit of the preceding section and specialize to power-law potentials. For s-wave bound states of nonsingular potentials of the form (6.7), eq. (6.24) can be integrated by elementary means to yield:

$$E_n \propto (n - 1/4)^{2\nu/(2+\nu)}, \tag{6.25}$$

where with an eye toward the intended applications I have suppressed the dependence on constituent mass and coupling strength given in (6.15).

For singular potentials additional care is required near the origin. A simple modification of the usual procedure leads to

$$E_n \propto [n - \gamma(\nu)]^{2\nu/(2+\nu)}, \quad -2 < \nu < 0, \tag{6.26}$$

where

$$\gamma(\nu) = \frac{1}{2}\left(\frac{1+\nu}{2+\nu}\right). \tag{6.27}$$

Similar expressions may be obtained for orbitally-excited states.

By evaluating the expectation value in eq. (6.18) with JKWB wavefunctions, it is also straightforward to derive

$$|\Psi_n(0)|^2 \propto \begin{cases} (n - 1/4)^{2(\nu - 1)/(2 + \nu)}, & \nu > 0, \tag{6.28a} \\ [n - \gamma(\nu)]^{(\nu - 2)/(2 + \nu)}, & 0 > \nu > -2. \tag{6.28b} \end{cases}$$

A more general result is derived in problem 23. Generalizations of these results to $\ell \neq 0$ have also been made, but we shall not require them here. Let us now see what can be learned by comparing the simple results of this section with experimental information.

6.2. Inferences from experiment

6.2.1. Data

The spectrum of $(c\bar{c})$ bound states is summarized in table 19 and fig. 30. In addition to refinements in the branching ratios for radiative decays, there are two recent developments of note [Scharre 1981, Schamberger 1981].

The candidate U(2980) for the 1^1S_0 hyperfine partner of the ψ now appears firmly established [Partridge et al. 1980, Himel et al. 1980b]. I

Table 19
Some properties of the $^3S_1$ ψ states [from Particle Data Group 1980]

Level	$\Gamma(\psi \to e^+e^-)$ (keV)	Γ_{tot}
$\psi(3097)$	4.60 ± 0.42	63 ± 9 keV
$\psi'(3685)$	2.05 ± 0.23	215 ± 40 keV
$\psi(4029)$	0.75 ± 0.15	52 ± 10 MeV
$\psi(4159)$	0.77 ± 0.23	78 ± 20 MeV
$\psi(4415)$	0.49 ± 0.13	42 ± 10 MeV

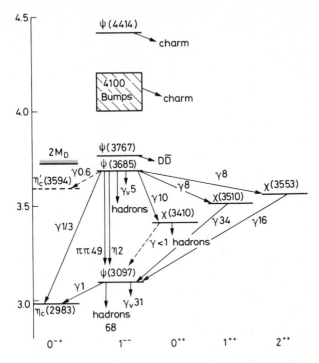

Fig. 30. The spectrum of charmonium (c̄c). Branching fractions (in percent) are shown for the important classes of decays [Particle Data Group 1980, Himel et al. 1980a, Oreglia et al. 1980, Schamberger 1981, Scharre 1981]. Charm threshold is indicated at twice the D meson mass.

therefore designate it as $\eta_c(2983 \pm 5)$ which corresponds to a hyperfine mass splitting of

$$M_\psi - M_{\eta_c} = 114 \pm 5 \text{ MeV}/c^2. \tag{6.29}$$

The state has a total width $\Gamma(\eta_c \to \text{all}) = 8\text{–}19$ MeV, considerably larger than that of the ψ. This is in qualitative accord with the ortho–parapositronium analogy. The η_c is observed in the decays $\psi \to \gamma\eta_c$ and $\psi' \to \gamma\eta_c$, with branching ratios

$$\Gamma(\psi \to \gamma\eta_c)/\Gamma(\psi \to \text{all}) = 0.7 \pm 1.5\%, \tag{6.30}$$

which implies:

$$\Gamma(\psi \to \gamma\eta_c) = 0.44\text{–}0.95 \text{ keV}, \tag{6.31}$$

Table 20
Some properties of the $^3S_1$ Υ states [from the review by Schamberger 1981]

Level	$\Gamma(\Upsilon \to e^+e^-)$ (keV)	Γ_{tot} (keV)
$\Upsilon(9433)$	1.17 ± 0.05	35.5^{+8}_{-6}
$\Upsilon'(9993)$	0.54 ± 0.03	$\approx \Gamma(\Upsilon)$
$\Upsilon''(10\,323)$	0.37 ± 0.03	
$\Upsilon'''(10\,546)$	0.27 ± 0.02	~ 15 MeV

and

$$\Gamma(\psi' \to \gamma \eta_c)/\Gamma(\psi' \to \text{all}) = 0.32 \pm 0.05 \pm 0.05\%, \tag{6.32}$$

which implies:

$$\Gamma(\psi' \to \gamma \eta_c) = 0.69 \pm 0.23 \pm 0.11 \text{ keV}. \tag{6.33}$$

Where two errors are shown, the first is statistical and the second is systematic.

There is now also a candidate or the radial excitation of η_c, which would be the 2^1S_0 hyperfine partner of ψ'. This state, which I provisionally designate as η_c' (3592 ± 5), has a total width smaller than 9 MeV. The hyperfine splitting is

$$M_{\psi'} - M_{\eta_c'} = 92 \pm 5 \text{ MeV}/c^2, \tag{6.34}$$

which is, as we shall see below in section 6.4, a reasonable value. The state is observed in the inclusive radiative decay $\psi' \to \gamma \eta_c'$ with a branching ratio of

$$\Gamma(\psi' \to \gamma \eta_c')/\Gamma(\psi' \to \text{all}) = 0.6 \times 2^{\pm 1}\%, \tag{6.35}$$

which implies a decay rate

$$\Gamma(\psi' \to \gamma \eta_c') = 1.3 \times 2^{\pm 1} \text{ keV}. \tag{6.36}$$

No exclusive channels have yet been identified.

In the upsilon spectrum, only vector states have been observed. Their properties are summarized in table 20 and fig. 31.

6.2.2. Consequences

The strategy embodied in section 6.1 has been pursued explicitly by several authors [including Quigg and Rosner 1977, 1978a, 1979, Quigg 1980, Martin 1980, Rosner 1980a, 1981a] and implicitly by many others.

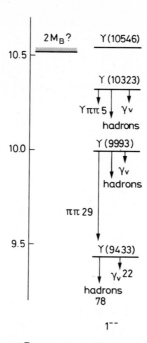

Fig. 31. The spectrum of upsilon ($b\bar{b}$) states. Branching fractions (in percent) are shown for the important classes of identified decays [Schamberger 1981]. Beauty threshold is indicated schematically.

The conclusion to be drawn from the data is that a potential of the form

$$V(r) = A + Br^v, \tag{6.37}$$

with $v \approx 0.1$ gives a good representation of the ψ and Υ spectra. This is based upon four distinct kinds of evidence:

(i) We may note by comparing figs. 30 and 31 that the level spacings are quite similar in the ψ and Υ families. Indeed, the observation that

$$M_{\Upsilon'} - M_{\Upsilon} \approx M_{\psi'} - M_{\psi}, \tag{6.38}$$

provided an early motivation for the logarithmic potential [Quigg and Rosner 1977]. A more detailed look at the intervals indicates that

$$\Delta E(\Upsilon) \approx 0.95 \Delta E(\psi). \tag{6.39}$$

Assuming that the potential strength does not vary between the ψ and Υ systems, this implies a small positive power for the effective potential.

The precise value of the exponent depends upon the ratio of quark masses, which is imperfectly known.

(ii) The principal-quantum-number dependence of observables within one quarkonium system is free from the assumption that the potential strength λ is the same for different quark flavors. Effective powers may be inferred independently from the ψ and Υ levels and compared for consistency. The level structures $(E_3 - E_2)/(E_2 - E_1)$ etc. are characteristic of the potential shape. These ratios of intervals are the same for ψ and Υ states, and are again compatible with $\nu \approx 0.1$.

(iii) Similarly, the 2S–2P spacing, known only for the ψ family, implies a small positive power.

(iv) Finally, the principal-quantum-number dependence of wavefunctions at the origin, or equivalently of the reduced leptonic widths (4.51), is approximately given by

$$|\Psi_n(0)|^2 \sim 1/(n - 1/4), \tag{6.40}$$

for both ψ and Υ. This behavior again corresponds to an effective potential which is a small positive power. It was this observation for the ψ family that led Machacek and Tomozawa (1978) to investigate softer-than-linear confining potentials, including logarithmic forms. Taken together, these results on principal-quantum-number dependence would seem to exclude the bizarre possibility that the nearly equal spacing in the ψ and Υ families results from a potential strength which varies approximately as

$$\lambda \propto \mu^{\nu/2}. \tag{6.41}$$

Martin (1980) has shown that careful attention to hyperfine effects does not change the conclusions of this analysis, namely that the inter-quark potential is flavor independent (as QCD would have it) and characterized by an effective power-law potential with a small positive exponent. This is also in agreement with the conclusions of all other analyses and fits: In the region of space between 0.1 fm and 1 fm, the interaction between heavy quarks is flavor independent, and roughly logarithmic in shape [Buchmüller and Tye 1981, Quigg and Rosner 1981].

6.3. Theorems and near-theorems

An excellent review of statements about bound-state properties which may be proved rigorously in nonrelativistic potential theory has been given by Grosse and Martin (1980). Many results have been deduced which pertain to the order of levels, inequalities for wavefunctions at the origin, bounds on quark mass differences and so forth. The value of such statements is not only that they are true, but also that they provide a context for computations based upon explicit potentials. It is of great value to understand what must be true for any reasonable potential, or for any potential of a particular class, in order to distinguish the consequences that may be peculiar to a specific model. I shall cite two examples that bear directly upon experimental results.

Considerer a quarkonium potential which is monotonic,

$$dV/dr \geqslant 0, \tag{6.42}$$

and concave downward,

$$d^2V/dr^2 \leqslant 0. \tag{6.43}$$

The first property is motivated by simplicity, and the second by the expectation that the confining potential rises no faster than linearly. Both are satisfied by the effective power-law potentials just discussed. Then if $m > \mu$ are masses of the constituents of two $Q\overline{Q}$ systems, one may prove [Rosner et al. 1978, Leung and Rosner 1979] that:

$$|\Psi_m(0)|^2 \geqslant (m/\mu)|\Psi_\mu(0)|^2. \tag{6.44}$$

This result holds for the ground state under the assumptions stated, for all levels in power-law potentials [compare eqn. (6.14)], and for all levels in a general potential satisfying the assumptions, in WKB approximation [Grosse and Martin 1980]. It implies a lower bound on leptonic widths in the more massive system as, in the case at hand,

$$\Gamma(\Upsilon_n \to e^+e^-) \geqslant \frac{m_b}{m_c} \cdot \frac{e_b^2}{e_c^2} \cdot \frac{M(\psi_n)^2}{M(\Upsilon_n)^2} \Gamma(\psi_n \to e^+e^-). \tag{6.45}$$

The lower bounds on upsilon leptonic widths are plotted in fig. 32, together with the experimental measurements. A b-quark charge of $2/3$ is seen to be incompatible with the bound. The conclusion that $|e_b| = 1/3$ is substantiated by the measurements of $R = \sigma(e^+e^- \to \text{hadrons})/\sigma(e^+e^- \to \mu^+\mu^-)$ shown in fig. 22.

A semiclassical near-theorem relates the number of levels below flavor threshold to the mass of the constitutents. This would seem to be a question ill-suited to a nonrelativistic approach because it is necessary to compute both quarkonium $(Q\overline{Q})$ masses and the mass of the lightest flavored $(Q\overline{q})$ state. The latter is unlikely to be governed by a potential theory description. However, a key simplifying observation was made by

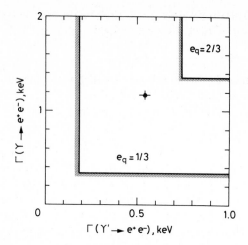

Fig. 32. Lower bounds for leptonic decays of Υ and Υ' [after Rosner et al. 1978] together with the data cited in table 20. The bounds are computed from eq. (6.45) using ψ leptonic widths 1σ below the central values and assuming $m_b/m_c \geq 3$.

Eichten and Gottfried (1977), who noted that the mass of the light-quark–heavy-quark state can be written as

$$M(Q\bar{q}) = M(Q) + M(q) + \text{binding} + \text{hyperfine}. \qquad (6.46)$$

Although the binding energy may not be calculable, it is reasonable to suppose that it depends upon the reduced mass of the constituents, which tends to $M(q)$ as $M(Q) \to \infty$. Thus the binding energy must become independent of the heavy quark mass. Furthermore, the hyperfine splitting of the 0^{-+} and 1^{--} $(Q\bar{q})$ levels must certainly (see section 5.3) vary as $1/M(Q)$. It therefore vanishes as $M(Q) \to \infty$. Hence in the limit of infinite quark mass, the difference

$$\delta(M(Q)) \equiv 2M(Q\bar{q}) - 2M(Q) \to \delta_\infty, \qquad (6.47)$$

independent of the heavy-quark mass.

In the regime in which $\delta(M(Q)) \approx \delta_\infty$ is a good approximation, the number of levels below flavor threshold is easily calculated [Quigg and Rosner 1978b]. Consider any confining potential. In semiclassical approximation the number of levels bound below $E = 2M(Q) + \delta_\infty$ is specified by the quantization condition

$$\int_0^{r_\delta} dr\{M(Q)[\delta_\infty - V(r)]\}^{1/2} = (n - 1/4)\pi, \qquad (6.48)$$

where to save writing the zero of energy has been set at $2M(Q)$. The classical turning point r_δ, defined through

$$V(r_\delta) = \delta_\infty, \tag{6.49}$$

is independent of $M(Q)$, so we have by inspection the result that

$$(n - 1/4) \propto \sqrt{M(Q)} . \tag{6.50}$$

It is likely that the limit (6.47) is already approached within 10% in the charmonium system, in which two $^3S_1$ levels lie below charm threshold. Thus there should be slightly less than four bound levels in the upsilon family, in agreement with the observation of three narrow vector states. The success of this prediction provides another verification of flavor independence, which was the principal assumption. Many narrow levels are thus to be expected for the next quarkonium family, when it is found, since the next quark mass certainly exceeds 18 GeV$/c^2$ [Wiik, 1981].

A corollary to the conclusion that the classical turning point of the last narrow level has become independent of quark mass is that the single-channel analysis cannot be extended past about 1 fm. Heavier $(Q\overline{Q})$ systems will extend our knowledge of the interaction to shorter distances [for some specifics see Moxhay et al. 1981], but are unlikely to address the nature of the confining potential.

6.4. Remarks on spin-singlet states

Experimental progress toward establishing the properties of the $^1S_0$ charmonium levels prompts some elementary comments. First let us examine the hyperfine splittings. If they are determined by the mechanism described in section 5.3 for the light hadrons, which should in any case be more trustworthy for quarkonium, we expect

$$M_\psi - M_{\eta_c} \approx \frac{32\pi\alpha_s}{9m_c^2}|\Psi(0)|^2 \approx \frac{M_\psi^2\alpha_s\Gamma(\psi \to e^+e^-)}{2m_c^2\alpha^2}, \tag{6.51}$$

where the second equality follows from the Van Royen–Weisskopf formula (6.22). The leptonic width quoted in table 19 leads to the numerical estimate

$$\alpha_s \approx 1/2, \tag{6.52}$$

but this is reduced by QCD radiative corrections [Buchmüller et al. 1981,

Barbieri et al. 1981] to

$$\alpha_s \approx 0.3, \tag{6.53}$$

a plausible value. The hyperfine splitting between the radial excitations may then be estimated as

$$M_{\psi'} - M_{\eta_c'} \approx \left(M_\psi - M_{\eta_c} \right) \frac{M_{\psi'}^2 \Gamma(\psi' \to e^+ e^-)}{M_\psi^2 \Gamma(\psi \to e^+ e^-)} \approx 75 \pm 20 \text{ MeV}/c^2.$$

$$\tag{6.54}$$

This is in reasonable agreement with the experimental suggestion (6.34). We have thus encountered no mysteries. The expression (6.51), applied mutatis mutandis to the upsilon ground state, yields the order of magnitude estimate

$$M_T - M_{\eta_b} \approx M_\psi - M_{\eta_c}. \tag{6.55}$$

To the extent that they are known, the M1 transition rates to the $^1S_0$ states also seem reasonable. In complete analogy to the treatment of subsection 3.4.4, we may compute

$$\Gamma(\psi \to \gamma \eta_c) = 4\alpha e_c^2 \omega^3 / 3m_c^2. \tag{6.56}$$

Bearing in mind the uncertainty in the charmed quark mass, but choosing $m_c = 1.5 \text{ GeV}/c^2$ for illustration, we expect:

$$\Gamma(\psi \to \eta_c \gamma) \approx 2.7 \text{ keV} \tag{6.57}$$

and

$$\Gamma(\psi' \to \eta_c' \gamma) \approx 1.5 \text{ keV} \tag{6.58}$$

for the observed photon energies. The predicted ground-state rate is somewhat too high, but the 2S prediction agrees well with the preliminary data. [Joy over the latter success is tempered by the discovery of a confusion between per cent and fractions in the ordinate of fig. 11b of Quigg (1980)]. It will be important to check, as data improve, the ω^3 dependence embodied in (6.56). At the present stage of quarkonium theory, one can only guess at the hindered $M1$ rate for the transition $\psi' \to \gamma \eta_c$. The standard estimates for the branching ratio,

$$\Gamma(\psi' \to \gamma \eta_c)/\Gamma(\psi' \to \text{all}) \approx 10^{-2} - 10^{-3}, \tag{6.59}$$

do not disagree with the measurement (6.32).

6.5. *Relation to light-quark spectroscopy*

Several authors, especially Lipkin (1978a, b, 1979b), Cohen and Lipkin (1980), Martin (1981), and Richard (1981), have attempted to extend the successful description of the quarkonium spectra to light mesons and baryons. This may be done either by abstracting the scaling laws from the ψ and Υ states or by transplanting the quarkonium potential to what would seem a manifestly relativistic regime. This activity requires more courage than I can summon, but the resulting numerical correlations are suggestive indeed. As food for thought I present in fig. 33 a highly speculative spectrum of $(s\bar{s})$ states. Many of the assignments are uncertain, but the resemblance to the ψ and Υ spectra is remarkable, as Martin (1981) and Close (1981) have also commented. Whether this spectrum (if correct !) shows that nonrelativistic analysis has a wider-than-expected range of validity, or that a deeper principle of hadron dynamics awaits

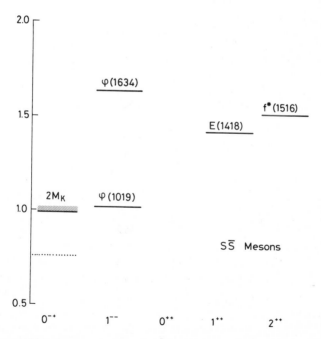

Fig. 33. A possible spectrum of strangeonium (s$\bar{s}$) levels. Identification of E(1418) and φ(1634) as pure s$\bar{s}$ states may be disputed. The dotted 0^{-+} entry is impressionistic, having been invented from the φ mass and the $\pi - \rho$ splitting, appropriately rescaled.

recognition, I do not know. The parallel between figs. 30, 31, and 33 and the nuclear level schemes shown in figs. 1–3 is haunting.

7. Toward the bag

We have already mentioned in subsection 3.4.2 some consequences of quark confinement, in the context of an extremely stylized description of confinement: the boundary condition that the Dirac wavefunctions vanish on a static spherical surface. The static cavity approximation, as it is called, is a principal technical assumption in the formulation of the MIT bag model [Chodos et al. 1974a, b, De Grand et al. 1975]. Of the bag model itself, which has been extremely influential in hadron spectroscopy, there exist several fine reviews, including those by Johnson (1975, 1977, 1979), DeTar (1980), Hasenfratz and Kuti (1978), Jaffe and Johnson (1977), and Jaffe (1977d, 1979) as well as the summary in Close (1979). A different but related picture of quark confinement, known as the SLAC bag, was put forward by Bardeen et al. (1975); see also Giles (1976). Our interest here is much more restricted: to understand how the mechanism of quark confinement [see Wilson 1974, Nambu 1976, Mandelstam 1980, 't Hooft 1980, Adler 1981, Bander 1981] thought to operate in QCD may give rise to hadronic bags. In the absence of a compelling argument, I follow the usual practice of giving two incomplete arguments. The heuristic discussions are themselves quite standard, and can be found in similar form in many places, including Kogut and Susskind (1974), Lee (1980) and Gottfried and Weisskopf (1981).

7.1. An electrostatic analog

It is typical in field theories that the coupling constant depend upon the distance scale. This dependence can be expressed in terms of a dielectric constant ε. We define

$$\varepsilon(r_0) \equiv 1, \tag{7.1}$$

and write

$$g^2(r) = g^2(r_0)/\varepsilon(r). \tag{7.2}$$

We assert that the implication of asymptotic freedom [Gross and Wilczek 1973a, Politzer 1973, see also 't Hooft 1973a, b, and Khriplovich 1969] is that in QCD the effective color charge decreases at short distances and increases at large distances. In other words, the dielectric "constant" will obey

$$\varepsilon(r) > 1, \quad \text{for } r < r_0, \tag{7.3a}$$

$$\varepsilon(r) < 1, \quad \text{for } r > r_0. \tag{7.3b}$$

Indeed, to second order in the strong coupling we may write

$$\varepsilon(r) = \left[1 + \frac{1}{2\pi}\frac{g^2(r_0)}{4\pi}(11 - 2n_f/3)\ln(r/r_0) + O(g^4)\right]^{-1} \tag{7.4}$$

in QCD, where n_f is the number of active quark flavors.

Let us now consider an idealization based upon electrodynamics. In Quantum ElectroDynamics, we choose

$$\varepsilon_{\text{vacuum}} = 1, \tag{7.5}$$

and can show [e.g., Landau and Lifshitz 1960] that physical media have $\varepsilon > 1$. The displacement field is

$$D = E + 4\pi P, \tag{7.6}$$

and atoms are polarizable with P parallel to the applied field E, so that $|D| \geqslant |E|$. Since the dielectric constant is defined through

$$D = \varepsilon E$$

in these simple circumstances, we conclude that $\varepsilon > 1$. For a thorough treatment, see Dolgov et al. (1981).

Now let us consider, in contrast to the familiar situation, the possibility of a dielectric medium with

$$\varepsilon_{\text{medium}} = 0, \tag{7.7}$$

a perfect dia-electric, or at least

$$\varepsilon_{\text{medium}} \ll 1, \tag{7.8}$$

a very effective dia-electric medium. We can easily show that if a test charge is placed within the medium, a hole will develop around it.

To see this, consider the arrangement depicted in fig. 34a, a positive charge distribution ρ_+ placed in the medium. Suppose that a hole is formed. Then because the dielectric constant of the medium is less than unity, the induced charge on the inner surface of the hole will also be positive. The test charge and the induced charge thus repel, and the hole is stable against collapse. In normal QED, the induced charge will be negative, as indicated in fig. 34b, and will attract the test charge. The hole is thus unstable against collapse.

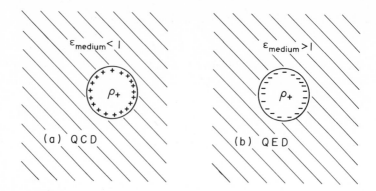

Fig. 34. Charge induced by a positive test charge placed at the center of a hole in a dielectric medium. (a) Dia-electric case $\varepsilon_{\text{medium}} < 1$ hoped to resemble QCD. (b) Dielectric case $\varepsilon_{\text{medium}} > 1$ of normal electrodynamics.

The radius R of the hole can be estimated on the basis of energetics. Within the hole the electrical energy W_{in} is finite and independent of the dielectric constant of the medium. The displacement field is radial and hence continuous across the spherical boundary. Thus it is given outside the hole by

$$\boldsymbol{D}_{\text{out}}(r > R) = \hat{r}Q/r^2, \tag{7.9}$$

where Q is the total test charge. The induced charge density on the surface of the hole is

$$\sigma_{\text{induced}} = (1 - \varepsilon)|\boldsymbol{D}(R)|/4\pi\varepsilon = (1 - \varepsilon)Q/4\pi\varepsilon R^2, \tag{7.10}$$

which has the same sign as Q, as earlier asserted. Outside the hole, the electric field is determined by the total interior charge

$$Q + (1 - \varepsilon)Q/\varepsilon = Q/\varepsilon, \tag{7.11}$$

so that

$$\boldsymbol{E}_{\text{out}}(r > R) = \hat{r}Q/\varepsilon r^2. \tag{7.12}$$

The energy stored in electric fields outside the hole is then

$$W_{\text{out}} = \frac{1}{8\pi}\int d^3r \boldsymbol{D}_{\text{out}}(r) \cdot \boldsymbol{E}_{\text{out}}(r) = \tfrac{1}{2}\int_R^\infty r^2\,dr\,Q^2/\varepsilon r^4 = Q^2/2\varepsilon R. \tag{7.13}$$

As the dielectric constant of the medium approaches zero, W_{out} becomes large compared to W_{in}, so that the total electric energy

$$W_{el} \equiv W_{in} + W_{out} \rightarrow W_{out}, \quad \text{as } \varepsilon \rightarrow 0. \tag{7.14}$$

One must consider as well the energy required to hew such a hole out of the medium. For a hole of macroscopic size, it is reasonable to suppose that

$$W_{hole} = \tfrac{4}{3}\pi R^3 v + 4\pi R^2 s + \cdots, \tag{7.15}$$

where v and s are non-negative constants. The total energy of the system,

$$W = W_{el} + W_{hole}, \tag{7.16}$$

can now be minimized with respect to R. In the regime where the volume term dominates W_{hole}, the minimum occurs at

$$R = \left(\frac{Q^2}{2\varepsilon} \cdot \frac{1}{4\pi v} \right)^{1/4} \neq 0, \tag{7.17}$$

for which

$$W_{el} \approx \left(\frac{Q^2}{2\varepsilon} \right)^{3/4} (4\pi v)^{1/4}, \tag{7.18}$$

and

$$W_{hole} \approx \tfrac{1}{3}\left(\frac{Q^2}{2\varepsilon} \right)^{3/4} (4\pi v)^{1/4}, \tag{7.19}$$

so that

$$W \approx \tfrac{4}{3}\left(\frac{Q^2}{2\varepsilon} \right)^{3/4} (4\pi v)^{1/4}. \tag{7.20}$$

Thus, in a very effective dia-electric medium, a test charge will induce a bubble or hole of finite radius. Notice, however, that in the limit of a perfect dia-electric medium

$$W \rightarrow \infty \quad \text{as} \quad \varepsilon \rightarrow 0. \tag{7.21}$$

An isolated charge in a perfect dia-electric thus has infinite energy. This is the promised analog of the argument used in subsection 5.2.1 to wish away isolated colored objects.

If instead of an isolated charge, we place a test dipole within the putative hole in the medium, we can again show that the minimum energy configuration occurs for a hole of finite radius about the test dipole. In this case, however, the field lines need not extend to infinity, so the hole radius remains finite as $\varepsilon \to 0$, and so does the total energy of the system. The analogy between the exclusion of chromoelectric flux from the QCD vacuum and the exclusion of magnetic flux from a superconductor is now obvious. To separate the dipole charges to $\pm \infty$ requires an infinite amount of work, as shown in the previous example. This is the would-be analog of quark confinement. For a recent attempt to deduce an effective dia-electric theory from QCD, see Nielsen and Patkós (1982).

Two issues arise in this line of reasoning. One is the question of quark (or as we have phrased it here, charge) confinement. The other is, what form does the sourceless QCD vacuum take if it is analogous to a perfect, or very effective, dia-electric medium? Is the QCD vacuum unstable against the formation of domains containing dipole pairs in the electrostatic model, corresponding to gluons in color-singlet spin-singlet configurations?

7.2. A string analog

Suppose, as discussed in subsection 5.2.2, that color-electric flux lines are squeezed into a flux tube. This effect can be parametized by the statement that a region of space of volume

$$V = \sigma r, \tag{7.22}$$

containing color-electric flux, contributes a term

$$W_{bag} = BV = B\sigma r \tag{7.23}$$

to the total energy of the world, where B is a positive constant. The effect of the "bag pressure" B will be to compress the flux lines as much as possible.

The region of color-electric field emanating from a source of charge Q contains an energy density $E^2/8\pi$, where the electric field strength is

$$E = 4\pi Q/\sigma \tag{7.24}$$

if the flux lines are confined within an area σ. The energy stored in the field is thus

$$W_{field} = \frac{E^2}{8\pi} V = \left(\frac{4\pi Q}{\sigma} \right)^2 \frac{\sigma r}{8\pi} = 2\pi Q^2 r/\sigma. \tag{7.25}$$

The total bag plus field energy is

$$W = W_{bag} + W_{field} = (B\sigma + 2\pi Q^2/\sigma)r, \tag{7.26}$$

which can be minimized with respect to the area σ, whereupon

$$\sigma_0 = Q(2\pi/B)^{1/2}. \tag{7.27}$$

At this minimum, the energy density per unit length is

$$k = B\sigma_0 + 2\pi Q^2/\sigma_0 = 4\pi Q^2/\sigma_0. \tag{7.28}$$

For a quark–antiquark pair, the replacement

$$Q^2 \rightarrow 4\alpha_s/3 \tag{7.29}$$

[compare (5.32a)] leads to

$$k = 16\pi\alpha_s/3\sigma_0, \tag{7.30}$$

or

$$\sigma_0 = 16\pi\alpha_s/3k \equiv \pi d^2. \tag{7.31}$$

Recalling from (5.22) that

$$k^{-1/2} \approx 1/2 \text{ fm}, \tag{7.32}$$

we find that:

$$d \approx 2(\alpha_s/3)^{1/2} \text{ fm}. \tag{7.33}$$

For a strong coupling constant $\alpha_s \approx 1$, the radius of the flux tube is

$$d \approx 1 \text{ fm}, \tag{7.34}$$

a reasonable hadronic dimension. We have therefore contrived a situation in which a flux tube of finite radius is a stable configuration. It remains to show that this situation actually obtains in QCD.

Both in the present discussion and in subsection 5.3.2 we have neglected quark masses. Their inclusion is interesting as a matter of principle and is of some practical importance for particles composed of heavy quarks. Within the framework of an extended bag, the problem has been addressed by Johnson and Thorn (1976) and by Johnson and Nohl (1979); see also Chodos and Thorn (1974). Their work suggests that the Regge trajectories of particles composed of massive quarks should be shallow at low spins, but should approach a universal slope as $J \rightarrow \infty$. Some evidence for the first half of this statement was noted in section 4.1, in connection with fig. 19.

If chromoelectric confinement is indeed the origin of the string picture, we also gain an understanding of the equality of the Regge slopes of the mesons and baryons, which is apparent from figs. 19 and 20. In an elongated bag, both mesons:

$$q\text{———}\bar{q}, \tag{7.35}$$

and baryons:

$$q\text{———}(qq) \tag{7.36}$$

are [3]–[3*] color configurations. They must therefore have the same chromoelectric flux density, hence the same amount of stored energy per unit area, hence the same Regge slope. It will thus be of considerable interest to learn the Regge trajectories of baryons containing several strange quarks or a heavy quark.

7.3. Quark nonconfinement?

If we assume, in view of the heuristic arguments reviewed above, that unbroken QCD is indeed a confining theory, how might we accommodate the observation of free quarks? At first sight it seems straightforward to consider a spontaneously broken color symmetry which endows gluons with small masses and permits quark liberation. This has been explored by De Rújula et al. (1978), for example. Georgi (1980) has countered that a small mass term in the Lagrangian need not, in the face of strong quantum corrections, lead to a spontaneous symmetry breakdown. This possibility is open to discussion [De Rújula et al. 1980]. Okun and Shifman (1981) have argued that this style of partial confinement is incompatible with the known evidence for asymptotic freedom and with the absence of fractionally charged hadrons. A different pattern of spontaneous symmetry breaking has been advocated by Slansky et al. (1981). Evidently the experimental search for fractionally charged matter and the theoretical search for proofs or evasions of confinement are research topics of no little importance.

8. Glueballs and related topics

The possibility of quarkless states, composed entirely of gluons, would seem to be unique to a non-Abelian field theory such as QCD—as opposed to the elementary quark model. In this short introduction to glueballs I shall try to explore the four important questions:
  (i) Should glueballs exist in QCD?
  (ii) What are their properties?

(iii) How can they be found?

(iv) Are they found in nature?

Since most of what we believe to be the solution to QCD is abstracted from the elementary quark model, and because the quark model provides no guidance for quarkless states, the answers given to all of these questions will be partial and frustratingly vague. In the course of explaining these partial answers, one naturally encounters some other issues of significance: violations of the Zweig rule, deviations from ideal mixing, and the continuing problem of the pseudoscalar masses. A common thread will be seen to run through all these topics, and to tie them to the properties of glueballs.

The search for quarkless states has become intense, and several candidate states have appeared. I am not prepared to endorse any of these claims, at least not yet, but I shall have a little bit to say about the experimental situation. This will include some general and specific suggestions for experimental studies.

The subject of glueballs is a newly active one, which remains to be distinctly defined by experimental observations and by theoretical predictions of greater clarity. The modest aim of this section is merely to underline the importance of the topic, and to introduce some of the issues involved. As for multiquark states, understanding the role of quarkless states in hadron spectroscopy remains in the future.

8.1. The idea of glueballs

If color is confined, color-singlet states composed entirely of glue may exist as isolated hadron resonances. This is in essence the argument for the existence of quarkless states, as emphasized quite early by Fritzsch and Gell-Mann (1972). If one assumes the existence of gluons, the gauge interaction among gluons, and the confinement of color, this conclusion cannot easily be challenged. After the recognition of asymptotic freedom and the increasingly explicit formulation of QCD [among them Fritzsch et al. 1973, Gross and Wilczek 1973b, Weinberg 1973], many authors have analyzed, in one or another framework, the possibilities for glueballs. A partial bibliography includes the papers by Freund and Nambu (1975), Fritzsch and Minkowski (1975), Bolzan et al. (1976), Jaffe and Johnson (1976), Willemsen (1976), Kogut et al. (1976), Veneziano (1976), Robson (1977), Roy and Walsh (1978), Koller and Walsh (1978), Ishikawa (1979a, b), Bjorken (1979, 1980), Novikov et al. (1979, 1980a, b, c, d, 1981), Zakharov (1980a, b), Suura (1980), Donoghue (1980, 1981), Roy

(1979, 1980), Soni (1980), Berg (1980), Coyne et al. (1980), Carlson et al. (1980, 1981), Bhanot and Rebbi (1981), Bhanot (1981), Shifman (1981), Barnes (1981).

Bjorken (1979, 1980) has emphasized the apparent inevitability of color-singlet, flavor-singlet multigluon states within QCD. In pure (sourceless) QCD, with no fermions, the existence of glueballs follows at once from our assumptions stated above. This may be argued in any of the pictures we have discussed before. A "most attractive channel" analysis is implicit in the work of Barnes (1981). Bag arguments, of the sort given in section 7, lead to the conclusion that the color-singlet configuration is energetically favored, whereas colored states require infinite energy. The string picture of subsection 5.2.2 is also easily transplanted, with gluon sources replacing quark sources and glueballs replacing $(q\bar{q})$ mesons. The larger flux density between octet sources (cf. table 12) than between triplets implies flatter Regge trajectories and hence a smaller level density for glueballs than for $(q\bar{q})$ states. In pure QCD there will be among the glueballs a lightest glueball state which, it is reasonable to expect, must be stable.

The introduction of massive quarks [stage II of Bjorken, 1980] does little but provide new sources of glueball production. Quarkonium states may now decay according to

$$^1S_0(Q\bar{Q}) \rightarrow gg, \tag{8.1}$$

a colorless, $J^{PC} = 0^{-+}$ final state, and

$$^3S_1(Q\bar{Q}) \begin{cases} ggg, & \text{(8.2a)} \\ \\ gg\gamma, & \text{(8.2b)} \end{cases}$$

colorless hadronic states with $J^{PC} = 1^{--}$ for the three-gluon semifinal state and $J^{PC} = 0^{++}, 0^{-+}, 2^{++}$, etc. for the $gg\gamma$ semifinal state. Again the lightest gluon will be stable, because all $(Q\bar{Q})$ states are—by assumption —extremely massive.

Extending QCD to the light-quark sector raises two questions that go directly to the heart of the matter: what is the mass scale for glueballs, and how prominently will they appear in the spectrum of hadrons? Given the small mass of the pion it is essentially a certainty that the lightest glueball will be unstable. We must then ask whether the quarkless states will become so broad as to be lost in a general continuum, whether they will mix so strongly with $(q\bar{q})$ and $(q\bar{q}g)$ states as to lose their identity, or whether they will remain relatively pure glue states of modest width. Until definite theoretical predictions can be given, we may conclude only that the observation of glueballs would support the notion that gluons exist and interact among themselves. Not finding glueballs, at the present level of understanding of QCD, has a less obvious significance.

8.2. The properties of glueballs

Some characteristics of quarkless states such as their flavor properties are unambiguous, but many others including masses and decay widths are predicted rather indecisively. It is reasonable to attempt to enumerate few-gluon states by analogy with Landau's (1948) classification of two-photon states, which incorporates the restrictions of Yang's (1950) theorem. This has been done by Fritzsch and Minkowski (1975), Barnes (1981), and within the bag model by Donoghue (1980). A pair of massless vector particles can be combined to yield states with

$$J^{PC} = \begin{cases} (\text{even} \geqslant 0)^{++} & (8.3a) \\ (\text{even} \geqslant 0)^{-+}, & (8.3b) \\ (\text{even} \geqslant 2)^{++}, & (8.3c) \\ (\text{odd} \geqslant 3)^{++}. & (8.3d) \end{cases}$$

Many papers [e.g. Robson 1977, Coyne et al. 1980] treat the gluons as massive vectors and arrive at longer lists of two-gluon states. Similarly, extra states may arise in the bag model unless spurious modes associated with the empty bag are eliminated [Donoghue et al. 1981]. The lowest-lying two-gluon configurations should therefore include $J^{PC} = 0^{++}$, 2^{++}, 0^{-+}, and 2^{-+} states.

A variety of estimates of varying degrees of sophistication have been made for the masses of these states. Keeping in mind that the scalar ground state has precisely the quantum numbers of the vacuum and may therefore be appreciably mixed or even subsumed into the vacuum, let us list some representative predictions. The bag model [Donoghue 1981] suggests that

$$M(0^{++}) \approx M(2^{++}) \approx 1 \text{ GeV}/c^2, \qquad (8.4)$$

neglecting hyperfine effects, and that

$$M(0^{-+}) \approx M(2^{-+}) \approx 1.3 \text{ GeV}/c^2, \qquad (8.5)$$

again neglecting hyperfine effects. The QCD sum rules of the ITEP

Group [Novikov et al. 1979, 1980a, b, c, d, 1981, Zakharov 1980a, b, Shifman 1981] lead to slightly larger values:

$$M(0^{++}) \approx M(2^{++}) \approx 1.2\text{--}1.4\,\text{GeV}/c^2, \tag{8.6}$$

and

$$M(0^{-+}) \approx 2\text{--}2.5\,\text{GeV}/c^2, \tag{8.7}$$

but with an important gluon component in η' (958). the effective potential calculations of Suura (1980) and Barnes (1981) lead to degenerate pseudoscalar and scalar states, with masses supposed to be on the order of 1–2 GeV/c^2. Barnes (1981) concludes that

$$M(2^{++})/M(0^{++}) \approx 1.2, \tag{8.8}$$

with his description of hyperfine forces.

Three-gluon bound states are more complicated to analyze, especially in terms of the dynamics. It will suffice to give one estimate [Donoghue 1981] of the masses, obtained in the bag model upon neglecting spin–spin forces:

$$M(0^{++}) \approx M(1^{+-}) \approx 1.45\,\text{GeV}/c^2, \tag{8.9}$$

and

$$M(0^{-+}) \approx M(1^{-+}) \approx M(1^{--}) \approx M(2^{-+})$$
$$\approx M(2^{--}) \approx 1.8\,\text{GeV}/c^2. \tag{8.10}$$

The general conclusion is that a host of states is to be expected, and that it is plausible that many exist in the region between 1 and 2 GeV/c^2. However all calculations have at least some degree of arbitrariness in the overall mass scale.

A simple lattice argument has also been presented [Kogut et al. 1976] for glueball masses in the 1–2 GeV/c^2 region. Figure 35a shows the minimal lattice configuration for a meson: a single link. On the other hand, on a rectangular lattice the minimal quarkless state consists of a closed loop made up of four links, as shown in fig. 35b. Consequently

Fig. 35. (a) A single link between two quarks in lattice gauge theory. (b) The smallest closed loop, corresponding to a quarkless excitation.

one may suppose that the mass of a typical ground-state glueball is approximately four times the mass of a typical ground-state meson and thus on the order of $1-2\ GeV/c^2$.

With respect to quantum numbers let us note that apart from the 1^{-+} level, which cannot occur as a $(q\bar{q})$ state, all of the glue states resemble ordinary mesons. Their distinctive property is that pure glue states must be flavor singlets. With this restriction, the allowed decay modes follow from standard selection rules, although branching ratios may be strongly influenced by phase-space effects and by the preëminence of quasi-two body final states.

It is quite possible, as we shall now discuss, that glueballs may be narrower structures than $(q\bar{q})$ mesons of comparable mass. This suspicion is tied up with the validity of the so-called Zweig rule and the mixing of glueballs with ordinary mesons, to which we now turn.

8.3. Gluons and the Zweig rule

In section 3.3.2 we concluded on the basis of simple mass formulas, that $\varphi(1019)$ is essentially a pure $(s\bar{s})$ state. This conclusion is sustained by an examination of the decay modes of φ, which are collected in table 21. The total width is

$$\Gamma(\varphi) = 4.1 \pm 0.2\ MeV. \tag{8.11}$$

Decays into $K\bar{K}$ are inhibited by the limited phase space available, and the relative rates for the charged and neutral final states are understood in terms of p-wave kinematics. For the suppression of the $3\pi(\pi\rho)$ mode, however, a dynamical explanation must be sought.

Table 21
Decay modes of the (s$\bar{s}$) state, $\varphi(1020)$ [Particle Data Group 1980]

Channel	Branching Fraction (%)	q_{max} (MeV/c)
K^+K^-	48.6 ± 1.2	127
$K_L K_S$	35.2 ± 1.2	111
$\pi^+\pi^-\pi^0$	14.7 ± 0.7	462
$\eta\gamma$	1.5 ± 0.2	362
$\pi^0\gamma$	0.14 ± 0.05	501
e^+e^-	0.031 ± 0.001	510
$\mu^+\mu^-$	0.025 ± 0.003	499

The suppression of nonstrange decays can be accounted for, if not explained from first principles, by the rule [Okubo 1963, Zweig 1964a, b, Iizuka et al. 1966, Iizuka 1966] that decays which correspond to connected quark-line diagrams are allowed, but those which correspond to disconnected diagrams are not. This is made concrete for the case of φ decay in fig. 36. The dissociation and subsequent dressing of the $(s\bar{s})$ pair is allowed (a), but the quarkless semifinal state reached by $(s\bar{s})$ annihilation is not. One may attribute the small observed rate for $\varphi \to 3\pi$ either to light-quark impurities in the φ wavefunction or to violations of the Zweig rule.

Additional evidence in favor of the rule comes from the remarkable metastability of ψ (3097), for which [Particle Data Group 1980]:

$$\Gamma(\psi \to \text{hadrons}) = 45 \text{ keV}, \tag{8.12}$$

and of $\Upsilon(9433)$, for which [Schamberger, 1981]:

$$\Gamma(\Upsilon \to \text{hadrons}) = 28 \text{ keV}. \tag{8.13}$$

The Zweig rule thus provides a notable mnemonic for forbidden decays. It is of interest to ask whether there is a dynamical basis for the rule within QCD, and whether there may be other manifestations of violations of the rule.

To this end, recall the outstanding failure of our description of meson masses: the problem of the η and η' masses. In the language of singlet and octet mixing we found it possible in subsection 3.3.2 to parametrize

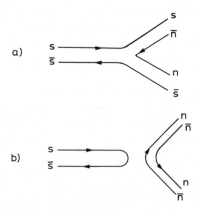

Fig. 36. The Okubo–Zweig–Iizuka rule applied to φ decay. The connected diagram (a) is allowed; the disconnected diagram (b) is forbidden.

$M(\eta)$ and $M(\eta')$ in terms of two free parameters: a flavor-singlet mass M_1 and a mixing angle θ. The resulting wave-functions imply relations between decay and reaction rates that are imperfectly respected by the data, as noted in subsection 5.5.1. In the otherwise successful quark language we were not able to understand the $\pi^0-\eta-\eta'$ splitting or the high mass of the η'. If we interpret the failure as pertaining only to isoscalar states, it is sensible to consider the possibility that virtual annihilations into glue states

$$q\bar{q} \rightarrow \text{glue} \rightarrow q'\bar{q}' \tag{8.14}$$

may influence the masses of $(q\bar{q})$ states. [See among others De Rújula et al. 1975, Isgur 1976]. Such transitions of course cannot affect flavored states.

In the $u\bar{u}$, $d\bar{d}$, $s\bar{s}$ basis the mass matrix of the pseudoscalar mesons can be written as

$$\mathfrak{M} = \begin{pmatrix} 2m_u - 3\delta M_{c.m.} + A & A & A \\ A & 2m_u - 3\delta M_{c.m.} + A & A \\ A & A & 2m_s - 3(m_u/m_s)^2 \delta M_{c.m.} + A \end{pmatrix} \tag{8.15}$$

in the notation of eqs. (5.52), (5.53), (5.57), (5.58), where A represents the flavor-independent amplitude for the process (8.14). Recognizing that virtual annihilations cannot affect the π^0 mass, we recast (8.15) in a basis of $(u\bar{u} \mp d\bar{d})/\sqrt{2}$, $s\bar{s}$ as

$$\mathfrak{M} = \begin{pmatrix} 2m_u - 3\delta M_{c.m.} & 0 & 0 \\ 0 & 2m_u - 3\delta M_{c.m.} + 2A & A\sqrt{2} \\ 0 & A\sqrt{2} & 2m_s - 3(m_u/m_s)^2 \delta M_{c.m.} + A \end{pmatrix}, \tag{8.16}$$

which retains the expected result $M(\pi^\pm) = M(\pi^0)$.

The remaining two-by-two isoscalar mass matrix suggests a common origin for the $\pi^0-\eta-\eta'$ splitting and the deviations from ideal mixing. With the parameters of section 5.3, the sum of η and η' masses is reproduced with the choice

$$A = 198 \text{ MeV}/c^2, \tag{8.17}$$

for which

$$M(\eta) = 408 \text{ MeV}/c^2, \tag{8.18a}$$

and

$$M(\eta') = 1109 \text{ MeV}/c^2. \tag{8.18b}$$

The wavefunctions implied are

$$|\eta'\rangle = 0.44(u\bar{u} + d\bar{d})/\sqrt{2} + 0.81\, s\bar{s}, \tag{8.19a}$$

and

$$|\eta\rangle = 0.81(u\bar{u} + d\bar{d})/\sqrt{2} - 0.44 s\bar{s}, \tag{8.19b}$$

which are similar to those (3.112) given by the linear mass formula in the singlet–octet picture. The masses (8.18) are considerably improved over those produced in section 5.3, but they are still not perfect. At any rate, we have succeeded in raising the $\eta-\eta'$ center of gravity by invoking vitual annihilations, and have thus been able to begin to reconcile the constituent picture with the symmetry approach. Note also that if physical glue states do exist, the mass matrix must be enlarged and the mixing pattern may be considerably more complicated.

The success of our earlier description of the vector meson masses argues that no appreciable annihilation is required there. For heavy mesons such as ψ and Υ, the analogy with ortho- and para-positronium seems apt. A coupling constant argument then suggests that in the asymptotically free regime:

$$\Gamma(^3S_1 \to \text{glue})/\Gamma(^1S_0 \to \text{glue}) = \alpha_s \times \text{numerical factors.} \tag{8.20}$$

A power of small coupling constant may inhibit mixing of vector states with gluons in quarkonium, but this is a tenuous argument for the light mesons. In the following section 9 we shall review an argument in favor of the Zweig rule that does not depend upon powers of the coupling constant.

Among the orbitally excited mesons, there is also room for virtual annihilations. One should in general be alert for the possibility whenever a breakdown of ideal mixing is signalled by the nondegeneracy of the isovector and would-be $(u\bar{u} + d\bar{d})/\sqrt{2}$ states. For a recent look at the 1^{++} and 2^{++} nonets, see Schnitzer (1981a, b).

If virtual transitions such as (8.14) occur, they may account for violations of the Zweig rule. This mechanism for Zweig-rule violations naturally suggests the pattern

$$\Gamma(Q\overline{Q} \to \text{hadrons}) < \Gamma(\text{glue} \to \text{hadrons}) < \Gamma(q\bar{q} \to \text{hadrons}),$$

where the Zweig-inhibited quarkonium decay rate is of order A^2, the decay rate of a glueball into light quarks is of order A^1, and the rate for Zweig-allowed dissociation of a light quark pair is of order A^0. The possibility therefore exists that a pure glue state will be relatively more stable than a light-quark meson of comparable mass.

8.4. Searching for glueballs

As strongly-interacting particles, glueballs should be produced routinely in hadronic collisions, where they may be sought out using the techniques of traditional meson spectroscopy. Special kinematic selections may enhance the glueball signal over ordinary mesonic background. An obvious choice with the CERN p$\bar{\text{p}}$ collider at hand is an investigation of "Double-Pomeron events", which yield hadronic states in the central region of rapidity with vacuum quantum numbers. If, as Freund and Nambu (1975) and others have suggested, there is a deep connection between the Pomeron and quarkless states, such a selection may be of more than merely kinematical benefit.

Another favorable situation may be in the decays of heavy quarkonium according to (8.2b), leading to transitions of the form

$$\psi \to \gamma + G, \tag{8.21}$$

where G denotes a glueball. This is not only a case in which the general arguments of section 8.1 lead us to expect that glueballs may be produced, but also one that permits inclusive as well as exclusive searches and lends itself to comparison with

$$\psi \to (\omega, \varphi) + \text{anything}, \tag{8.22}$$

in which the anything is presumably composed of quarks.

Interest in quarkonium decay has been increased by the recent observation of suggestive structures in

$$\psi \to \gamma + \text{hadrons}. \tag{8.23}$$

According to Scharre (1981), there is evidence in the Crystal Ball Experiment at SPEAR for two new states. The first, named iota (1440) is seen in the cascade decay

$$\psi \to \gamma + \iota(1440),$$
$$ \hookrightarrow \text{K}\bar{\text{K}}\pi \neq \text{KK*} \tag{8.24}$$

with a mass $M_\iota = 1440^{+10}_{-15}$ MeV/c^2 and a width of $\Gamma_\iota = 50^{+30}_{-20}$ MeV. The state has $J^{PC} = 0^{-+}$ and the combined branching ratio for the cascade is:

$$B(\psi \to \gamma\iota)B(\iota \to K\bar{K}\pi) \approx 4 \times 10^{-3}. \tag{8.25}$$

The second is seen in

$$\psi \to \gamma + \theta(1640),$$
$$\rightarrow \eta\eta \tag{8.26}$$

with a mass $M_\theta = 1650 \pm 50$ MeV/c^2 and a width of $\Gamma_\theta = 220^{+100}_{-70}$ MeV. The decay angular distribution favors $J^{PC} = 2^{++}$, and the product of branching ratios is

$$B(\psi \to \gamma\theta)B(\theta \to \eta\eta) \approx 5 \times 10^{-4}. \tag{8.27}$$

An upper limit exists for the decay of θ into $\pi^0\pi^0$:

$$B(\theta \to \pi\pi) \lesssim B(\theta \to \eta\eta). \tag{8.28}$$

We have seen above that scalar, pseudoscalar, and tensor glueballs are to be expected in this mass range. In addition, an analysis [Billoire et al. 1979] of the spin–parity content of the gluon pair in (8.2b) suggests that 2^{++} formation is favored with equal but smaller probabilities for 0^{++} and 0^{-+} configurations. At the same time, radial excitations of the low-lying mesons are to be expected in precisely this region [Cohen and Lipkin 1979]. Thus it is easily possible that any new states be traditional ($q\bar{q}$) states, or mixtures of ($q\bar{q}$) with glueballs, or other exotic possibilities [Close 1981], as well as states of pure glue. How can these possibilities be distinguished?

Without going into details, let us note that Chanowitz (1981), Ishikawa (1981), Donoghue et al. (1981), Lipkin (1981b), and Cho et al. (1981) have examined the case that $\iota(1440)$ is a glueball. Opinion is divided. Chanowitz (1981) has shown that a large number of seemingly contradictory experiments may be reconciled if, in addition to the 1^{++} E(1420) there is a nearby pseudoscalar state for which $\iota(1440)$ is the obvious candidate. He further argues that $\iota(1440)$ has the characteristics of a glueball, and that $\eta'(1400)$—see table 18—is distinct from $\iota(1440)$. If we accept the spin–parity assignments, then there are at least two isoscalar states around 1400 MeV/c^2. The conclusion that $\iota(1440)$ is pseudoscalar and not axial [as E(1420)] removes a potential embarrassment for the two-gluon glueball interpretation. Lipkin (1981), on the other hand, argues that the absence of an appreciable $\iota \to \eta\pi\pi$ signal is inconsistent with a flavor singlet assignment. Obviously there are many experimental questions to settle, among them the spin–parity assignments and the relationship between η' and ι.

Another obvious test may be available in two-photon reactions

$$e^+e^- \rightarrow e^+e^- + \text{hadrons,} \tag{8.29}$$

at least for the pseudoscalar state(s). A $(q\bar{q})$ state should decay into two photons, whereas a pure glue state should not, in lowest order. This inspires a search for the reactions

$$e^+e^- \rightarrow e^+e^-\, \eta'(1400), \tag{8.30a}$$
$$\quad\ \ \ \boxed{}\!\!\!\!\rightarrow \eta\pi\pi$$

$$e^+e^- \rightarrow e^+e^-\, \iota(1440). \tag{8.30b}$$
$$\quad\ \ \ \boxed{}\!\!\!\!\rightarrow K\bar{K}\pi$$

Given an estimate for the two-photon decay rate, standard techniques [described in Quigg 1980] lead to the two-photon production cross section. The rate for production of a $J^{PC} = 0^{-+}$ $\iota(1440)$ is shown in fig. 37 under the assumption that

$$\Gamma(\iota \rightarrow \gamma\gamma) = 1 \text{ keV.} \tag{8.31}$$

At the energies accessible at PEP and PETRA, an ample cross section is to be expected.

If a prominent signal is observed, one may conclude that the hadronic state is not an axial vector meson and that it is not dominantly a glueball. But if no signal is found, what then? I see four possibilities:

(i) the hadron is an axial state,

(ii) the hadron is a glueball,

(iii) the hadron is a $(q\bar{q})$ state with a small width for two-photon decay,

(iv) the hadron is a mixed $(q\bar{q})$–glueball state.

The first and second points are self-evident. The third is more problematic. I believe a two-photon width of 0.1–1 keV is reasonable for a radially-excited pseudoscalar, but I cannot convince myself that this represents the full range of "reasonable" possibilities. If glueballs exist, the fourth possibility seems the most reasonable to me. It has been studied in some detail by Donoghue et al. (1981), by Rosner (1981b), and by Cho et al. (1981).

In general we may expect some degree of mixing between nearby (or overlapping) hadrons. The simplest case of one glueball and one $(q\bar{q})$

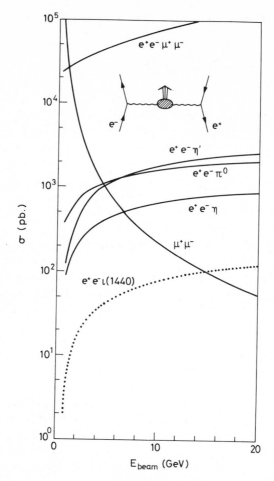

Fig. 37. Cross sections for the two-photon reactions $e^+e^- \to e^+e^- +$ hadrons. The cross section for excitation of $\iota(1440)$ is computed under the assumption that $\Gamma(\iota \to \gamma\gamma) = 1$ keV, and so should be multiplied by $\Gamma(\iota \to \gamma\gamma)/(1$ keV$)$. The cross section for the reaction $e^+e^- \to \mu^+\mu^-$ ("one unit of R") is shown for reference.

state can be parametrized as

$$|h_1\rangle = |q\bar{q}\rangle\cos\theta + |G\rangle\sin\theta, \tag{8.32a}$$

$$|h_2\rangle = -|q\bar{q}\rangle\sin\theta + |G\rangle\cos\theta, \tag{8.32b}$$

in an obvious notation. The decay $\psi \to \gamma + $ glue would then lead to a line shape characteristic of

$$|h_1\rangle \sin\theta + |h_2\rangle \cos\theta, \tag{8.33}$$

whereas two-photon collisions would excite

$$|h_1\rangle \cos\theta - |h_2\rangle \sin\theta. \tag{8.34}$$

Thus we are led to ask whether, for example, the $f^0(1270)$ seen in the decay $\psi \to \gamma + f^0$ is identical with that observed in two-photon collisions. To answer such questions requires careful measurements of line shapes and branching ratios in both kinds of reactions, as well as in peripheral and central hadron collisions. There is much to be learned here, but the experimental work called for is demanding and meticulous.

Before leaving the subject of glue, let us note that there may be other manifestations of degrees of freedom beyond those of quark and antiquark. The specific possibility of "vibrational modes" has been raised by Giles and Tye (1977, 1978) and by Buchmüller and Tye (1980). States with constituent gluons ($q\bar{q}g$ states) have been examined by Horn and Mandula (1978) and by deViron and Weyers (1981); see also Close (1981). Glue-bearing baryons ($qqqg$) have been considered by Bowler et al. (1980).

9. The idea of the $1/N$ expansion in QCD

The search for small parameters which can play the part of expansion parameters is a central element of the process of approximation and model making that is theoretical physics. In many physical situations, extremes of energy or distance suggest highly accurate and readily improved approximation schemes. In classical electrodynamics the indispensable far-field approximation is applicable when the size of a radiator is negligible compared to the distance between the radiator and receiver. The Born approximation for the scattering of charged-particle beams from atomic electrons is trustworthy for beam energies greatly in excess of the atomic binding energy. In Quantum ChromoDynamics, a perturbative treatment (which is to say an expansion in powers of the strong coupling parameter $\alpha_s(Q^2)$) is expected to be reliable when the invariant momentum transfer Q^2 is large compared to a characteristic mass scale denoted by Λ^2.

For the problem of hadron structure, no similar expansion is applicable. All of the relevant energies of the problem are on the order of the naturally occurring scale. In a typical hadron, the separation of the quarks is simply the hadronic size of approximately 1 fm—hardly a regime in which perturbative QCD is likely to make any sense. We may, of course, simply await the day when a very heavy quarkonium family is found, and then happily apply conventional perturbative measures. That insouciant course however leaves untouched the problem of the structure of all the hadrons not known, so other actions are called for.

The strategy of the $1/N$ expansion is a familiar one. When confronted with a problem we cannot solve, we invent a related problem that we can solve. If this is done adroitly, the new problem will not only be simpler but will also capture the physical essence of the original one. More specifically, the $1/N$ expansion represents an attempt to introduce a parameter that permits a simplification of the calculation at hand. Problem 24 introduces an elementary example.

For QCD, this simplification is achieved ['t Hooft 1974a, b] by generalizing the color gauge group from $SU(3)_c$ to $SU(N)_c$ and considering the limit in which N becomes very large. Although $SU(N)$ is in general more complicated than $SU(3)$, the hadron structure problem is simplified by two observations:

(i) At any order in the strong coupling constant, some classes of diagrams are found to be combinatorially negligible.

(ii) The remaining diagrams have common consequences, in large-N perturbation theory.

This technique does not entirely free us from the constraints of perturbative analysis. Since we shall find, by inspection, that entire classes of combinatorially favored diagrams have common features to all orders in the coupling constant, we shall have to assume that the content of the theory is accurately represented by the set of all diagrams. For QCD, the reliability of the $1/N$ expansion is inferred from the fact that $SU(N)_c$ QCD seems to resemble the world we observe. Clear introductions to the method, with allusions to other physical situations, are given by Coleman (1980), and by Witten (1979b, 1980a, b). See also Lipkin (1969).

The combinatorial analysis of $SU(N)_c$ QCD is most transparent in terms of the double-line notation introduced for this purpose by 't Hooft (1974a), which is illustrated in fig. 38. Several examples will suffice to make the main points.

Consider first the lowest-order vacuum polarization contributions to the gluon propagator, the quark loop illustrated in fig. 39a and the gluon loop pictured in fig. 39b, in conventional notation. These are redrawn in the double-line notation in fig. 39c, d. For an initial gluon of type $i\bar{j}$, only

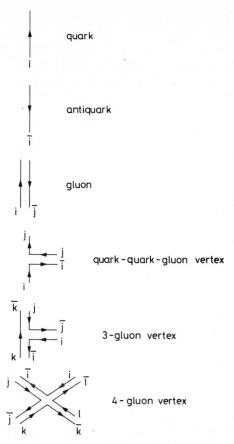

Fig. 38. Double-line notation for quarks, gluons, and their interactions, useful for $1/N_c$ analyses.

a single color configuration is possible for the quark loop intermediate state: a quark of color i and an antiquark of color $\bar{j}$. For the gluon loop, however, the index k is free to take on any value $1, 2,...,N$. Thus the gluon loop diagram has a combinatoric factor N associated with it. This illustrates the general rule that gluon loops dominate over quark loops by a factor of N, as $N \to \infty$.

The presence of the factor N would seem to imply that the gluon loop diagram diverges as $N \to \infty$. This can be cured by choosing the coupling

C. Quigg

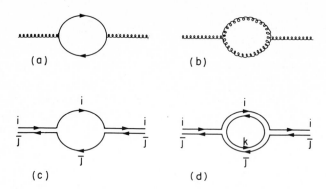

Fig. 39. Lowest order vacuum polarization contributions to the gluon propagator. (a) quark loop; (b) gluon loop; (c) quark loop in the double-line notation; (d) gluon loop in the double-line notation.

constant to be $g/\sqrt{N}$, with g fixed as $N \to \infty$. Then for any value of N, the contribution of the gluon loop goes as

$$\left(g/\sqrt{N}\right)^2 \times N \to g^2, \tag{9.1}$$

a smooth limit.

That this device solves the divergence problem in general is indicated by an analysis of diagrams with more than one loop. The two-loop diagram depicted in fig. 40 in (a) standard and (b) double-line notation is immediately seen to be proportional to

$$\left(g/\sqrt{N}\right)^4 \times N^2 \to g^4. \tag{9.2}$$

Similarly, the three-loop diagram of fig. 41 obviously goes as

$$\left(g/\sqrt{N}\right)^6 \times N^3 \to g^6. \tag{9.3}$$

The situation is different for nonplanar graphs, however. The simplest such graph is shown in fig. 42. The double-line notation makes it apparent that this graph contains but a single, tangled color loop, and

Fig. 40. A two-loop diagram in (a) conventional and (b) double-line notation.

Fig. 41. A three-loop diagram in (a) conventional and (b) double-line notation.

therefore goes as

$$\left(g/\sqrt{N} \right)^6 \times N \to g^6/N^2, \tag{9.4}$$

and is therefore suppressed by $1/N^2$ compared to its planar counterpart at the same order in g^2. It is generally the case that nonplanar graphs are reduced by $1/N^2$, as $N \to \infty$.

These combinatorial arguments select planar graphs as an important subclass. To evaluate and sum all the graphs thus selected is no trivial task. Instead, we may identify their common features and speculate that these survive confinement. It is possible in this way to establish the following results in the large-N limit:

(i) Mesons are free, stable, and noninteracting. For each allowed combination of J^{PC} and flavor quantum numbers, there are an infinite number of resonances.

(ii) Zweig's rule is exact. Singlet–octet mixing (through virtual annihilations) and meson–glue mixing are suppressed. Mesons are pure $(q\bar{q})$ states, with no quark–antiquark sea.

(iii) Meson–meson bound states, which would include particles with exotic quantum numbers, are absent.

(iv) Meson decay amplitudes are proportional to $1/\sqrt{N}$, so mesons are narrow structures.

Fig. 42. A nonplanar graph in (a) conventional and (b) double-line notation.

(v) The meson–meson elastic scattering amplitude is proportional to $1/N$ and is given, as in Regge theory, by an infinite number of one-meson exchange diagrams.

(vi) Multibody decays of unstable mesons are dominated by resonant, quasi-two body channels whenever they are open. The partial width of an intrinsically k-body final state goes as $1/N^{k-1}$.

(vii) For each allowed J^{PC} there are infinitely many glueball states, with widths of order $1/N^2$. They are thus more stable than (q$\bar{\text{q}}$) mesons, interact feebly with (q$\bar{\text{q}}$) mesons, and mix only weakly with (q$\bar{\text{q}}$) states.

Until QCD is actually solved, we will not know how closely the $N \to \infty$ limit of SU(N)$_c$ resembles the case of interest, which is color SU(3). The preceding list of large-N results does bear, however, a quite striking resemblance to the world described earlier in these lectures. To the extent that the $1/N$ expansion faithfully represents the consequences of QCD, much of the foregoing phenomenology is explained, and many of the model approximations are justified.

To see how conclusions (i)–(vii) may be reached, let us consider the $1/N$ derivation of the Zweig rule. A possible mechanism for the Zweig-forbidden decay of (q$\bar{\text{q}}$) state is shown in fig. 43, the process

$$(\text{q}\bar{\text{q}}) \to \text{gg} \to \text{q}'\bar{\text{q}}' \to \text{mesons}. \tag{9.5}$$

This is shown in standard notation in fig. 43a, and in double-line notation in fig. 43b. In the latter case I have tied together the ends of the quark and antiquark lines in mesons to emphasize that the mesons are color singlets. The Zweig-forbidden decay amplitude contains a single color loop. It therefore goes as

$$\left(g/\sqrt{N}\right)^4 \times N \sim g^4/N. \tag{9.6}$$

At the same order in the strong coupling constant, the allowed decay is illustrated in fig. 44. In the double-line representation, it is seen to

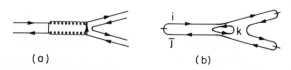

$$(a) \qquad\qquad\qquad (b)$$

Fig. 43. A mechanism for OZI-forbidden decay, at order g^4, in (a) conventional and (b) double-line notation.

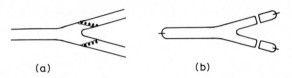

(a)   (b)

Fig. 44. OZI-allowed decay of a meson, at order g^4, in (a) conventional and (b) double-line notation.

contain two color loops. The allowed amplitude is therefore proportional to

$$\left(g^2/\sqrt{N} \right)^4 \times N^2 \sim g^4. \tag{9.7}$$

Thus at each order in perturbation theory, the Zweig-forbidden decay is down by a power of $1/N$ in amplitude compared with the Zweig-allowed decay. Since this reasoning does not rely upon the smallness of the strong coupling constant, which may well be appropriate for the ψ and Υ families, it is an appealing argument for the inhibition of $\varphi \rightarrow \rho\pi$. The $1/N$ expansion has also been applied to the problem of baryon structure by Witten (1979b).

To close this brief section on the $1/N$ expansion, let us briefly return to the difficulty of understanding the η' mass. A clear statement of the puzzle of the flavor-singlet pseudoscalar meson, which is known as the U(1) problem, was given by Gell-Mann et al. (1968) and by Weinberg (1975). What seems a promising phenomenological explanation is the influence of virtual states composed of glue alone, as described in section 8.3. A formal solution to the U(1) problem was given by 't Hooft (1976), who argued that the U(1) current has an anomaly which leads to a physical non-conservation of the U(1) charge. This removes the raison d'être for a ninth light pseudoscalar. The relationship between the intuitive and formal approaches was exhibited in the context of the $1/N$ expansion by Witten (1979a, 1980c), Di Vecchia (1979), and Veneziano (1979b).

10. Regrets

One cannot reach the end of a course such as this without contemplating what might have been, or what should have been. There are a number of subjects that I have been forced by the pressure of time to omit. Here I attempt to make amends by providing a brief bibliography for some of the topics I had hoped to discuss.

10.1. The masses of quarks

At various points in the analysis of hadron masses we have had occasion to refer to the effective masses of confined quarks. Several important issues have thus been swept under the rug, or at best talked around. One is how QCD behaves in the limit of vanishing quark masses, for which the Lagrangian will have an exact $SU(n) \otimes SU(n)$ chiral symmetry operating independently on the left-and right-handed parts of the quark fields for the n massless flavors. That this is approximately so in nature is evidenced by the success of soft-pion theorems [see Adler and Dashen 1968, Renner 1968, Lee 1972]. The Lagrangian will also have the chiral $U(1)$ symmetry which leads to the puzzle of the η' mass dealt with in section 9.

In the limit of zero up-, down-, and strange-quark masses, QCD possesses an octet of exactly conserved axial currents. It is believed that the corresponding chiral symmetry must be spontaneously broken along the lines described by Nambu and Jona-Lasinio (1961a, b). Accordingly, in the world of three massless quark flavors there should be eight massless Goldstone (1961) bosons which we identify with the pseudoscalar octet. See also Nambu (1960). The pattern of the spontaneous breaking of the chiral symmetry of QCD has been discussed from the point of view of the $1/N$ expansion by Coleman and Witten (1980).

Nonzero values of the quark masses which appear in the Lagrangian are thought to arise from the spontaneous breakdown of the $SU(2) \otimes U(1)$ gauge symmetry of the electroweak interactions by means of the Higgs (1964a, b) mechanism [for an elementary discussion, see Quigg 1981], or through dynamical symmetry breaking [Weinberg 1976, 1977, Susskind 1979, Farhi and Susskind 1981]. It then follows that the π, K, and η are only approximately massless, although they are presumed to retain some memory of their chiral origin. The Lagrangian ("current quark") masses have been studied by Leutwyler (1974a, b), Pagels (1975), and Langacker and Pagels (1979), among others.

If the masses of the up and down quarks are not identical—a possibility we have entertained in connection with electromagnetic mass differences of hadrons—there may be a number of observable violations of isospin symmetry. The effect upon $\rho - \omega$ mixing was mentioned in passing in section 3.1.3, and many other applications are discussed by Gross et al. (1979), Isgur et al. (1979), Langacker (1979, 1980), and Shifman, et al. (1979d).

For recent attempts to understand the chiral nature of the pion within the framework of QCD and confinement, consult Pagels (1979), Pagels and Stokar (1979), Donoghue and Johnson (1980), Goldman and Haymaker (1981), and Haymaker and Goldman (1981).

10.2. Decays and interactions of hadrons

Important support for flavor-SU(3) symmetry and for specific multiplet assignments derives from the systematic study of hadron decay rates and hadron–hadron reaction rates. The quark model, with or without specific dynamical assumptions, makes many predictions that are sharper than those of SU(3) alone. Entry to the extensive literature on these subjects may be gained via the lecture notes by Rosner (1981a) and the book by Close (1979).

10.3. QCD sum rules

A very different and extremely provocative approach to hadron spectroscopy has been pioneered by a group from the Institute for Theoretical and Experimental Physics in Moscow. I regard my omission of their method of analysis as particularly unfortunate. For the students at Les Houches, although not for posterity, this void was filled by informative seminars by John Bell and Eduardo de Rafael [but see in part Bourrely et al. 1981]. A short course is provided by the following articles: Shifman et al. (1979a, b, c, d); Voloshin (1979); Leutwyler (1981); Reinders et al. (1981); Bell and Bertlmann (1981); and Ioffe (1981).

10.4. Relation to other pictures of hadrons

Finally, and still more telegraphically, I wish to note a few articles which pertain to other approaches to hadron structure and their connections with the schemes I have discussed. Renormalization group techniques for quarks and strings are reviewed by Kadanoff (1977). The theory of dual models and strings is summarized by Scherk (1975). Parallels between QCD, especially in the $1/N_{color}$ expansion, dual theories, and the Reggeon calculus are drawn by Veneziano (1976, 1979a).

Acknowledgements

Many individuals and institutions have contributed to the existence of these lecture notes. Mary K. Gaillard invited me to teach a course on hadrons at Les Houches. I thank her and Raymond Stora for their warm encouragement and for the pleasant atmosphere I found there. Jonathan Rosner has taught me much about spectroscopy; in addition I have benefited enormously in planning this course from hearing his St. Croix lectures on quark models. Students at Les Houches and in a related course at the University of Chicago in the Spring of 1981 have asked many provocative questions, not all of which have been answered here. The writing of these notes was facilitated by the hospitality of Jacques Prentki and his colleagues in the Theoretical Studies Division at CERN and by that of Jean-Loup Gervais and other members of the Laboratoire de Physique Théorique de l'Ecole Normale Supérieure. No less important has been the generosity of Leon Lederman and the Trustees of Universities Research Association in granting the leave of absence during which this work was completed. Special thanks are owed to Valérie Lecuyer and Yolande Ruelle for typing the lengthy manuscript under unusual conditions and to Nicole Ribet for her assistance. Finally, I thank my family for a long season of tolerant support.

Problems

1. Consider bound states composed of fundamental scalar particles (denoted σ). The quantum numbers of σ are $J^{PC} = 0^{++}$. For $(\sigma\sigma)$ composites,

   a) Show that a bound state with angular momentum L (i.e. an orbital excitation) must have quantum numbers

   $$C = (-1)^L, \qquad P = (-1)^L.$$

   b) Allowing for both orbital and radial excitations, construct a schematic mass spectrum of $(\sigma\sigma)$ bound states. Label each state with its quantum numbers J^{PC}.

   c) Now suppose that the fundamental scalars have isospin I. Compute C, P, and G for $(\sigma\sigma)$ bound states, and redo part b).

2. Consider bound states composed of fundamental spin-$1/2$ particles (denoted f), with isospin $= 1/2$. For $(f\bar{f})$ composites,

a) Show that a bound state with angular momentum L must have quantum numbers

$$C = (-1)^{L+s}; \qquad P = (-1)^{L+1}; \qquad G = (-1)^{L+s+I},$$

where s is the spin of the composite system, and I is its isospin.

b) Allowing for both orbital and radial excitations, construct a schematic mass spectrum of $(f\bar{f})$ bound states. Label each state with its quantum numbers J^{PC}.

3. The η-meson (550 MeV/c^2) has quantum numbers $J^{PC} = 0^{-+}$ and isospin zero. Its principal decay modes, and branching fractions, are

$$\gamma\gamma, 38\%; \quad \pi^0\pi^0\pi^0, 30\%; \quad \pi^+\pi^-\pi^0, 24\%.$$

We wish to understand the surprising competition of photonic and hadronic decay modes. Show that the hadronic decays are isospin violating. Analyze the $3\pi^0$ and the $\pi^+\pi^-\pi^0$ decays separately. What qualitative explanation can be offered for the relative decay rates?

4. The permutation group on three objects admits three representations: symmetric (S), antisymmetric (A), and mixed (M). For the first two, the group elements are

	Element					
Representation	I	(12)	(13)	(23)	(123)	(132)
S	1	1	1	1	1	1
A	1	-1	-1	-1	1	1

When baryon wavefunctions are constructed from three isospin quarks, $|I = 1/2, I_z = \pm 1/2\rangle$, the antisymmetric representation cannot be formed. Consider the M representation of $I = 1/2$ final states, which may be built by first coupling quarks 1 and 2 to isospin 0 or 1, and then coupling the third quark. Use as a basis the two states $|1/2, 1/2\rangle_1$ [symmetric in (12)] and $|1/2, 1/2\rangle_0$ [antisymmetric in (12)]. Denote these states as $|1\rangle$ and $|0\rangle$, and use $\begin{pmatrix} |1\rangle \\ |0\rangle \end{pmatrix}$ as a basis vector for M. Find the 2×2 matrices representing the action of the permutations listed in the table for the M representation.

5. The flavor symmetry $SU(2)_{\text{isospin}}$ and the rotational symmetry $SU(2)_{\text{spin}}$ may be combined systematically in the group $SU(4)$. In nuclear

physics, this symmetry group provides the basis for classification into "Wigner supermultiplets". The fundamental representation of SU(4) is

$$\mathbf{4} \equiv \begin{pmatrix} u_{\uparrow} \\ d_{\uparrow} \\ u_{\downarrow} \\ d_{\downarrow} \end{pmatrix}.$$

Using the notation $(2I+1, 2s+1)$ for the isospin $\times$ spin decomposition of SU(4) representations, we may write

$$\mathbf{4} = (\mathbf{2,2}),$$

which shows that the **4** of SU(4) transforms as a doublet under isospin rotations and as a doublet under spin rotations.

a) Using the techniques for SU(N) computations developed for example in Ch. 3 of Close (1979) or in Bacry (1967), work out the SU(4) content of the product

$$\mathbf{4 \otimes 4 \otimes 4}.$$

Characterize each SU(4) representation in the product by its Young tableau, symmetry properties, and dimension.

b) Give the $(2I+1, 2s+1)$ content of each of the SU(4) representations in your expansion of $\mathbf{4 \otimes 4 \otimes 4}$.
Reference: Lipkin (1966).

6. Compute the magnetic moments of $\Delta^{++}, \Delta^{+}, \Delta^{0}, \Delta^{-}$. Assume that the magnetic moment of a quark is given by

$$\mu_i = e_i \hbar / 2 m_i c,$$

where $i = $ u, d and e_i is the quark charge in units of $|e|$. Further assume that $m_u = m_d$.
References: Close (1979) ch. 4; Kokkedee (1969) ch. 11.

7. Work out the explicit SU(6) wavefunctions for the strange members of the baryon octet. The expressions will be the analogs of

$$|p\uparrow\rangle = (1/\sqrt{18})(2u_{\uparrow}d_{\downarrow}u_{\uparrow} - u_{\downarrow}d_{\uparrow}u_{\uparrow} - u_{\uparrow}d_{\uparrow}u_{\downarrow}$$
$$- d_{\uparrow}u_{\downarrow}u_{\uparrow} + 2d_{\downarrow}u_{\uparrow}u_{\uparrow} - d_{\uparrow}u_{\uparrow}u_{\downarrow}$$
$$- u_{\uparrow}u_{\downarrow}d_{\uparrow} - u_{\downarrow}u_{\uparrow}d_{\uparrow} + 2u_{\uparrow}u_{\uparrow}d_{\downarrow}),$$

and

$$|n\uparrow\rangle = \left(-1/\sqrt{18}\right)\left(2d_\uparrow u_\downarrow d_\uparrow - d_\downarrow u_\uparrow d_\uparrow - d_\uparrow u_\uparrow d_\downarrow\right.$$
$$- u_\uparrow d_\downarrow d_\uparrow + 2u_\downarrow d_\uparrow d_\uparrow - u_\uparrow d_\uparrow d_\downarrow$$
$$\left.- d_\uparrow d_\downarrow u_\uparrow - d_\downarrow d_\uparrow u_\uparrow + 2d_\uparrow d_\uparrow u_\downarrow\right).$$

8. Using your explicit wavefunctions, express the magnetic moments of the strange baryons in terms of μ_u, μ_d, μ_s.

9. In the SU(3) symmetry limit, the quark magnetic moments are proportional to quark charge:

$$\mu_d = \mu_s = -\tfrac{1}{2}\mu_u.$$

Using the proton moment,

$$\mu_p = 2.793 \text{ n.m.},$$

as input, predict the numerical values of the magnetic moments of the octet baryons. Compare with the measured values given by the Particle Data Group (1980).

Reference for problems 7–9: Thirring (1966).

10. Consider the weak decay of a Λ-hyperon (with four-momentum p_Λ) into a proton (with four-momentum p) and a π^- (with four-momentum q). In general, the Feynman amplitude for the decay will have vector and axial vector terms. We write the general form for the amplitude as

$$\mathfrak{M} = \bar{u}_p(p)\left(A + B\gamma_5\right)\gamma_\mu u_\Lambda(p_\Lambda) \cdot q^\mu.$$

Work in the rest frame of the Λ ($p_\Lambda = 0$) and let the proton momentum lie along the $\hat{z}$ direction. Compute the decay angular distribution for a Λ with net polarization P_Λ along an arbitrary direction $\hat{n}$. Show that it takes the form

$$d\sigma/d\Omega = \text{const.} \times \left(1 + \alpha P_\Lambda \hat{q} \cdot \hat{n}\right),$$

so a measurement of the decay angular distribution determines αP_Λ. Express the asymmetry parameter α in terms of A, B, M_p, and M_Λ.

11. Now consider the decay of an unpolarized Λ. Show that a measurement of the proton's helicity leads to a determination of the asymmetry parameter α.

References for problems 10 and 11: Gasiorowicz (1966) ch. 33; Cronin and Overseth (1963); Okun (1965).

12. The magnetic dipole transitions among charmed mesons may be relatively immune from recoil effects, because of the large masses and small mass differences.

    a) Neglecting the small phase space difference, and approximating $\mu_c = 0$, calculate the ratio $\Gamma(D^{*0} \to D^0\gamma)/\Gamma(D^{*+} \to D^+\gamma)$.

    b) Redo your calculation assuming $\mu_c = -\frac{2}{3}\mu_s = 0.41$. Continue to use $\mu_u = -2\mu_d = \frac{2}{3}\mu_p$.

    c) Now using the masses, branching ratios, and momenta given in the Particle Data Group (1980) meson table, compare your predictions with experiment. You will need to use isospin invariance for the strong decay amplitudes, and to correct the strong decay rates for phase space differences.

    d) Assume the masses of the charmed-strange mesons are F$^+$: 2030 MeV/c^2 and F^{*+}: 2140 MeV/c^2. Using μ_c and μ_s as in part b), estimate the absolute width for the decay F$^* \to$ Fγ. What branching ratio do you expect?

13. Derive the connection between $|\Psi(0)|^2$ and the leptonic decay rate of a (q$\bar{\text{q}}$) vector meson. It is convenient to proceed by the following steps:

    a) Compute the spin-averaged cross section for the reaction q$\bar{\text{q}} \to$ e$^+$e$^-$. Show that it is

$$\sigma = \frac{\pi\alpha^2 e_q^2}{12E^2} \cdot \frac{\beta_\ell}{\beta_q}\left(3 - \beta_\ell^2\right)\left(3 - \beta_q^2\right),$$

where E is the c.m. energy of a quark and β is the speed of a particle.

    b) The annihilation rate in a $^3S_1$ vector meson is the density $\times$ relative velocity $\times 4/3$ (to undo the spin average) $\times \sigma$, or

$$\Gamma = |\Psi(0)|^2 \times 2\beta_q \times \tfrac{4}{3} \times \sigma.$$

    c) How is the result modified if the vector meson wavefunction is

$$|V^0\rangle = \sum_i c_i |q_i\bar{q}_i\rangle?$$

d) Now neglect the lepton mass and the quark binding energy and assume the quarks move nonrelativistically. Show that

$$\Gamma(V^0 \to e^+e^-) = \frac{16\pi\alpha^2}{3M_V^2}|\Psi(0)|^2 \left(\sum_i c_i e_i\right)^2.$$

e) How is the result modified if quarks come in N_c colors and hadrons are color singlets?
References: Van Royen and Weisskopf (1967); Pietschmann and Thirring (1966); Jackson (1976).

14. The Han–Nambu (1965) model is an integer-charge alternative to the fractional-charge quark model, with quark charges assigned as

color	flavor		
	u	d	s
R	0	-1	-1
G	1	0	0
B	1	0	0

a) Show that below the threshold for color liberation, the ratio

$$R \equiv \sigma(e^+e^- \to \text{hadrons})/\sigma(e^+e^- \to \mu^+\mu^-)$$

is $R = 2$, as in the fractional-charge model, and that $R = 4$ if color can be liberated.

b) Consider the reaction $\gamma\gamma \to$ hadrons, viewed as $\gamma\gamma \to q\bar{q}$. Show that with fractionally charged quarks

$$\sigma(\gamma\gamma \to \text{hadrons}) \propto \sum_i e_i^4 = 2/3,$$

and that in the Han–Nambu model

$$\sigma(\gamma\gamma \to \text{hadrons}) \propto \begin{cases} 2, & \text{below color threshold,} \\ 4, & \text{above color threshold.} \end{cases}$$

References: Close (1979) ch. 8; Chanowitz (1975); Lipkin (1979a); see also Okun et al. (1979).

15. Consider the electromagnetic interaction of two classical charged particles, with charges of q_1 and q_2, masses m_1 and m_2, and positions r_1 and r_2. In the static limit the interaction Lagrangian is the familiar Coulomb Lagrangian,

$$\mathcal{L}_{\text{int;NR}} = -q_1 q_2/r,$$

where $r \equiv r_1 - r_2$ is the relative coordinate. Derive the interaction Lagrangian through order $(v/c)^2$, and show that it may be written in the form obtained by Darwin in 1920:

$$\mathcal{L}_{\text{int}} = -\frac{q_1 q_2}{r} \left\{ 1 - \frac{1}{2c^2} \left[v_1 \cdot v_2 + (v_1 \cdot \hat{r})(v_2 \cdot \hat{r}) \right] \right\}.$$

The derivation is most gracefully carried out in the Coulomb gauge. Reference: Jackson (1975) ch. 12.

16. a) Show that the magnetic field due to a classical particle with magnetic dipole moment μ at the origin of coordinates is

$$B(r) = \frac{8\pi}{3} \mu \delta^3(r) + \frac{3\hat{r}(\hat{r} \cdot \mu) - \mu}{r^3}$$

b) Now consider the (classical) interaction of a static nucleus with magnetic moment μ_N, fixed at the origin, with an electron (with magnetic moment μ_e and electric charge e) orbiting about it with angular momentum L. Show that the interaction energy is given by the hyperfine Hamiltonian

$$\mathcal{H}_{\text{HFS}} = -\frac{8\pi}{3} \mu_e \cdot \mu_N \delta^3(r)$$

$$+ \frac{1}{r^3} \left[\mu_e \cdot \mu_N - 3(\hat{r} \cdot \mu_e)(\hat{r} \cdot \mu_N) - \frac{e}{mc} L \cdot \mu_N \right].$$

Discuss the origin of each term.
References: Fermi (1930), Jackson (1975) ch. 5.

17. The Darwin Lagrangian for two charged particles is given by the interaction Lagrangian $\mathcal{L}_{\text{int}}$ of problem 15 plus the free-particle Lagrangian expanded to order $1/c^2$,

$$\mathcal{L}_{\text{free}} = \frac{1}{2} \left(m_1 v_1^2 + m_2 v_2^2 \right) + \frac{1}{8c^2} \left(m_1 v_1^4 + m_2 v_2^4 \right).$$

a) Introduce relative coordinates $r = r_1 - r_2$ and $v = v_1 - v_2$ and c.m. coordinates. Write out the Lagrangian $\mathcal{L}_{\text{Darwin}} = \mathcal{L}_{\text{free}} + \mathcal{L}_{\text{int}}$ in the reference frame in which the velocity of the center of mass vanishes and evaluate the canonical momentum components $p_x = \partial \mathcal{L}/\partial v_x$, etc.

b) Compute the Hamiltonian to first order in $1/c^2$ and show that it is

$$\mathcal{H} = \frac{p^2}{2}\left(\frac{1}{m_1} + \frac{1}{m_2}\right) + \frac{q_1 q_2}{r} - \frac{p^4}{8c^2}\left(\frac{1}{m_1^3} + \frac{1}{m_2^3}\right)$$

$$+ \frac{q_1 q_2}{2m_1 m_2 c^2}\left(\frac{p^2 + (\boldsymbol{p}\cdot\hat{\boldsymbol{r}})^2}{r}\right).$$

Compare with the various terms in eq. (42.1) on p. 193 of Bethe and Salpeter (1957). Discuss the agreements and disagreements.
References: Jackson (1975), problem 12.12; Berestetskii et al. (1971), pp. 280–284; Breit (1930); Heisenberg (1926).

18. By coupling together first the quarks and antiquarks separately, show that the colorspin for a collection of n constituents is given by

$$\sum_{i<j}\langle\boldsymbol{\lambda}_i\cdot\boldsymbol{\lambda}_j\boldsymbol{\sigma}_i\cdot\boldsymbol{\sigma}_j\rangle = -4G_{\text{tot}}^{(6)2} + \tfrac{4}{3}s_{\text{tot}}(s_{\text{tot}} + 1)$$

$$+ 8\left[G_{\text{quarks}}^{(6)2} + G_{\text{antiq}}^{(6)2}\right] - 4\left[G_{\text{quarks}}^{(3)2} + G_{\text{antiq}}^{(3)2}\right]$$

$$- \tfrac{8}{3}\left[s_{\text{quarks}}(s_{\text{quarks}} + 1) + s_{\text{antiq}}(s_{\text{antiq}} + 1)\right] - 8n,$$

where the labels (quarks, antiquarks) refer to the collective representations of quarks and antiquarks. Verify that for a state composed only of quarks you recover (5.81).

19. Consider quark–antiquark states. Using SU(6) techniques, identify the colorspin representations containing color singlets, and compute the expectation value of the colorspin operator. Compare with the results in table 15 for $0^- - 1^-$ splitting.

20. Enumerate the SU(6)$_{\text{colorspin}}$ representations that can be formed out of two quarks and two antiquarks. Give the SU(3)$\otimes$SU(2) decomposition of each. Compute $G^{(6)2}$ for each representation.

Reference for problems 18–20: Jaffe (1977a, b).

21. Show that quarkonium level spacings independent of the constituent mass occur in a logarithmic potential, $V(r) = \lambda \log(r/r_0)$.
Reference: Quigg and Rosner (1977).

22. Using the Schrödinger equation (6.3), prove the identity

$$\frac{|u'(0)|^2}{4\pi} = |\Psi(0)|^2 = \frac{\mu}{2\pi\hbar^2}\left\langle\frac{dV}{dr}\right\rangle$$

for a system with reduced mass μ.

Reference: This result is apparently due to Fermi and to Schwinger, in unpublished work. A general derivation appears in section 2.2 of Quigg and Rosner (1979).

23. By evaluating the identity just derived in semiclassical approximation, show that for a general nonsingular potential

$$|\Psi_n(0)|^2 \simeq \left(\frac{2\mu}{\hbar^2}\right)^{3/2}\frac{E_n^{1/2}}{4\pi^2}\frac{\partial E_n}{\partial n}.$$

References: Krammer and Léal Ferreira (1976); Quigg and Rosner (1978c); Bell and Pasupathy (1979).

24. Consider the Schrödinger equation for s-wave bound states of a $1/r$ potential in N space dimensions:

$$\left[\nabla^2 + 2\mu(E + \alpha/r)\right]\Psi(r) = 0. \tag{1}$$

(a) Show that the radial equation is

$$\left[\frac{d^2}{dr^2} + \frac{(N-1)}{r}\frac{d}{dr} + 2\mu(E + \alpha/r)\right]\Psi(r) = 0. \tag{2}$$

(b) Now take the limit of large N, so that $(N-1) \to N$. Introduce a reduced radial wavefunction

$$u = r^{N/2}\Psi, \tag{3}$$

and a scaled radial coordinate

$$R = r/N^2. \tag{4}$$

Show that the Schrödinger equation becomes

$$\left[\frac{1}{N^2}\frac{d^2u}{dR^2} - \frac{u}{4R^2} + 2\mu(N^2E + \alpha/R)u\right] = 0. \tag{5}$$

(c) Apart from the factor N^2 which sets the scale of E, this equation describes a particle with effective mass μN^2 moving in an effective potential

$$V_{\text{eff}} = 1/8\mu R^2 - \alpha/R. \tag{6}$$

Find the energy of the ground state in the limit as $N \to \infty$, for which the kinetic energy vanishes. Show that it is given by the absolute minimum of V_{eff}, so that

$$E_{N \to \infty} = -2\mu\alpha^2/N^2. \tag{7}$$

Corrections to (7) may be computed by expanding V_{eff} about the minimum and treating the additional terms as perturbations.

(d) The exact solution to the exact eigenvalue problem (2) is easily verified to be

$$E_{\text{exact}} = -2\mu\alpha^2/(N-1)^2. \tag{8}$$

Show that the exact eigenvalue can be recast in the form of an expansion in powers of $1/N$ as

$$E_{\text{exact}} = -\frac{2\mu\alpha^2}{N^2} \sum_{j=1}^{\infty} jN^{1-j} = E_{N \to \infty}\left(1 + \sum_{j=2}^{\infty} jN^{1-j}\right),$$

so that the $N \to \infty$ result may form the basis for a systematic approximation scheme. How many terms must be retained to obtain a 1% approximation for $N = 3$?
Reference: Mlodinow and Papanicolaou (1980).

References

Abrams, G.S., et al. (1974), Phys. Rev. Lett. 33, 1453.
Abrams, G.S., et al. (1979), Phys. Rev. Lett. 43, 477.
Ademollo, M., and R. Gatto (1964), Phys. Rev. Lett. 13, 264.
Adler, S. L. (1981), Phys. Rev. D23, 2905; D24, 1063E.
Adler, S.L., and R. Dashen (1968), Current Algebras (Benjamin, New York).
Ajzenberg-Selove, F. (1975), Nucl. Phys. A248, 1.
Ajzenberg-Selove, F. (1976), Nucl. Phys. A268, 1.
Ajzenberg-Selove, F. (1979), Nucl. Phys. A320, 1.
Akhiezer, A.I., and V.B. Berestetskii (1965), Quantum Electrodynamics (Wiley, New York), § 11.
Allen, E. (1975), Phys. Lett. 57B, 263.
Amaldi, U., M. Jacob and G. Matthiae (1976), Ann. Rev. Nucl. Sci. 26, 385.
Amirzadeh, J., et al. (1979), Phys. Lett. 89B, 125.
Appelquist, T. (1978), Chromodynamic Structure and Phenomenology, in: Particles and Fields, 1977, Banff Summer Institute, eds. D.H. Boal and A.N. Kamal (Plenum, New York) p. 33.

Appelquist, T., and H.D. Politzer (1975), Phys. Rev. Lett. 34, 43.

Appelquist, T., R.M. Barnett and K. Lane (1978), Ann. Rev. Nucl. Part. Sci. 28, 387.

Armstrong, T., et al. (1982), Phys. Lett. 110B, 77.

Aston, D., et al. (1981), Phys. Lett. 106B, 183.

Aubert, J.J., et al. (1974), Phys. Rev. Lett. 33, 1404.

Augustin, J.E., et al. (1974), Phys. Rev. Lett. 33, 1406.

Bacry, H. (1967), Leçons sur la Théorie des Groupes et les Symétries des Particules Elémentaires (Gordon and Breach/Dunod, Paris and New York).

Bander, M. (1981), Phys. Rep. 75C, 205.

Bander, M., T.-W. Chiu, G.L. Shaw and D. Silverman (1981), Phys. Rev. Lett. 47, 549.

Barbieri, R., R. Gatto and E. Remiddi (1981), Phys. Lett. 106B, 497.

Bardeen, J., L.N. Cooper and J.R. Schrieffer (1957), Phys. Rev. 108, 1175.

Bardeen, W.A., H. Fritzsch and M. Gell-Mann (1973), in: Scale and Conformal Symmetry in Hadron Physics, ed. R. Gatto (Wiley, New York) p. 139.

Bardeen, W.A., M. Chanowitz, S.D. Drell, M. Weinstein and T.-M. Yan (1975), Phys. Rev. D11, 1094.

Barnes, T. (1981), Z. Phys. C10, 275.

Bell, J.S., and R. Bertlmann (1981), Nucl. Phys. B187, 285.

Bell, J.S., and J. Pasupathy, (1979), Phys. Lett. 83B, 389.

Berard, R.W., et al. (1973), Phys. Lett. 47B, 355.

Berestetskii, V.B., E.M. Lifshitz and L.P. Pitaevski (1971), Relativistic Quantum Theory, part 1, transl. J.B. Sykes and J.S. Bell (Pergamon, Oxford, 1971, 1979).

Berg, B. (1980), Phys. Lett. 97B, 401.

Berg, D., et al. (1980a), Phys. Rev. Lett. 44, 706.

Berg, D., et al. (1980b), in: High Energy Physics 1980, XX Int. Conf., Madison, eds. L. Durand and L.G. Pondrom (American Institute of Physics, New York) p. 537.

Berg, D., et al. (1981), Phys. Lett. 98B, 119.

Berkelman, K. (1980), in: High Energy Physics 1980, XX Int. Conf., Madison, eds. L. Durand and L.G. Pondrom (American Institute of Physics, New York) p. 1499.

Bethe, H., and E.E. Salpeter (1957), Quantum Mechanics of One- and Two-Electron Atoms (Springer Verlag, Berlin).

Bhanot, G. (1981), Phys. Lett. 105B, 95.

Bhanot, G., and C. Rebbi (1981), Nucl. Phys. B180, 469.

Biagi, S.F., et al. (1981), Z. Phys. C9, 305.

Billoire, A., R. Lacaze, A. Morel and H. Navelet (1979), Phys. Lett. 80B, 381.

Binnie, D.M., et al. (1979), Phys. Lett. 83B, 141.

Bjorken, J.D. (1979), in: Proc. of the Eur. Phys. Soc. Int. Conf. on High Energy Physics, Geneva, 1979 (CERN, Geneva) p. 245.

Bjorken, J.D. (1980), Elements of Quantum Chromodynamics, in: Quantum Chromodynamics, Proc. 7th SLAC Summer Inst. on Particle Physics, 1979, ed. A. Mosher (SLAC, Stanford) p. 219.

Blatt, J.M., and V.F Weisskopf (1952), Theoretical Nuclear Physics (Wiley, New York).

Bogoliubov, P.N. (1967), Ann. Inst. Henri Poincaré 8, 163.

Bohm, A., and R.B. Teese (1981), Lett. Nuovo Cim. 32, 122.

Bohr, A., and B.R. Mottelson (1969), Nuclear Structure, vol. 1 (Benjamin, New York).

Bolzan, J.F., W.F. Palmer and S.S. Pinsky (1976) Phys. Rev. D14, 3202.

Bonesini, M., et al. (1981), Phys. Lett. 103B, 75.

Borkowski, F., et al. (1974) Nucl. Phys. A222, 269.

Bouchiat, C., and Ph. Meyer (1964), Nuovo Cim. 34, 1122.

Bourrely, C., B. Machet and E. de Rafael (1981), Nucl. Phys. B189, 157.

Bowler, K.C., P.J. Corvi, A.J.G. Hey and P.D. Jarvis (1980), Phys. Rev. Lett. 45, 97.

Brandenburg, G., et al. (1976), Phys. Rev. Lett. 36, 1239.

Breit, G. (1930), Phys. Rev. 36, 383.

Breit, G., E.U. Condon and R.D. Present (1936), Phys. Rev. 50, 846.

Brower, R.C. (1981) Course 5, this volume.

Brown, G.E., and M. Rho (1979), Phys. Lett. 82B, 177.

Brown, G.E., M. Rho and V. Vento (1979), Phys. Lett. 84B, 383.

Buchmüller, W., and S.-H.H. Tye, (1980), Phys. Rev. Lett. 44, 850.

Buchmüller, W., and S.-H.H. Tye (1981), Phys. Rev. D24, 132.

Buchmüller, W., G. Grunberg and S.-H.H. Tye (1980), Phys. Rev. Lett. 45, 103, 587(E).

Buchmüller, W., Y.J. Ng and S.-H.H. Tye (1981), Phys. Rev. D24, 3003.

Buras, A.J., (1980), Rev. Mod. Phys. 52, 199.

Buras, A.J. (1981), Proc. 1981 Int. Symp. on Lepton and Photon Interactions at High Energies, Bonn, ed. W. Pfeil (Univ. of Bonn) p. 636.

Cabibbo, N. (1963), Phys. Rev. Lett. 10, 531.

Cahn, R.N., and M.B. Einhorn (1971), Phys. Rev. D4, 3337.

Carlitz, R., and D. Creamer (1979), Ann. Phys. (NY) 118, 429.

Carlitz, R.D., S.D. Ellis and R. Savit (1977), Phys. Lett. 68B, 443.

Carlson, C., J. Coyne, P. Fishbane, F. Gross and S. Meshkov (1980), Phys. Lett. 98B, 110.

Carlson, C., J. Coyne, P. Fishbane, F. Gross and S. Meshkov (1981), Phys. Lett. 99B, 353.

Carroll, A.S., et al. (1978), Phys. Rev. Lett. 41, 777.

Carruthers, P. (1966), Introduction to Unitary Symmetry (Interscience, New York).

Cashmore, R.J. (1980), in: Proc. 1980 Conf. on Experimental Meson Spectroscopy, Brookhaven, ed. S.-U. Chung (American Institute of Physics, New York) p. 1.

Cassen, B., and E.U. Condon (1936), Phys. Rev. 50, 846.

Celmaster, W., and F. Henyey (1978), Phys. Rev. D18, 1688.

Chanowitz, M. (1975), Color and Experimental Physics, in: Particles and Fields 1975, eds. H.J. Lubatti and P.M. Mockett (Univ. of Washington, Seattle) p. 448.

Chanowitz, M.S. (1980), Phys. Rev. Lett. 44, 59.

Chanowitz, M.S. (1981), Phys. Rev. Lett. 46, 981.

Chao, K.-T., N. Isgur and G. Karl (1981), Phys. Rev. D23, 155.

Chew, G.F. (1965), The Analytic S Matrix, in: High Energy Physics, 1965 Les Houches Lectures, eds. C. DeWitt and M. Jacob (Gordon and Breach, New York) p. 187.

Chew, G.F., and S.C. Frautschi (1961), Phys. Rev. Lett. 7, 394.

Chew, G.F., and F.E. Low (1959), Phys. Rev. 113, 1640.

Cho, Y.M., J.L. Cortes and X.Y. Pham (1981), Univ. Pierre et Marie Curie preprint PAR/LPTHE 81/08.

Chodos, A., and C.B. Thorn (1974), Nucl. Phys. B72, 509.

Chodos, A., R.L. Jaffe, K. Johnson, C.B. Thorn and V.F. Weisskopf, (1974a), Phys. Rev. D9, 3471.

Chodos, A., R.L. Jaffe, K. Johnson and C.B. Thorn (1974b), Phys. Rev. D10, 2599.

Chou, T.T., and C.N. Yang (1968), Phys. Rev. 170, 1591.

Chounet, L.-M., J.-M. Gaillard and M.K. Gaillard (1972), Phys. Rep. 4C, 199.

Chung, S.-U., ed. (1980), Proc. 1980 Conf. on Experimental Meson Spectroscopy, Brookhaven (American Institute of Physics, New York).

Cleland, W.E., et al. (1980a), Observation of a Spin-6 Isospin-1 Boson Resonance in the Charged K$\overline{\text{K}}$ System, Univ. de Genève preprint UGVA-DPNC 1980/07-101.

Cleland, W.E., et al (1980b), Phys. Lett. 97B, 465.

Close, F.E. (1979), An Introduction to Quarks and Partons (Academic, New York).

Close, F.E. (1980), The First Lap in QCD, Rutherford Lab. preprint RL-80-063.

Close, F.E. (1981), Glueballs, Hermaphrodites, and QCD Problems for Baryon Spectroscopy, Proc. Eur. Phys. Soc. Conf. on High Energy Physics, Lisbon, eds. J. Dias de Deus
and J. Soffer (Eur. Phys. Soc., 1983) p. 549.

Cohen, I., and H.J. Lipkin (1979), Nucl. Phys. B151, 16.

Cohen, I., and H.J. Lipkin (1980), Phys. Lett. 93B, 56.

Cohen, I., and H.J. Lipkin (1981), Phys. Lett. 106B, 119.

Coleman, S. (1980), 1/N, in: Pointlike Structures Inside and Outside Hadrons, ed. A.
Zichichi (Plenum, New York) p.11.

Coleman, S., and S.L. Glashow (1961), Phys. Rev. Lett. 6, 423.

Coleman, S., and S.L. Glashow (1964), Phys. Rev. 134, B671.

Coleman, S., and E. Witten (1980), Phys. Rev. Lett. 45, 100.

Collins, P.D.B. (1977), An Introduction to Regge Theory and High Energy Physics
(Cambridge Univ. Press, Cambridge).

Copley, L.A., N. Isgur and G. Karl (1979), Phys. Rev. D20, 768.

Cox, P.T. (1981), seminar related to this course.

Cox, P.T., et al. (1981), Phys. Rev. Lett. 46, 877.

Coyne, J., P. Fishbane and S. Meshkov (1980), Phys. Lett. 91B, 259.

Cronin, J., and O. Overseth (1963), Phys. Rev. 129, 1795.

Dalitz, R.H. (1965), Quark Models for the Elementary Particles, in: High Energy Physics,
1965 Les Houches Lectures, eds. C. DeWitt and M. Jacob (Gordon and Breach, New
York) p. 251.

Dalitz, R.H. (1977), in: Fundamentals of Quark Models, Scottish Univ. Summer School in
Physics, 1976, eds. I.M. Barbour and A.T. Davies (SUSSP, Edinburgh) p. 151.

Dalitz, R.H., and D.G. Sutherland (1966), Phys. Rev. 146, 1180.

Dally, E.B., et al. (1977) Phys. Rev. Lett. 39, 1176.

Dally, E.B., et al. (1980a) Elastic Scattering Comparison of the Negative Pion Form Factor,
contribution to the XX Int. Conf. on High Energy Physics, Madison.

Dally, E.B., et al. (1980b), Direct Comparison of the π^- and K^- Form Factors, contribution to the XX Int. Conf. on High Energy Physics, Madison.

Dally, E.B., et al. (1980c), Phys. Rev. Lett. 45, 232.

Dankowych, J.A., et al. (1981), Phys. Rev. Lett. 46, 580.

DeGrand, T. (1980), Thinking about Baryon Excited States with Bags, in: Baryon 1980,
IVth Int. Conf. on Baryon Resonances, ed. N. Isgur (Univ. of Toronto, Toronto) p. 209.

DeGrand, T., R.L. Jaffe, K. Johnson and J. Kiskis (1975), Phys. Rev. D12, 2060.

De Rújula, A., H. Georgi and S.L. Glashow (1975), Phys. Rev. D12, 147.

De Rújula, A., R. Giles and R. Jaffe (1978), Phys. Rev. D17, 285.

De Rújula, A., R. Giles and R. Jaffe (1980), Phys. Rev. D22, 227.

DeTar, C. (1978), Phys. Rev. D17, 302, 323 [Erratum D19 (1979) 1028].

DeTar, C. (1979), Phys. Rev. D19, 1451.

DeTar, C. (1980), The MIT Bag Model, in: Quantum Flavordynamics, Quantum Chromodynamics, and Unified Theories, eds K.T. Mahanthappa and J. Randa, NATO Adv.
Study Inst. Ser. B, Physics, vol. 54 (Plenum, New York) p. 393.

DeTar, C.E. (1981a), Phys. Rev. D24, 752.

DeTar, C.E. (1981b), Phys. Rev. D24, 762.

DeViron, F., and J. Weyers (1981), Nucl. Phys. B185, 391.

Di Vecchia, P. (1979), Phys. Lett. 85B, 357.

Dolgov, A.D., L.B. Okun, I.Ya. Pomeranchuk and V.V. Soloviev (1965), Phys. Lett. 15, 84.

Dolgov, O.V., D.A. Kirzhnits and E.G. Maksimov (1981), Rev. Mod. Phys. 53, 81.

Donoghue, J.F. (1980), in: High Energy Physics 1980, XX Int. Conf., Madison, eds. L. Durand and L.G. Pondrom (American Institute of Physics, New York) p. 35.

Donoghue, J.F. (1981), Glueballs in the Bag, in: Gauge Theories, Massive Neutrinos, and Proton Decay, ed. A. Perlmutter (Plenum, New York) p. 85.

Donoghue, J.F., and K. Johnson (1980), Phys. Rev. D21, 1975.

Donoghue, J.F., E. Golowich and B.R. Holstein (1975), Phys. Rev. D12, 2875.

Donoghue, J.F., K. Johnson and Li Bing-An (1981), Phys. Lett. 99B, 416.

Dorsaz, P.-A. (1981), Univ. de Genève Thesis UGVA-Thèse-1994.

Dothan, Y. (1981), Linear Symmetric Quark Theories of Baryon Magnetic Moments, Univ. of Minnesota preprint.

Drell, S.D., and T.-M. Yan (1970), Phys. Rev. Lett. 25, 316.

Drell, S.D., and T.-M. Yan (1971), Ann. Phys. (NY) 66, 578.

Dremin, I.M., and C. Quigg (1978), Science 199, 937.

Dydak, F., et al. (1977), Nucl. Phys. B118, 1.

Eichten, E. (1978), in: New Results in High Energy Physics, Vanderbilt Conf., eds. R.S. Panvini and S.E. Csorna (American Institute of Physics, New York) p. 252.

Eichten, E., and F. Feinberg (1979), Phys. Rev. Lett. 43, 1205.

Eichten, E., and F. Feinberg (1981), Phys. Rev. D23, 2724.

Eichten, E., and K. Gottfried (1977), Phys. Lett. 66B, 286.

Eichten, E., K. Gottfried, T. Kinoshita, J. Kogut, K.D. Lane and T.-M. Yan, (1975), Phys. Rev. Lett. 34, 369.

Einhorn, M.B. (1975), Introduction to SU(4) and the Properties of Charmed Particles, Fermilab-Lectures-75/1-THY.

Eisenbud, L., and E.P. Wigner (1941), Proc. Nat. Acad. Sci. (USA) 27, 281.

Ellis, J., M.K. Gaillard, D.V. Nanopoulos and S. Rudaz (1981), Phys. Lett. 99B, 101.

Estabrooks, P. (1977), $K\pi$ and $\pi\pi$ Data and Partial Wave Analyses, in: Experimental Meson Spectroscopy 1977, 5th Int. Conf., Boston, eds. E. von Goeler and R. Weinstein (Northeastern Univ. Press, Boston) p. 185.

Fairley, G.T., and E.J. Squires (1975), Nucl. Phys. B93, 56.

Farhi, E., and L. Susskind (1981), Phys. Rep. 74C, 277.

Feld, B. (1969), Models of Elementary Particles (Blaisdell, Waltham MA.).

Fermi, E. (1930), Mem. Accad. d'Italia I, (Fis.) 139; reprinted in: E. Fermi, Collected Papers, vol. 1, eds. E. Segrè et al. (Univ. of Chicago Press, Chicago, 1962) paper 57a.

Fermi, E., and C.N. Yang (1949), Phys. Rev. 76, 1739.

Feynman, R.P. (1972), Photon–Hadron Interactions (Benjamin, Reading, MA.).

Feynman, R.P., (1974), Science 183, 601.

Feynman, R.P. (1977), Gauge Theories, in: Weak and Electromagnetic Interactions at High Energy, ed. R. Balian and C.H. Llewellyn Smith (North-Holland, Amsterdam) p. 120.

Fiarman, S., and S.S. Hanna (1975), Nucl. Phys. A251, 1.

Field, R.D., and C. Quigg (1975), Estimates of Associated Charm Production Cross Sections, Fermilab-75/15-THY.

Fishbane, P.M., and M.T. Grisaru (1978), Phys. Lett. 74B, 98.

Fogleman, G., D.B. Lichtenberg and J.G. Wills (1979), Lett. Nuovo Cim. 26, 369.

Fox, G.C., and C. Quigg (1973), Ann. Rev. Nucl. Sci. 23, 219.

Frauenfelder, H., and E.M. Henley (1974), Subatomic Physics (Prentice-Hall, Englewood Cliffs, NJ).

Frazer, W. R. (1966), Elementary Particles (Prentice-Hall, Englewood Cliffs, NJ).

Freund, P.G.O., and Y. Nambu (1975), Phys. Rev. Lett. 34, 1645.

Fritzsch, H. (1979) Chromodynamics, in: Proc. 1978 Int. Meeting on Frontier of Physics, Singapore, eds. K. K. Phua, C.K. Chew and Y.K. Lim (Singapore National Academy of Science, Singapore) p. 1005.

Fritzsch, H., and M. Gell-Mann (1972), in: Proc. XVI Int. Conf. on High Energy Physics, eds. J.D. Jackson, A. Roberts and R. Donaldson (Fermilab, Batavia) p. II-135.

Fritzsch, H., and P. Minkowski (1975), Nuovo Cim. 30A, 393.

Fritzsch, H., M. Gell-Mann and H. Leutwyler (1973), Phys. Lett. 47B, 365.

Fujii, Y., and K. Mima (1978), Phys. Lett. 79B, 138.

Gaillard, M.K., B.W. Lee and J.L. Rosner (1975), Rev. Mod. Phys. 47, 277.

Gasiorowicz, S. (1966), Elementary Particle Physics (Wiley, New York).

Gavela, M.B., et al. (1979), Phys. Lett. 82B, 431.

Geffen, D.A., and W. Wilson (1980), Phys. Rev. Lett. 44, 370.

Gell-Mann, M. (1953), Phys. Rev. 92, 833.

Gell-Mann, M. (1961), CalTech Synchrotron Lab. Report CTSL-20, reprinted in Gell-Mann and Ne'eman (1964) p. 11.

Gell-Mann, M. (1962), Phys. Rev. 125, 1067.

Gell-Mann, M. (1964), Phys. Lett. 8, 214.

Gell-Mann, M., and S.L. Glashow (1961), Ann. Phys. (NY) 15, 437.

Gell-Mann, M., and Y. Ne'eman (1964), The Eightfold Way (Benjamin, New York).

Gell-Mann, M., R.J. Oakes and B. Renner (1968), Phys. Rev. 175, 2195.

Georgi, H. (1980), Phys. Rev. D22, 225.

Gerasimov, S.B. (1966), Zh. Eksp. Teor. Fiz. 50, 1559; Engl. transl.: Sov. Phys. JETP 23, 1040.

Gidal, G., G. Goldhaber, J.G. Guy, et al. (1981), Phys. Lett. 107B, 153.

Giles, R.C. (1976), Phys. Rev. D13, 1670.

Giles, R.C., and S.-H.H. Tye (1977), Phys. Rev. D16, 1079.

Giles, R.C., and S.-H.H. Tye (1978), Phys. Lett. 73B, 30.

Gilman, F., and I. Karliner (1974), Phys. Rev. D10, 2194.

Gilman, F., and D.H. Miller (1978), Phys. Rev. D17, 1846.

Glashow, S.L. (1961), Phys. Rev. Lett. 7, 469.

Goebel, C. (1958), Phys. Rev. Lett. 1, 337.

Goldhaber, A.S., G.C. Fox and C. Quigg (1969), Phys. Lett. 30B, 249.

Goldhaber, G. (1970), Experimental Results on the $\omega - \rho$ Interference Effect, in: Experimental Meson Spectroscopy, eds C. Baltay and A.H. Rosenfeld (Columbia Univ. Press, New York and London) p. 59.

Goldman, T., and R.W. Haymaker (1981), Phys. Rev. D24, 724.

Goldstone, J. (1961), Nuovo Cim. 19, 154.

Golowich, E. (1975), Phys. Rev. D12, 2108.

Golub, R., et al. (1979), Scientific American 240, no. 6 (June) p. 106.

Gopal, G.N. (1980), in: Baryon 1980, IVth Int. Conf. on Baryon Resonances, ed. N. Isgur (Univ. of Toronto, Toronto) p. 159.

Gottfried, K. (1977), in: Proc. 1977 Int. Symp. on Lepton and Photon Interactions at High Energies, Hamburg, ed. F. Gutbrod (DESY, Hamburg) p. 667.

Gottfried, K. (1980), in: High Energy e^+e^- Interactions, Vanderbilt Conf., eds. R.S. Panvini and S.E. Csorna (American Institute of Physics, New York) p. 88.

Gottfried, K., and V.F. Weisskopf (1981), textbook in preparation.

Gourdin, M. (1967), Unitary Symmetries and Their Applications to High Energy Physics (North-Holland, Amsterdam).

Gourdin, M. (1974), Phys. Rep. 11C, 29.

Greenberg, O.W. (1964), Phys. Rev. Lett. 13, 598.

Greenberg, O.W. (1978), Ann. Rev. Nucl. Part. Sci. 28, 327.

Greenberg, O.W., and H.J. Lipkin (1981), Nucl. Phys. A370, 349.

Greenberg, O.W., and C.A. Nelson (1977), Phys. Rep. 32C, 69.

Greenberg, O.W., and M. Resnikoff (1967), Phys. Rev. 163, 1844.

Greenberg, O.W., and D. Zwanziger (1966), Phys. Rev. 150, 1177.

Greenberg, O.W., S. Nussinov and J. Sucher (1977), Phys. Lett. 70B, 465; reprinted Phys. Lett. 72B (1977) 87.

Greene, G.L., et al. (1979), Phys. Rev. D20, 2139.

Gromes, D. (1977), Nucl. Phys. B131, 80.

Gross, D.J., and F. Wilczek (1973a), Phys. Rev. Lett. 30, 1343.

Gross, D.J., and F. Wilczek (1973b), Phys. Rev. D8, 3633.

Gross, D.J., S.B. Treiman and F. Wilczek (1979), Phys. Rev. D19, 2188.

Grosse, H., and A. Martin (1979), Nucl. Phys. B148, 413.

Grosse, H., and A. Martin (1980), Phys. Rep. 60C, 341.

Halprin, A., C.M., Anderson and H. Primakoff (1966), Phys. Rev. 152, 1295.

Hamermesh, M. (1963), Group Theory (Addison-Wesley, Reading, MA).

Han, M.-Y., and Y. Nambu (1965), Phys. Rev. 139B, 1006.

Handler, R., et al. (1980), in: High Energy Physics 1980, XX Int. Conf. Madison, eds. L. Durand and L.G. Pondrom (American Institute of Physics, New York) p. 539.

Hasenfratz, P., and J. Kuti (1978), Phys. Rep. 40C, 75.

Haymaker, R.W., and T. Goldman (1981), Phys. Rev. D24, 743.

Heisenberg, W. (1926), Z. Phys. 39, 499.

Heisenberg, W. (1932), Z. Phys. 77, 1.

Hendry, A.W., and D.B. Lichtenberg (1978), Rep. Prog. Phys. 41, 1707.

Henriques, A.B., B.H. Kellett and R.G. Moorhouse (1976), Phys. Lett. 64B, 85.

Herb, S.W., et al. (1977), Phys. Rev. Lett. 39, 252.

Hey, A.J.G. (1979), Particle Systematics, in: Proc. Eur. Phys. Soc. Conf. on High Energy Physics, Geneva, 1979 (CERN, Geneva) p. 523.

Hey, A.J.G. (1980a), Quark Models of Baryons, in: Baryon 1980, IVth Int. Conf. on Baryon Resonances, Toronto, ed. N. Isgur (University of Toronto, Toronto) p. 223.

Hey, A.J.G. (1980b), Multiquark States and Exotics, in: High Energy Physics 1980, XX Int. Conf., Madison, eds. L. Durand and L.G. Pondrom (American Institute of Physics, New York) p. 22.

Higgs, P.W. (1964a), Phys. Rev. Lett. 12, 132.

Higgs, P.W. (1964b), Phys. Rev. Lett. 13, 508.

Himel, T.M., et al. (1980a), Phys. Rev. Lett. 44, 920.

Himel, T.M., et al. (1980b), Phys. Rev. Lett. 45, 1146.

Hofstadter, R. (1963), Electron Scattering and Nuclear and Nucleon Structure (Benjamin, New York).

Horn, D., and J. Mandula (1978), Phys. Rev. D17, 898.

Igi, K. (1977), Phys. Rev. D16, 196.

Iizuka, J. (1966), Suppl. Prog. Theor. Phys. 37–38, 21.

Iizuka, J., K. Okada and O. Shito (1966), Prog. Theor. Phys. 35, 1061.

Ioffe, B.L. (1981), Nucl. Phys. B188, 317.

Isgur, N. (1976), Phys. Rev. D13, 122.

Isgur, N. (1978), Phys. Rev. D17, 369.

Isgur, N. (1980a), Baryons with Chromodynamics, in: High Energy Physics 1980, XX Int. Conf., Madison, eds. L. Durand and L.G. Pondrom (American Institute of Physics, New York) p. 30.

Isgur, N., ed. (1980b), Baryon 1980, IVth Int. Conf. on Baryon Resonances, Toronto (University of Toronto, Toronto).

Isgur, N. (1980c), Soft QCD: Low Energy Hadron Physics with Chromodynamics, in: New Aspects of Subnuclear Physics, 1978 Int. School of Physics "Ettore Majorana", ed. A. Zichichi (Plenum, New York).

Isgur, N., and G. Karl (1977), Phys. Lett. 72B, 109.

Isgur, N., and G. Karl (1978a), Phys. Lett. 74B, 353.

Isgur, N., and G. Karl (1978b), Phys. Rev. D18, 4187.

Isgur, N., and G. Karl (1979), Phys. Rev. D19, 2653.

Isgur, N., and G. Karl (1980), Phys. Rev. D21, 3175.

Isgur, N., G. Karl and R. Koniuk (1978), Phys. Rev. Lett. 41, 1269 [Erratum 45 (1980) 1738].

Isgur, N., H.R. Rubinstein, A. Schwimmer and H.J. Lipkin (1979), Phys. Lett. 89B, 79.

Isgur, N., G. Karl and D.W.L. Sprung (1981), Phys, Rev. D23, 163.

Ishikawa, K. (1979a), Phys. Rev. D20, 731.

Ishikawa, K. (1979b), Phys. Rev. D20, 2903.

Ishikawa, K. (1981), Phys. Rev. Lett. 46, 978.

Jackson, J.D. (1966), in: Proc. XIIIth Int. Conf. on High Energy Physics, Berkeley, ed. M. Alston-Garnjost (Univ. of California Press, Berkeley and Los Angeles) p. 149.

Jackson, J.D. (1975), Classical Electrodynamics, 2nd ed. (Wiley, New York).

Jackson, J.D. (1976), Lectures on the New Particles, in: Proc. 4th SLAC Summer Inst. on Particle Physics, ed. M.C. Zipf (SLAC-198).

Jackson, J.D. (1977a), The Nature of Intrinsic Magnetic Dipole Moments, CERN Yellow Report 77-17.

Jackson, J.D. (1977b), in: Proc. 1977 Eur. Conf. on Particle Physics, Budapest, eds. L. Jenik and I. Montvay (Central Research Inst. for Physics, Budapest) vol. 1, p. 603.

Jackson, J.D., C. Quigg and J.L. Rosner (1979), in: Proc. 19th Int. Conf. on High Energy Physics, Tokyo, 1978, eds. S. Homma, M. Kawaguchi and H. Miyazawa (Physical Society of Japan, Tokyo) p. 391.

Jaffe, R.L. (1977a), Phys. Rev. D15, 267.

Jaffe, R.L. (1977b), Phys. Rev. D15, 281.

Jaffe, R.L. (1977c), Phys. Rev. Lett. 38, 195.

Jaffe, R.L. (1977d), Nature 268, 201.

Jaffe, R.L. (1979), The Bag, in: Pointlike Structures inside and outside Hadrons, ed. A. Zichichi (Plenum, New York) p. 99.

Jaffe, R.L., and K. Johnson (1976), Phys. Lett. 60B, 201.

Jaffe, R.L., and K. Johnson (1977), Comm. Nucl. Part. Phys. 7, 107.

Jaffe, R.L., and F.E. Low (1979), Phys. Rev. D19, 2105.

Jensen, D.A., et al. (1980), in: High Energy Physics 1980, XX Int. Conf., Madison, eds. L. Durand and L.G. Pondrom (American Institute of Physics, New York) p. 364.

Jensen, T. (1980), Univ. of Rochester thesis, UR-747.

Jensen, T. (1981), seminar related to this course.

Johnson, K. (1975), Acta Phys. Polon. B6, 865.

Johnson, K. (1977), in: Fundamentals of Quark Models, Scottish Univ. Summer School in Physics, 1976, eds. I.M. Barbour and A.T. Davies (SUSSP, Edinburgh) p. 245.

Johnson, K. (1979), Scientific American 241, no. 1 (July) p. 112.

Johnson, K., and C. Nohl (1979), Phys. Rev. D19, 291.

Johnson, K., and C.B. Thorn (1976), Phys. Rev. D13, 1934.

Jones, L.W. (1977), Rev. Mod. Phys. 49, 717.

Kadanoff, L.P. (1977), Rev. Mod. Phys. 49, 267.

Kane, G.L. (1972), Acta Phys. Polon. B3, 845.

Karl, G. (1979), in:\ Proc. 19th Int. Conf. on High Energy Physics, Tokyo, 1978, eds. S. Homma, M. Kawaguchi and H. Miyazawa (Physical Society of Japan, Tokyo) p. 135.

Karl, G., and E. Obryk (1968), Nucl. Phys. B8, 609.

Kawamoto, N., and A.I. Sanda (1978), Phys. Lett. 76B, 446.

Kelly, R.L. (1980), in: Baryon 1980, IVth Int. Conf. on Baryon Resonances, Toronto, ed. N. Isgur (Univ. of Toronto, Toronto) p. 149.

Khriplovich, I.B. (1969), Yad. Fiz. 10, 409; Engl. transl.: Sov. J. Nucl. Phys. 10, 235 (1970).

Kinson, J.B. (1980), in: Baryon 1980, IVth Int. Conf. on Baryon Resonances, Toronto, ed. N. Isgur (Univ. of Toronto, Toronto) p. 263.

Koester, L., et al. (1976), Phys. Rev. Lett. 36, 1021.

Kogut, J.B., and L. Susskind (1974), Phys. Rev. D9, 3501.

Kogut, J., D. Sinclair and L. Susskind (1976), Nucl. Phys. B114, 199.

Kokkedee, J.J.J. (1969), The Quark Model (Benjamin, Reading, MA.).

Koller, K., and T.F. Walsh (1978), Nucl. Phys. B140, 449.

Krammer, M., and H. Krasemann (1979a), Quarkonium, in: New Phenomena in Lepton–Hadron Physics, NATO Adv. Study Inst. Series B, Physics, vol. 49, eds. D.E.C. Fries and J. Wess (Plenum, New York and London) p. 161.

Krammer, M., and H., Krasemann (1979b), Quarkonia, in: Quarks and Leptons as Fundamental Particles, ed. P. Urban, Acta Phys. Austr., Suppl. XXI (Springer, New York) p. 259.

Krammer, M., and P. Léal Ferreira (1976), Rev. Bras. Fis. 6, 7.

Krasemann, H., and S. Ono (1979), Nucl. Phys. B154, 283.

Krohn, V.E., and G.R. Ringo (1973), Phys. Rev. D8, 1305.

Landau, L.D. (1948), Dokl. Akad. Nauk SSSR 60, 207.

Landau, L.D., and E.M. Lifshitz (1960), Electrodynamics of Continuous Media (Addison-Wesley, Reading, MA) § 14.

Langacker, P. (1979), Phys. Rev. D20, 2983.

Langacker, P. (1980), Phys. Lett. 90B, 447.

Langacker, P. (1981), Phys. Rep. 72C, 185.

Langacker, P., and H. Pagels (1979), Phys. Rev. D19, 2070.

LaRue, G.S., J.D. Phillips and W.M. Fairbank (1981), Phys. Rev. Lett. 46, 967.

Lee, B.W. (1972), Chiral Dynamics (Gordon and Breach, New York).

Lee, B.W., C. Quigg and J.L. Rosner (1977), Phys. Rev. D15, 157.

Lee, T.D. (1980), QCD and the Bag Model of Hadrons, in: Proc. 1980 Guangzhou Conf. on Theoretical Physics (Science Press, Beijing) p. 32.

Leith, D.W.G.S. (1977), Progress in K* Spectroscopy, in: Experimental· Meson Spectroscopy 1977, Proc. 5th Int. Conf., Boston, eds. E. von Goeler and R. Weinstein (Northeastern Univ. Press, Boston) p. 207.

Leung, C.N., and J.L. Rosner (1979), J. Math. Phys. 20, 1435.

Leutwyler, H. (1974a), Phys. Lett. 48B, 45.

Leutwyler, H. (1974b), Nucl. Phys. B76, 413.

Leutwyler, H. (1981), Phys. Lett. 98B, 447.

Levine, R., and Y. Tomozawa (1979), Phys. Rev. D19, 1572.

Levine, R., and Y. Tomozawa (1980), Phys. Rev. D21, 840.

Levinson, C.A., H.J. Lipkin and S. Meshkov (1962), Phys. Lett. 1, 44.
LeYaouanc, A., et al. (1977), Phys. Rev. D15, 844.
Lichtenberg, D.B. (1978), Unitary Symmetry and Elementary Particles (Academic Press, New York).
Lichtenberg, D.B. (1981), Z. Phys. C7, 143.
Lipkin, H.J. (1966), Lie Groups for Pedestrians, 2nd Ed. (North-Holland, Amsterdam).
Lipkin, H.J. (1969), in: Nuclear Physics, eds. C. DeWitt and V. Gillet (Gordon and Breach, New York) p. 585.
Lipkin, H.J. (1973a), Phys. Rep. 8C, 173.
Lipkin, H.J. (1973b), Phys. Rev. D7, 846.
Lipkin, H.J. (1973c), Phys. Lett. 45B, 267.
Lipkin, H.J. (1978a), Phys. Lett. 74B, 399.
Lipkin, H.J. (1978b), Phys. Rev. Lett. 41, 1629.
Lipkin, H.J. (1979a), Nucl. Phys. B155, 104.
Lipkin, H.J. (1979b), A Quasi-Nuclear Colored Quark Model for Hadrons, in: Common Problems in Low- and Medium-Energy Nuclear Physics, Banff Summer School, eds. B. Castel, B. Goulard and F.C. Khanna (Plenum, New York) p. 173.
Lipkin, H.J. (1981a), Phys. Rev. D24, 1437.
Lipkin, H.J. (1981b), Phys. Lett. 106B, 114.
Lipkin, H.J., and A. Tavkhelidze, 1965, Phys. Lett. 17, 331.
Llewellyn Smith, C.H. (1980), Topics in QCD, in: Quantum Flavordynamics, Quantum Chromodynamics, and Unified Theories, NATO Adv. Study Inst. Ser. B, Physics, vol. 54, eds. K.T. Mahanthappa and J. Randa (Plenum, New York) p. 59.
Lyons, L. (1981), Current Status of Quark Search Experiments, in: Progress in Particle and Nuclear Physics, vol. 7, ed. D. Wilkinson (Pergamon Press, Oxford) p. 169.
Machacek, M., and Y. Tomozawa (1978), Ann. Phys. (NY) 110, 407.
Mandelstam, S. (1979), Phys. Rev. D19, 2391.
Mandelstam, S. (1980), Phys. Rep. 67C, 109.
Marciano, W.J., and H. Pagels (1978), Phys. Rep. 36C, 137.
Marciano, W.J., and H. Pagels (1979), Nature 279, 479.
Marshak, R., S. Okubo and E.C.G. Sudarshan (1957), Phys. Rev. 106, 599.
Martin, A. (1980), Phys. Lett. 93B, 338.
Martin, A. (1981), Phys. Lett. 100B, 511.
Matsuyama, S., and H. Miyazawa (1979), Prog. Theor. Phys. 61, 942.
McKay, W.G., and J. Patera (1981), Tables of Dimensions, Indices, and Branching Rules for Representations of Simple Lie Algebras, Lecture Notes in Pure and Applied Mathematics, vol. 69 (Dekker, New York and Basel).
Mistretta, C., et al. (1969), Phys. Rev. 184, 1487.
Mlodinow, L.D., and N. Papanicolaou (1980), Ann. Phys. (NY) 128, 314.
Molzon, W.R., et al. (1978), Phys. Rev. Lett. 41, 1213.
Montanet, L. (1980), in: High Energy Physics 1980, XX Int. Conf., Madison, eds. L. Durand and L.G. Pondrom (American Institute of Physics, New York) p. 1196.
Montanet, L., G.C. Rossi and G. Veneziano (1980), Phys. Rep. 63C, 149.
Morgan, D. (1981), Acta Phys. Polon. B12, 43.
Moxhay, P., J.L. Rosner and C. Quigg (1981), Phys. Rev. D23, 2638.
Nakano, T., and K. Nishijima (1955), Prog. Theor. Phys. 10, 581.
Nambu, Y. (1960), Phys. Rev. Lett. 4, 380.
Nambu, Y. (1966), A Systematics of Hadrons in Subnuclear Physics, in: Preludes in Theoretical Physics in Honor of V.F. Weisskopf, eds. A. De-Shalit, H. Feshbach and L. Van Hove (North-Holland, Amsterdam; Wiley, New York) p. 133.

Nambu, Y. (1974), Phys. Rev. D10, 4262.

Nambu, Y. (1976), Scientific American 235, no. 5 (November) p. 48.

Nambu, Y., and G. Jona-Lasinio (1961a), Phys. Rev. 122, 345.

Nambu, Y., and G. Jona-Lasinio (1961b), Phys. Rev. 124, 246.

Ne'eman, Y. (1961), Nucl. Phys. 26, 222.

Nefkens, B.M.K., et al. (1978), Phys. Rev. D18, 3911.

Nielsen, H.B., and P. Olesen (1973), Nucl. Phys. B61, 45.

Nielsen, H.B., and A. Patkós (1981), Nucl. Phys. B195, 137.

Novikov, V.A., L.B. Okun, M.A. Shifman, A.I. Vainshtein, M.B. Voloshin and V.I. Zakharov (1978), Phys. Rep. 41C, 1.

Novikov, V.A., M.A. Shifman, A.I. Vainshtein and V.I. Zakharov (1979), Phys. Lett. 86B, 347.

Novikov, V.A., M.A. Shifman, A.I. Vainshtein and V.I. Zakharov (1980a), Nucl. Phys. B165, 55.

Novikov, V.A., M.A. Shifman, A.I. Vainshtein and V.I. Zakharov (1980b), Nucl. Phys. B165, 67.

Novikov, V.A., M.A. Shifman, A.I. Vainshtein and V.I. Zakharov (1980c), ITEP, Moscow preprint ITEP-87.

Novikov, V.A., M.A. Shifman, A.I. Vainshtein and V.I. Zakharov (1980d), ITEP, Moscow preprint ITEP-88.

Novikov, V.A., M.A. Shifman, A.I. Vainshtein and V.I. Zakharov (1981), Acta Phys. Polon. B12, 399.

O'Donnell, P.J. (1981), Rev. Mod. Phys. 53, 673.

Ohshima, T. (1980), Phys. Rev. D22, 707.

Okubo, S. (1962a), Prog. Theor. Phys. 27, 949.

Okubo, S. (1962b), Prog. Theor. Phys. 28, 24.

Okubo, S. (1963), Phys. Lett. 5, 165.

Okun, L.B. (1965), Weak Interactions of Elementary Particles (Addison-Wesley, Reading, MA) ch. 13.

Okun, L.B., and M.A. Shifman (1981), Z. Phys. C8, 17.

Okun, L.B., M.B. Voloshin and V.I. Zakharov (1979), On the Price of Integer Charge Quarks, ITEP, Moscow preprint ITEP-79.

Oreglia, M., et al. (1980), Phys. Rev. Lett. 45, 959.

Pagels, H. (1975), Phys. Rep. 16C, 219.

Pagels, H. (1979), Phys. Rev. D19, 3080.

Pagels, H., and S. Stokar (1979), Phys. Rev. D20, 2947.

Particle Data Group (1980), Rev. Mod. Phys. 52, S1.

Partridge, R., et al. (1980), Phys. Rev. Lett. 45, 1150.

Patera, J., and D. Sankoff (1973), Tables of Branching Rules for Representations of Simple Lie Algebras (Univ. de Montréal, Montréal).

Perkins, D.H. (1982), Introduction to High-Energy Physics, 2nd Ed. (Addison-Wesley, Reading, MA).

Perl, M.L. (1974), High Energy Hadron Physics (Wiley, New York).

Pietschmann, H., and W. Thirring (1966), Phys. Lett. 21, 713.

Pilcher, J. (1980), in: Proc. 1979 Int. Symp. on Lepton and Photon Interactions at High Energies, eds. T.B.W. Kirk and H.D.I. Abarbanel (Fermilab, Batavia) p. 185.

Politzer, H.D. (1973), Phys. Rev. Lett. 30, 1346.

Preston, M.A. (1962), Physics of the Nucleus (Addison-Wesley, Reading, MA).

Primakoff, H. (1951), Phys. Rev. 81, 899.

Protopopescu, S., and N.P. Samios (1979), Ann. Rev. Nucl. Part. Sci. 29, 339.

Quigg, C. (1979), Lectures on Charmed Particles, in: Proc. XIth Int. School for Young Scientists on High Energy Physics and Relativistic Nuclear Physics, Gomel, Byelorussia, 1977 (JINR, Dubna) p. 203.

Quigg, C. (1980), Bound States of Heavy Quarks and Antiquarks, in: Proc. 1979 Int. Symp. on Lepton and Photon Interactions at High Energies, eds. T.B.W. Kirk and H.D.I. Abarbanel (Fermilab, Batavia) p. 239.

Quigg, C. (1981), Introduction to Gauge Theories of the Strong, Weak, and Electromagnetic Interactions, in: Techniques and Concepts of High Energy Physics, St. Croix, 1980, ed. T. Ferbel (Plenum, New York) p. 143.

Quigg, C. (1983), Gauge Theories of the Strong, Weak, and Electromagnetic Interactions (Addison–Wesley, Reading, MA).

Quigg, C., and J.L. Rosner (1976), Phys. Rev. D14, 160.

Quigg, C., and J.L. Rosner (1977), Phys. Lett. 71B, 153.

Quigg, C., and J.L. Rosner (1978a), Comm. Nucl. Part. Phys. 8, 11.

Quigg, C., and J.L. Rosner (1978b), Phys. Lett. 72B, 462.

Quigg, C., and J.L. Rosner (1978c), Phys. Rev. D17, 2364.

Quigg, C., and J.L. Rosner (1979), Phys. Rep. 56C, 167.

Quigg, C., and J.L. Rosner (1981), Phys. Rev. D23, 2625.

Quigg, C., H.B. Thacker and J.L. Rosner (1980), Phys. Rev. D21, 234.

Ramsey, N.F. (1956), Molecular Beams (Oxford Univ. Press, Oxford).

Reinders, L.J., H.R. Rubinstein and S. Yazaki (1981), Nucl. Phys. B186, 109.

Renner, B. (1968), Current Algebras and Their Applications (Pergamon Press, London).

Richard, J.M. (1981), Phys. Lett. 100B, 515.

Richardson, J.L. (1979), Phys. Lett. 82B, 272.

Roberts, B.L., et al. (1975), Phys. Rev. D12, 1232.

Robson, D. (1977), Nucl. Phys. B130, 328.

Rosner, J.L. (1968), Phys. Rev. Lett. 21, 950 (Erratum 1422).

Rosner, J.L. (1974a), Phys. Rep. 11C, 189.

Rosner, J.L. (1974b), in: Proc. XVIIth Int. Conf. on High Energy Physics, London, ed. J.R. Smith (Rutherford Lab., Chilton, UK) p. II-171.

Rosner, J.L. (1980a), in: Particles and Fields 1979, eds. B. Margolis and D.G. Stairs (American Institute of Physics, New York) p. 325.

Rosner, J.L. (1980b), in: High Energy Physics 1980, XX Int. Conf. Madison, eds. L. Durand and L.G. Pondrom (American Institute of Physics, New York) p. 540.

Rosner, J.L. (1981a), Quark Models, in: Techniques and Concepts of High Energy Physics, St. Croix, 1980, ed. T. Ferbel (Plenum, New York) p. 1.

Rosner, J.L. (1981b), Phys. Rev. D24, 1347.

Rosner, J.L., C. Quigg and H.B. Thacker (1978), Phys. Lett. 74B, 350.

Roy, P. (1979), Nature 280, 106.

Roy, P. (1980), The Glueball Trail, Rutherford Lab. report RL-80-007.

Roy, P., and T.F. Walsh (1978), Phys. Lett. 78B, 62.

Rutherglen, J. (1969), in: Proc. 4th Int. Symp. on Electron and Photon Interactions at High Energies, Liverpool, eds. D.W. Braben and R. E. Rand (Daresbury Nuclear Physics Lab., Daresbury) p. 163.

Sakata, S. (1956), Prog. Theor. Phys. 16, 686.

Sakharov, A.D. (1975), Zh. Eksp. Teor. Fiz. Pisma 21, 554; Engl. transl.: JETP Lett. 21, 258.

Sakharov, A.D. (1980), Zh. Eksp. Teor. Fiz. 78, 2112 [Sov. Phys. JETP 51 (1980) 1059].

Sakurai, J.J. (1964), Invariance Principles and Elementary Particles (Princeton Univ. Press, Princeton).

Samios, N.P., M. Goldberg and B.T. Meadows (1974), Rev. Mod. Phys. 46, 49.

Schachinger, L.C., et al. (1978), Phys. Rev. Lett. 41, 1348.

Schamberger, R.D. (1981), Proc. 1981 Int. Symp. on Lepton and Photon Interactions at High Energies, Bonn, ed. W. Pfeil (Univ. of Bonn) p. 217.

Scharre, D. (1981), Proc. 1981 Int. Symp. on Lepton and Photon Interactions at High Energies, Bonn, ed. W. Pfeil (Univ. of Bonn) p. 163.

Scherk, J. (1975), Rev. Mod. Phys. 47, 123.

Schnitzer, H.J. (1975), Phys. Rev. Lett. 35, 1540.

Schnitzer, H.J. (1976), Phys. Rev. D13, 74.

Schnitzer, H.J. (1978), Phys. Rev. D18, 3482.

Schnitzer, H.J. (1981a), in: New Flavors and Hadron Spectroscopy, ed. J. Tran Thanh Van (Editions Frontières, Dreux, France) vol. 2, p. 293.

Schnitzer, H.J. (1981b), Spin-Dependence and Gluon Mixing in Ordinary Meson Spectroscopy, Brandeis preprint.

Schonfeld, J.F., W. Kwong, J.L. Rosner, C. Quigg and H.B. Thacker (1980), Ann. Phys. (NY) 128, 1.

Settles, R., et al. (1979), Phys. Rev. D20, 2154.

Shaw, R. (1955), The Problem of Particle Types and Other Contributions to the Theory of Elementary Particles, Cambridge Univ. Thesis.

Shifman, M.A. (1981), Z. Phys. C9, 347.

Shifman, M.A., A.I. Vainshtein and V.I. Zakharov (1979a), Phys. Rev. Lett. 42, 297.

Shifman, M.A., A.I. Vainshtein and V.I. Zakharov (1979b), Nucl. Phys. B147, 385.

Shifman, M.A., A.I. Vainshtein and V.I. Zakharov (1979c), Nucl. Phys. B147, 448.

Shifman, M.A., A.I. Vainshtein and V.I. Zakharov (1979d), Nucl. Phys. B147, 519.

Shrock, R., and L.-L. Wang (1978), Phys. Rev. Lett. 41, 1692.

Slansky, R., T. Goldman and G.L. Shaw (1981), Phys. Rev. Lett. 47, 887.

Soni, A. (1980), Nucl. Phys. B168, 147.

Spinetti, M. (1980), in: Proc. 1979 Int. Symp. on Lepton and Photon Interactions at High Energies, eds. T.B.W. Kirk and H.D.I. Abarbanel (Fermilab, Batavia) p. 506.

Stanley, D.P., and D. Robson (1980), Phys. Rev. Lett. 45, 235.

Stanley, D.P., and D. Robson (1981), Phys. Rev. D23, 2776.

Stanton, N., et al. (1979), Phys. Rev. Lett. 42, 346.

Susskind, L. (1979), Phys. Rev. D20, 2619.

Suura, H. (1980), Phys. Rev. Lett. 44, 1319.

Tati, T. (1966), Prog. Theor. Phys. 35, 126.

Teese, R.B. (1981), Phys. Rev. D24, 1413.

Thacker, H.B., C. Quigg and J.L. Rosner (1978a), Phys. Rev. D18, 274.

Thacker, H.B., C. Quigg and J.L. Rosner (1978b), Phys. Rev. D18, 287.

Thirring, W. (1966), Acta Phys. Austr. Suppl. II, 205; reprinted in Kokkedee (1969), p. 183.

't Hooft, G. (1973a), Phys. Lett. 61B, 455.

't Hooft, G. (1973b), Phys. Lett. 62B, 444.

't Hooft, G. (1974a), Nucl. Phys. B72, 461.

't Hooft, G. (1974b), Nucl. Phys. B75, 461.

't Hooft, G. (1976), Phys. Rev. Lett. 37, 8.

't Hooft, G. (1980), Scientific American 242, no. 6 (June) p. 104.

Tsyganov, E. (1979), in: Proc. 19th Int. Conf. on High Energy Physics, Tokyo, 1978, eds. S. Homma, M. Kawaguchi, and H. Miyazawa (Physical Society of Japan, Tokyo) p. 315.

Van Royen, R., and V.F. Weisskopf (1967), Nuovo Cim. 50, 617; 51, 583.
Veneziano, G. (1976), Nucl. Phys. B117, 519.
Veneziano, G. (1979a), Dynamics of Hadron Reactions, in: Proc. 19th Int. Conf. on High
    Energy Physics, Tokyo, 1978, eds. S. Homma, M. Kawaguchi, and H. Miyazawa
    (Physical Society of Japan, Tokyo) p. 725.
Veneziano, G. (1979b), Nucl. Phys. B159, 213.
Voloshin, M.B. (1979), Nucl. Phys. B154, 365.
Weber, G. (1967), in: Proc. 1967 Int. Symp. on Electron and Photon Interactions at High
    Energies, Stanford, ed. S.M. Berman (SLAC, Stanford) p. 59.
Weinberg, S. (1973), Phys. Rev. Lett. 31, 494.
Weinberg, S. (1975), Phys. Rev. D11, 3583.
Weinberg, S. (1976), Phys. Rev. D13, 974.
Weinberg, S. (1977), The Problem of Mass, in: A Festschrift for I.I. Rabi, ed. L. Motz,
    Trans. NY Acad. Sci. Series II, 38, 185.
Wess, J. (1981), course 1, this volume.
Wicklund, A.B., et al. (1980), Phys. Rev. Lett. 45, 1469.
Wigner, E.P. (1937), Phys. Rev. 51, 106.
Wiik, B. (1981), course 8, this volume.
Willemsen, J.F. (1976), Phys. Rev. D13, 1327.
Willey, R.S. (1978), Phys. Rev. D18, 270.
Williams, P.K., and V. Hagopian, eds. (1973), $\pi\pi$ Scattering 1973, Tallahassee Conference
    (American Institute of Physics, New York).
Willis, W., and J. Thompson (1968), Leptonic Decays of Elementary Particles, in: Advances
    in Particle Physics, vol. 1, eds. R.L. Cool and R.E. Marshak (Interscience, New York) p.
    295.
Wilson, K. (1974), Phys. Rev. D10, 2445.
Witten, E. (1979a), Nucl. Phys. B156, 269.
Witten, E. (1979b), Nucl. Phys. B160, 57.
Witten, E. (1980a), Phys. Today 33, no. 7 (July) p. 38.
Witten, E. (1980b), The $1/N$ Expansion in Atomic and Particle Physics, in: Recent
    Developments in Gauge Theories, Cargèse, 1979, NATO Adv. Study Inst. Ser. B, Physics,
    vol. 59, eds. G. 't Hooft et al. (Plenum, New York) p. 403.
Witten, E. (1980c), Ann. Phys. (NY) 128, 363.
Yang, C.N. (1950), Phys. Rev. 77, 242.
Yang, C.N., and R.L. Mills (1954), Phys. Rev. 96, 191.
Zakharov, V.I. (1980a), in: High Energy Physics 1980, XX Int. Conf. Madison, eds. L.
    Durand and L.G. Pondrom (American Institute of Physics, New York) p. 1027.
Zakharov, V.I. (1980b), in: High Energy Physics 1980, XX Int. Conf., Madison, eds. L.
    Durand and L.G. Pondrom (American Institute of Physics, New York) p. 1234.
Zee, A. (1972), Phys. Rep. 3C, 127.
Zel'dovich, Ya. B., and A.D. Sakharov (1966), Yad. Fiz. 4, 395; Engl. transl.: Sov. J. Nucl.
    Phys. 4, 283(1967).
Zweig, G. (1964a), CERN Report 8182/TH.401.
Zweig, G. (1964b), CERN Report 8419/TH.412; reprinted in: Developments in the Quark
    Theory of Hadrons, vol. 1: 1964–1978, eds. D.B. Lichtenberg and S.P. Rosen (Hadronic
    Press, Nonantum, MA, 1980) p. 22.
Zweig, G. (1964c), Fractionally Charged Particles and SU(6), in: Symmetries in Elementary
    Particle Physics, 1964 Int. School of Physics "Ettore Majorana", ed. A. Zichichi (Academic
    Press, New York, 1965) p. 192.
Zweig, G. (1980), Origins of the Quark Model, in: Baryon 1980, IVth Int. Conf. on Baryon
    Resonances, Toronto, ed. N. Isgur (Univ. of Toronto, Toronto) p. 439.

Seminars related to Course 6

1. Higher twist and pion production, by Jonathan A. Bagger (Princeton Univ.).

2. On the ITEP approach to resonance physics, by John S. Bell (CERN).

3. Hyperon magnetic moments, by Tim Cox (CERN).

4. Higher twist in the MIT bag model, by Robert Jaffe (MIT).

5. M1 electromagnetic transitions in mesons, by Terry Jensen (CERN).

6. Sum rules for light quark masses in QCD, by Eduardo de Rafael (CNRS, Marseille).

COURSE 7

HIGH ENERGY EXPERIMENTS

Leon M. LEDERMAN

Fermi National Accelerator Laboratory
Batavia, IL 60510, USA

M.K. Gaillard and R. Stora, eds.
Les Houches, Session XXXVII, 1981
Théories de jauge en physique des hautes énergies / Gauge theories in high energy physics
© *North-Holland Publishing Company, 1983*

Contents

0. Preamble

This set of lectures is being given to an audience consisting largely of European trained theory students. The lectures are an attempt to correct the tendency of very early specialization, and lack of familiarity with experimental matters.

1. Introduction

Perhaps there are some theorists who view experiments as annoying complications in what should be a mathematically simple universe. In ancient times the Greeks saw harmony and simplicity in perfect circles; modern theorists study gauge symmetries looking for simple truths. It is the responsibility of an experimentalist to point out what actually exists. In fact, it is the responsibility of both theorists and experimentalists to create new ideas and to test their validity. Hopefully these lectures will provide some insight into modern experimental particle physics. Perhaps it might even be possible to make a few converts.

Recently I composed a list of some of the best experimentalists of our time and discovered that, in fact, many know quite a lot of theory. The list included people such as J. Steinberger, C. Rubbia, B. Richter, R. Schwitters, B. Wiik, S.C.C. Ting, V. Fitch and J. Cronin. All too often though, particle physicists become so specialized that theorists do not know what the experimentalists are doing and vice versa.

As an introduction to the philosophy of experimental physics we will first split the variety of possible experiments into two classifications. The first class may be termed *observations*. Some examples are the observation that parity is not conserved, the observation that ν_e is not the same as ν_μ, and the observation that $\Upsilon, \Upsilon', \Upsilon''$ states exist. The second class includes experiments that produce a *measurement* of a specific quantity such as the pion nucleon scattering cross section or the K_s lifetime.

In either case the experiments must be well designed with careful regard given to the physics goals, human engineering and economics. Furthermore, systems must be built in to provide independent cross checks that can give confidence in the results. Unfortunately some discoveries are not anticipated and can easily slip past the eyes of anxious physicists whose experiment lacks adequate resolution or adequate statistical validity. Furthermore during any experiment countless opportunities for disaster can and do arise. An essential ingredient which can help is to designate the most critical and hostile graduate student as the "Devil's Advocate". His assignment is to find bugs in the experiment during design. The more he hates the professor the more zest he will put into showing him up. I have fired more advocates in my time...

2. Measuring space and time

In order to perform an experiment certain basic tools are needed. It has been said that "all things in physics can be determined with a clock and a ruler". The tools we use are, in fact, technologically complex clocks and rulers. Armed with these tools, experimenters can attempt the impossible experiments, e.g. measure the proton lifetime or the quark structure function. We have, these days, at our disposal a literal arsenal of modern detectors large and small to measure quantities of space and time. Some of the detectors in today's experiments are listed in table 1 with their resolutions in space and time. Continuing the list are various Čerenkov counters, transition radiators, time projection chambers and calorimeters.

The nature of the expected physics, as well as economics determines the variety and the scale of the detectors built for experiments. It is

Table 1
Some of today's detectors

Detector	Spatial resolution Δx	Integration time (ns)	Resolution time (ns)
Scintillator	~ 1 cm	~ 5	~ 0.5
Proportional wire chambers	~ 1 mm	≥ 50	50
Drift chambers	~ 100 μm	~ 100	5
Streamer chamber	~ 50 μm	1000	1000
Bubble chamber	2mm → 50 μm (holography)	∞	→ ∞
Emulsion	~ 1 μm	∞	→ ∞
Silicon strips(?)	~ 10 μm	~ 1	1

always a struggle to balance the apparatus and beam design for maximum experimental sensitivity at a reasonable cost. Moreover, time is a pressing issue. The experiment must be built and then analyzed in a reasonable length of time in order to assure the topical nature of the physics and to maintain the creative spirit of the physicists involved.

The complicated equipment is linked together by sophisticated information processing systems. Each new generation of experiments requires faster processors in order to handle more intricate triggers and higher data rates. The computer systems, designed by such people as Sippach of Columbia have helped ease the conscience of the physicist looking for the social justification of high energy physics. The resulting engineering systems and computer architecture are finding ready application in the burgeoning information processing industry.

3. Can experiments be wrong?

One can invent several tests to aid in deciding if an experiment is correct. If the data confirm a popular theory, it "must be right". One can say that the results were refereed and published and therefore must be correct. Or one could assume correctness if an experiment received a large number of citations. Certainly these tests are incorrect. Many experiments have been performed, the published results producing numerous debates, sometimes baffling current theoretical work, sometimes suggesting new directions, but later these experiments where shown to be in error. It is important to note that experiments can be misleading in that they provide correct but incomplete information, or data that can be easily misinterpreted. Then of course, there is the class of experiments that have produced results that are simply wrong. Some observed or suggested phenomena that have since disappeared were the split A_2 [1], the R, S, T... [2], the 2.8 GeV η_c resonance at DESY [3], the "high y anomaly" [4], the 6.0 GeV upsilon by myself and collaborators at Fermilab [5], the failure to see $\pi \to e + \nu$ decay [6], and $K_L \to \mu\mu$ [7] are a few examples. Wrong measurements and missed opportunities for discoveries are the pitfalls of the unwary and/or unlucky experimentalist.

4. Missed opportunities

The following personal history of missed opportunities illustrates another class of difficulties facing experimental physicists.

4.1. The 3/2 – 3/2 πN resonance

In 1952 our group at Columbia was intent on extending the measurement of the π + p elastic cross begun by Fermi at the Chicago cyclotron [8].

Positive pions were extracted from the fringing field of the Chicago cyclotron only if they were produced near ~ 180° relative to the protons striking a target near the outside edge of the cyclotron field. This limited Fermi's energy to ≤ 135 MeV. Their data showed a steep rise, strongly suggesting a nearby resonance. My colleagues and I reasoned that forward-produced positive pions would have much higher energy. Unfortunately they were bent inwards towards the center of the accelerator. We designed a collimator and an emulsion camera to receive pions in two energy bins: 155 MeV and 188 MeV. The idea is that nuclear emulsion contains a small amount of hydrogen and it is not difficult to distinguish π –p collisions by its kinematics from π –nucleus collisions. After much trial and error, we succeeded in obtaining clean exposures and we observed very large yields of positive pions in these energy bins. Results were obtained with what seemed to be solid cross checks. We concluded that our data did not require that the π^+ –p cross section go through a resonance [9]. Some months later, Julius Ashkin, at Carnegie Tech, reported [10] a clear resonance at 180 MeV with $\sigma \sim 4\pi\lambda^2$. The Carnegie cyclotron was able to extract π^+ up to 200 MeV from their fringe field and do the experiment in a clean, simple manner. What had gone wrong? Well, for one, we were unlucky in our choice of energies and our two points missed the peak energy. The other problem, appreciated much later, was simply one of having a scanning efficiency which was significantly lower than we thought. We found a *systematic* error which was ~ 1 standard deviation but which led to a misleading interpretation of the data. Here, we cleverly (!) used small angle (Coulomb) scattering as a test for efficiency. When the scanners correctly obtained the Rutherford scattering cross section, we assumed we had verified their efficiency. Wrong! This test failed, because in a crowded emulsion with strong impetus to speed, small scatters are less easily missed than very large angle scatters. The successful exposure and observation of high energy π^+ mesons in collision with protons was a tour de force of experimental achievement. But obviously this is not enough.

A good experiment should include many independent checks to prove that the results obtained are correct and that they are interpreted properly.

Once again, early in 1956, we ignored important clues in our data, and missed the observation of parity violation in $\mu^+ \rightarrow e^+ + 2\nu$. We were at

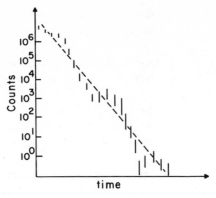

Fig. 1. Negative muons decaying in carbon (1956). Cyclotron fringing field was ~ 10 G.

that time measuring the apparent decrease in lifetime of negative muons in substances of increasing Z. One set of our data appears in fig. 1; and we fit the points to a straight line, ignoring the fluctuations. Shortly after (July, 1956), Lee and Yang [10a] predicted the occurrence of parity violation, and Wu et al [11] soon observed the phenomenon in the decay of ^{60}Co. Too late, we recognized our bumps as an asymmetry in the angular distribution of the electron coupled with a precession of our muons in the fringing field of the Nevis cyclotron, thereby substantiating the observation of parity violation and extending it to meson decay. By January 1957, we made up for our blunder [12]. We discovered parity violation in π decay, in muon decay, we measured the decay asymmetry and used the polarized muons for the first time to measure the g-value of the muon and to explore atomic magnetic fields.

Looking back, in the era before the Lee–Yang paper, an interpretation of the curious deviation from experimental decay would have required a level of inspiration which was just not available.

4.2. μ form factor and neutrino collisions

In 1963, in an experiment measuring the μ "form factor" $\mu + p \rightarrow \mu + p$ at BNL [13], we observed many unusual events that seemed to be inelastic. At the time we could not measure all the hadronic products, and the term inclusive reaction had not been invented. Ignorant as to how to deal with these events, we ignored them. Five years later scientists at SLAC were theoretically motivated to look at these events inclusively and plotted the inelastic cross section as a function of Q^2. They discovered that at large

Q^2, the cross section decreased as $1/Q^4$ rather than as $1/Q^{12}$ as originally expected [14]. Deep inelastic scattering, the first dynamical test of the quark model, was discovered.

Remember that at the time of this research, the question of μ–e universality was one of the most perplexing problems in particle physics. (It still is, but we've grown used to it!.) We were fascinated by the possibility of using the proton as a "ruler" to compare the muon and electron radii. In fact, to my knowledge, these early BNL experiments on proton elastic form factors as probed by muons still stand as the best measurements. However, to our present chagrin, inelastic events were irrelevant since there was good electron–proton elastic data and this seemed clear to us whereas inelastic stuff was "complex". How times have changed!

The following year, in an experiment examining neutrino–nucleon collision events with small hadron recoil, we saw many events we now know represent neutral currents. We were prone to blame "neutrons" as a background. There were many things to measure and neutral currents was not, at that time, a burning issue. It took another *ten years* before *neutral currents were discovered*.

In fairness, it is not clear that the experiment could have cleanly separated neutral currents from a neutron induced background but the point is that we didn't think to try. Now is there a common denominator to the above anecdotes, e.g. stupidity? Mitigating circumstance is that it was too easy to do experimental research in those days—one could almost immediately proceed to the next piece of research, leaving the more subtle aspects of the data untouched. But probably the main "lesson" is that we were always working at the frontier, where observation and measurement are as far from routine and programmatic as we can get. Our muon research was the first AGS-scale muon scattering experiment. We had initiated high energy neutrino physics in 1962 and so neutrinos in 1964 were not the industry they are today. Our thinking, our grasp of the crucial elements of the physics, were fuzzy. In short, yes! Dumb!

Lesson 1: "Try to be intelligent".

4.3. How to ask questions

This is a guide to asking questions at experimental seminars. It is true that experiments can be wrong, but it is difficult for even the most experienced observers who are not part of the research, to discern the fact from the fiction. Theorists must take advantage of opportunities to

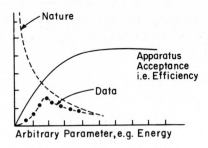

Fig. 2. Typical acceptance curve to be folded with "nature".

challenge and stimulate experimental colleagues. Since it is unreasonable to ask: "What exciting clues have you ignored?", one may as well concentrate on the most likely places for trouble. In evaluating an experiment the first question to ask the proud experimentalist is "Are you sure you understand the distortions introduced by your apparatus?" Things such as the *acceptance* of the apparatus must be examined in great detail. Acceptance is related to the size of the experimental apparatus, the experiment's trigger and the detector efficiencies as a function of the particle masses or energies. It is important to remember that all experimental apparatus will distort. Usually the acceptance takes a form sketched in fig. 2: What is actually recorded as data is the convolution of the two curves.

It is crucial to know the exact place where the acceptance rises, because it is possible to generate a false resonance in the region where the efficiency is changing rapidly, if one mistakenly assumes it to be flat. An experimental acceptance is estimated by a Monte Carlo program. Monte Carlos can never be completely correct and must be carefully examined. By making the experimental acceptance tunable, a check can be made of the simulated acceptance (fig. 3). For a cross section measurement, the total cross section,

$$\sigma_{\text{total}}(E) = \sigma_{\text{measured}}(E) \times \text{Efficiency}(E),$$

should be the same for each acceptance curve at a given energy.

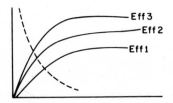

Fig. 3. Varying the efficiency to provide tests of the apparatus effect.

It can take years for an experimental team to understand their detector's acceptance. Particularly this has been true of the large colliding detectors at SPEAR, DORIS and the ISR.

4.4. Systematic and statistical errors

The second check of an experiment is to examine the resolution. Experimental errors are never exactly Gaussian. Misinterpretation of a non-Gaussian tail introduced by systematic errors could lead to an anomalous measurement. One must evaluate the statistical errors and ask if a 4.2σ bump is actually a bump, or merely a statistical fluctuation. The answer lies in the number of measurements you made; 4.2σ fluctuations *will* occur if you look often enough.

A third factor to consider is that it is sometimes necessary to examine an experimenter's psychological profile before blindly accepting his results. Pressures of old age, job security and the urge to publish can have great influence. Furthermore, it is safe to say that experiments of great significance should and will be doubted until they can be repeated independently.

5. Case history

5.1. Introduction

In order to give a more detailed description in which all the ingredients of experimental work are illustrated, I will present an experimental case study. This will be a personal history of a linked chain of experiments begun in 1964 and continuing up to the present. The subject of these experiments is dilepton production in hadronic collisions (also known as Drell–Yan processes). The actors come from an ever changing group at Columbia, in collaboration with a variety of institutions. In this case history I intend to emphasize key ingredients that influence every experiment:

    (i) Interplay between theory and experiment.
    (ii) Evolution of experiment.
    (iii) Intuition and choice.
    (iv) Missed opportunities.
    (v) Luck**.
    (vi) Statistical misfortune.
    (vii) Checks and rechecks.

5.2. Pre-history (addendum to the history)

The program started at about the time the first dynamical attacks on constituent physics were being formulated by Feynman and Bjorken on the West Coast. At the Nevis Laboratory (Columbia University) the motivations were based on a little known principle of Heisenberg:

$$\Delta Q \Delta x \geqslant h/2\pi, \tag{1}$$

which implied that large Q^2 would reveal small distance structure.

Back in 1964 we were already searching for the W^+ at the AGS at Brookhaven [15], a machine that accelerated protons to 30 GeV. The reaction explored was:

$$
\begin{array}{l}
p + N \rightarrow W^+ + \text{anything,} \\
\qquad\quad \llcorner\!\!\rightarrow \mu^+ + \nu_\mu
\end{array}
\tag{2}
$$

where we looked for a wide angle muon in a region where the π decay background would be small.

This experiment was hastily designed during our second generation neutrino experiment, one that followed the first observations of high energy neutrino scattering and the discovery of the muon neutrino [16].

We had acquired over 1000 ν-induced events in our second generation neutrino research. A study of these events convinced us that systematic effects, essentially the large number of "Z00" events, i.e. undecipherable events, would dominate over systematics and that we already had plenty of data. We switched over to the search for W's via wide angle muons and published a limit [15]:

$$M_W > 2 \, \text{GeV}.$$

Much to our chagrin, we soon read a critical paper by Yamaguchi [17], noting that even *if* we had seen muons, these would more probably have come from virtual photon decays. He noted that:

$$\frac{p + N \rightarrow \text{``}\gamma\text{''} \rightarrow \mu^+ + \mu^-}{p + N \rightarrow W \rightarrow \mu^+ + \nu} \geqslant 1,$$

if only a single muon is detected. The production processes of γ and W were related by a theorem (Conserved Vector Current) which is now incorporated in electroweak theory. (L. Okun made the same comment at the same time but published in Russian.)

5.3. Idea (Inspiration)

With this idea in mind, it became clear that if we looked at events of the type

$$p + N \rightarrow \text{``}\gamma\text{''} + \text{anything},$$
$$\hookrightarrow \mu^+ + \mu^+ \tag{3}$$

it would be possible to probe the high-Q^2 behavior of hadronic matter to study the resonances such as the ρ, ϕ and ω and in general to probe excited nuclear matter at small distances.

In January 1967 the Columbia–BNL group (now numbering six physicists) proposed our first dilepton experiment with hadronic collisions. Theorists were skeptical, because they felt that the $e^-e^+ \rightarrow$ hadrons data from SPEAR would be much simpler to analyze, and thus much better suited to search for these broad resonances. We, however, felt that we could use the full power of the proton beam at the AGS to conduct a clean search with a signal of about ten events per day assuming an experimental efficiency of 10%.

5.4. Design, Phase I

The proposed experiment involved the collision $p + {}^{238}U \rightarrow \mu^+ + \mu^- +$ anything. Later, this would be classified as an inclusive experiment. We intended to maintain the same general design as the W^+ experiment, but certain aspects required careful consideration. The experiment required a 10% overall acceptance to provide an adequate data rate. Furthermore, we had to build the target to withstand 10^{12} interactions per pulse, or 10^{16} nuclear collisions in total. This corresponds to 10^{17} pions per day produced in the target, which, by decay, results in a substantial muon background. The apparatus had to be designed so as to measure p_{μ^+} and p_{μ^-} with reasonable resolution and with minimal background.

The basic plan of an experiment examining hadronic production of muon pairs is as follows (fig. 4):

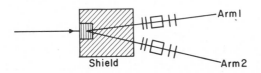

Fig. 4. Typical dilepton (dimuon) experiment.

An incoming hadron beam strikes a large target and interacts. The target also acts as a beam dump to stop the outgoing hadrons before they decay. Downstream is a double arm spectrometer consisting of magnets, counters, and track chambers that can distinguish two muons and measure their momentum. Finally, the data is analyzed and a mass for the muon pair determined.

$$M_x = 2p^+ p^- (1 - \cos\theta), \tag{4}$$

where θ is the angle between the two muons. The cross section as a function of dimuon mass (M_x) is plotted by dividing the events into mass bins and plotting the number of events in each bin versus the M_x. The major background in all of these experiments is the accidental coincidence of two muons, each arising from the decay of a pion or a kaon (see fig. 5).

Some specific comments can be made about the Columbia–BNL experiment of 1967–8. The muon energy was measured in four bins by observing the range of the particles in four blocks of steel and the angle was determined by hodoscopes of ten scintillation counters. Since the multiple Coulomb scattering in the steel would be high, good angular resolution was not terribly important.

The success of the experiment lay in a few tricks we used to reduce the backgrounds.

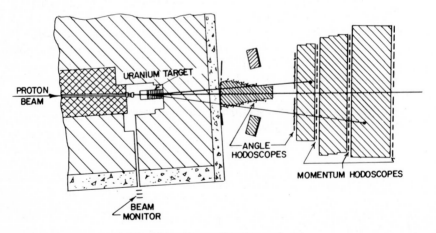

Fig. 5. Brookhaven dimuon experiment.

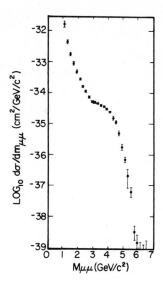

Fig. 6. BNL results.

Trick 1: The logic simultaneously counted in time and out of time muon rates and subtracted these for each mass bin. This gave an automatic background subtraction as a function of energy.

Trick 2. The ^{238}U target was constructed as a series of 1″ thick plates with variable spacing. This permitted us to vary the ratio of absorbed pions to pions that decay to muons. An extrapolation to infinite density was made to find the limit of zero π decays.

In designing the apparatus the group had to choose between high luminosity which would enable us to see smaller cross sections, and high resolution which would give a small $\Delta m/m$. The collaboration was split, but finally decided to aim for high luminosity. Ultimately we achieved an astonishing regime for hadronic cross sections of $\sigma \sim 4 \times 10^{-39}$ cm^2 with very low background, unfortunately at the expense of the mass resolution [18]. Since our resolution $\Delta m/m$ was only 14% the world was forced to wait six more years for the J/ψ ([19], see fig. 6).

5.5. Results

In interpreting the results we could not determine if the shoulder apparent at approximately 3.5 GeV was the sum of two continuum distributions or a resolution-broadened resonance, superimposed on the continuum. With our 14% mass resolution we could not *prove* beyond any doubt that a resonance hypothesis was correct. It would have been impossible to improve the resolution with the same apparatus since the design was carefully balanced; many features contributed equally to the resolution. Instead we finally decided to build new experiments for the higher energy accelerators that were soon to operate at the ISR and at FNAL. This would, we reasoned, allow more valid tests of the continuum soon to be named "Drell–Yan Process", and search for resonances over a wider kinematic domain.

Some direct results of the BNL dilepton experiment were "137" theoretical papers on various topics such as light cone algebra, multiperipheral models and quark bremsstrahlung. In 1970 the continuum behavior was accurately described by Drell and Yan [20]. The Columbia group moved on to two other collaborations, one at CERN (CCR) and one at FNAL (CFS), which would continue to study these hadronic processes through the 1970s. Other dilepton experiments were performed by other groups, most notably by Ting's group from MIT/BNL.

5.6. J/ψ digression

The MIT/BNL collaboration looked at e^+e^- coming from the reaction [19]

$$p + Be \rightarrow e^+e^- + \text{anything.}$$

In 1974, it was possible for them to achieve a mass resolution of $\leq 3\%$ with a magnetic spectrometer and new Charpak wire chambers. With only 242 events in the peak they were able to clearly identify the $c\bar{c}$ resonance at 3.1 GeV (see fig. 7).

5.7. Digression on the Drell–Yan model

Although by 1970 the dynamical quark–parton model had been employed to analyze e–p Deep Inelastic Scattering (DIS) it wasn't until "Paper 105" that S. Drell and T.M. Yan [20] outlined the generation of $\mu^+\mu^-$ pairs via parton–antiparton annihilation (see fig. 8).

In general two particles are observed in the final state. Therefore, five variables are needed to describe this state in the center of mass system:

$$M, x_F, p_T, \theta^*, \phi^*.$$

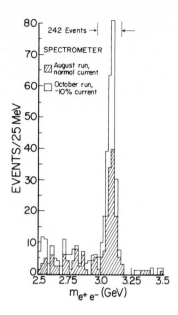

Fig. 7. MIT–BNL results in J/ψ region

Thus, the differential cross section can be written as:

$$\frac{E \, d^5\sigma}{dx_F \, dM \, dp_T \, d\Omega^*} = W_T(1 + \cos^2\theta^*) + W_L \sin^2\theta^*$$

$$+ W_\Delta \sin 2\theta^* \cos\phi^* + W_{\Delta\Delta} \sin^2\theta^* \cos 2\phi^*. \quad (5)$$

The four structure functions W depend only on M, x_F, p_T and p, the momentum of the individual particles in the initial state.

In the Drell–Yan model [20] the annihilating partons (quarks) are treated as free spin-1/2 objects in the impulse approximation; that is, in the limit as $M^2 \to \infty$ and $s \to \infty$, while $\tau \equiv M^2/s$ does not become "too

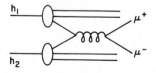

Fig. 8. Diagram of Drell–Yan process.

small". In the spirit of the model, $W_L = W_\Delta = W_{\Delta\Delta} = 0$ and:

$$p_T = 0, \qquad d\sigma/d\Omega^* = 1 + \cos^2\theta^*. \tag{6}$$

Integration gives the result

$$\frac{d\sigma}{dM^2 dx_F} = \frac{4\pi\alpha^2}{9M^4} F(x_1, x_2), \tag{7}$$

where $F(x_1, x_2)$ is the sum of the products of the quark and antiquark distribution functions.

An extra factor of three in the denominator of the rhs has been attributed to color degree of freedom. Looking at the kinematics yields

$$M^2 = (q_1 + q_2)^2 = (x_1 p + x_2 p)^2 = -2mq + 2x_1 x_2 p^2. \tag{8}$$

In the impulse approximation it is possible to neglect the quark mass and, since $s = 2p^2$:

$$x_1 - x_2 = x_F, \qquad x_1 x_2 = M^2/s \equiv \tau. \tag{9}$$

Rewriting in terms of the rapidity $y = \frac{1}{2}\ln[(E + p)/(E - p)]$ gives:

$$x_1 = \sqrt{\tau}\, e^y, \qquad x_2 = \sqrt{\tau}\, e^{-y}. \tag{10}$$

Note that three sets of variables can be used to describe these processes: x_1 and x_2, x_F and M and τ and y.

5.8. Inclusive processes

Inclusive processes are conventionally characterized by the variables S and Q^2. In these *models* a summation is carried out over the final states of a given total hadron energy and momentum. The derived parton distributions and details of the electromagnetic and weak forces are independent of the details of these non-observed final states. The processes described in this manner include charged lepton–nucleon inelastic scattering, neutrino–nucleon inelastic scattering, and of course dilepton production in nucleon–nucleon collisions.

In the mode of inclusive interactions, if the interaction time is short enough, the collision between the probe and a quark inside the hadron is free. The dynamical quark-parton model formulated in 1967 by Bjorken, Feynman and others predicts that the constituents of nucleons scatter as if they are free. The constituents carry a momentum xp, where x is the Bjorken variable, the fraction of the parent momentum p carried by the parton.

5.9. Scaling and structure

Quark distributions in a nucleon are made of two parts. The *valence* [$V(x)$] structure is given by a static quark model. The *sea* [$S(x)$] is made up of virtual quark and antiquark pairs.

Originally the Drell–Yan picture assumed $S(x) = V(x)$. Since that time more information on the sea structure has been obtained from neutrino scattering.

The model also gives a differential cross section, e.g. (7). In the simple kinematic region where $x_F = 0$, the Drell–Yan model predicts that

$$M^3 \left. \frac{d^2\sigma}{dMdy} \right|_0 = \frac{8\pi\alpha^2}{9} \tau S(\sqrt{\tau}) V(\sqrt{\tau}).$$

By measuring the cross section, $S(\tau)$ and $V(\tau)$ can be determined and scaling can be tested [21].

Phase II: CCR

In 1971, the Columbia group, in collaboration with CERN and Rockefeller University physicists, moved to the CERN ISR. This time the group focused on electron pair production. They set up a lead glass electromagnetic calorimeter on either side of a beam crossing region. Each side of the two-arm calorimeter could detect e, π^0 and γ over a solid angle of 1 sr. (see figure 9).

Lead glass blocks ($PbO + SiO_2$) were originally used as visual ports in radiation shields. An e, π^0, or γ will create an electromagnetic cascade, part of the energy of the shower goes into Čerenkov light, which is collected by photomultiplier tubes. The blocks must be carefully calibrated by an electron beam, and the calibration monitored. In the ISR, monitoring of the lead glass calibration is very difficult. Our group adopted the idea of embedding an α-emitter (^{124}Am) in a small crystal of sodium iodide to enhance the light output. We bonded this to the glass by optical cement. The ~ 100 counts/s were distributed with a 6% full width Gaussian, and provided us with a reliable pulse height equivalent to a 2 GeV electron shower.

The results of the experiment were unexpected. We found a background of π^0's at 90°, that was seven orders of magnitude larger than expected! Earlier data from Cosmic Rays, the AGS and the PS had given rise to a formula:

$$d\sigma/dp_T \sim A e^{-6p_T} \tag{11}$$

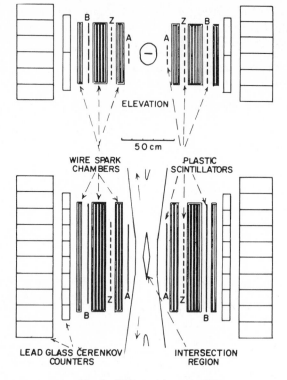

Fig. 9. CCR experiment at ISR.

which at $p_T = 5$ GeV would yield 10^{-12} of the total cross section. However, we saw many events at $p_T \approx 5$ GeV.

We published our discovery of the wide angle π^0,s in our background. The experimental law, e.g. (11), was invalid for $p_T > 1$ GeV/c:

$$Ae^{-6p_T} \rightarrow Bp_T^{-n}e^{-26p_T/\sqrt{s}}. \tag{12}$$

We found $n = 8$ [22]. This 1972 result was the start of a new particle physics industry: high transverse momentum physics. However, for the second time we missed the J/ψ. The experiment was plagued with problems. The calibration of the lead glass via NaI, kept telling us that something was wrong. We later learned that the new accelerator (ISR) dumped huge amounts of radiation into the lead glass blocks, slowly turning them yellow. To compensate for the decreased signal and the enormous yield of π^0's we kept raising the trigger threshold on the glass. The threshold was finally set at 1.5 GeV on each side, thus desensitizing us to the J/ψ! Therefore we observed no e^+e^- pairs.

Lesson 2: "If you are going to miss a discovery, always find something else to publish".

Lesson 3: "New accelerators are hostile environments".

5.10. Digression on high p_T physics

The outgrowth of the CCR experiment was 137 additional theoretical papers on the $n = 8$ result which differed from the predictions of Bjorken and coworkers [23] who predicted pointlike behavior

$$\frac{d\sigma}{dp_T^2} = \frac{1}{p_T^4}. \tag{13}$$

High p_T physics is clearly a consequence of hard collisions with pointlike constituents and its pursuit dominates the present experimental programs at Fermilab, SPS and ISR. CCOR, the successor to CCR, found that as the ISR luminosity increased, and good data was obtained to 14 GeV/c, $n = 8$ changed to $n = 5$, i.e. the pointlike value seems to be approached [24]. I believe that $n = 8$ can be understood in QCD perturbative calculations. Recently Levinthal (CCOR) has shown [25] that $\pi^0 - \pi^0$ pairs with each π^0 having high p_T are a very good signature for elastic q–q scattering. A single high p_T π^0 dominates the jet kinematics and gives a reasonable account of quark (gluon)–quark (gluon) scattering. JETS AHEAD!

5.11. The Fermilab era 1973 →

5.11.1. The beginning
The Fermilab collaboration CFS (Columbia–Fermilab–Stony Brook) built a magnetic spectrometer and lead glass calorimeter to look at final state single $e^\pm$ and e^+e^- pairs. This proposal (E70) was submitted in 1969 in order to follow up on the Brookhaven results. Arriving on the experimental floor in 1973, we made a decision to look at the production of single electrons first (in expectation of a hostile environment in a new accelerator) and postponed the study of electron pairs until later. Simultaneously Cronin and coworkers [26] looked at single final state μ's, π's,

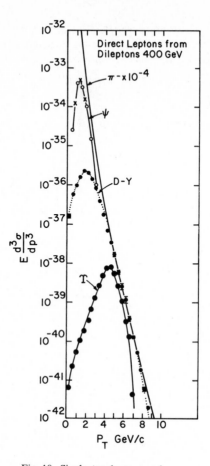

Fig. 10. Single arm leptons and sources.

or K's versus p_T. Both groups made the discovery that the lepton to pion ratio versus p_T was 1.0×10^{-4} independent of p_T or s [27]. Another 137 theoretical papers appeared explaining direct lepton production in hadronic collisions. It became important to test many theoretical hypotheses and our single arm experiment stretched on [28] (see fig. 10). The latest word shows that the fixed ratio $e/\pi = 10^{-4}$ is purely an accident, there being a number of independent sources of direct leptons each contributing over a different range of p_T. At the time, however, the known sources of leptons, the ρ^0, ϕ^0, ω^0 and Drell–Yan continuum did not account for the data.

During the summer of 1974, at the London HEP conference I concluded with great wisdom that there was something curious going on above 3.0 GeV. In the fall the CFS collaboration started to set up to study pair production. That was the fall of the "November Revolution". Ting and Richter announced the discovery of the J/ψ at BNL [19] and at SLAC [29]. We saw our first e^+e^- pairs in 1975, but it was too late. It was our third miss! Recall that the first miss, BNL in 1968, discovered dilepton continuum (what is now known as the Drell–Yan process!) and the second miss resulted in the discovery of high p_T hadrons. However, this time the "direct lepton" discovery turned out to be of minor consequence: a "cocktail" of vector mesons and continuum pairs.

5.11.2. The upsilon (E-288)

The dilepton experiments continued and the size of the collaborations continually became larger and larger. By this time the CFS collaboration had grown to sixteen people. The experiment was designed to observe electron pairs produced in hadronic collisions. Electrons rather than muons were the leptons of choice because muons would require an absorber to absorb the π's and K's in flight. This would generate multiple scattering and destroy our knowledge of the pair opening angle. Recall that we were trying to observe very small cross sections and keep very good resolution. The only background to electrons are converted γ's from π's and Dalitz pairs. These were considered manageable.

The design consisted of a 0.2 mm wide Be target, about 10 cm long to reduce conversion of π^0 γ's. Downstream there was a conventional double-arm spectrometer. No detectors were placed upstream of the magnet in order to permit a maximum amount of beam on target. The beam we could take was 10^9 protons per pulse at 400 GeV. The resolution was quite good, $\Delta m/m \approx 1\%$. The combination of that proton flux and acceptance yielded about 50 events above 5 GeV of mass, at that time a record number. The results appear in fig. 11. We announced that a suggestion of a bump was visible at 6 GeV and provisionally named it the Υ [30].

In 1976 we tried a "quickie" muon run by adding a hadron absorber to each arm. We wanted a rapid check of the 6 GeV signal and some experience with muon problems. The resolution decreased to 4%, but with the increased rate we were able to lay to rest the 6 GeV bump as a fluctuation. We *did* see a small bump near 9 GeV [31] (see inset on above figure). John Yoh of our group labelled a bottle of champagne "9.5" and placed it in the group refrigerator. With great anticipation we repeated

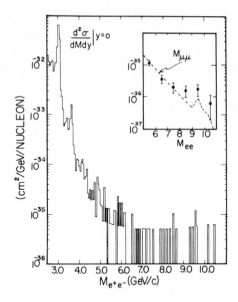

Fig. 11. Early dielectron and dimuon (inset) data.

the e^+e^- run, but found that the rates were unsatisfactory for exploring the higher mass region.

5.11.3. New idea

Muon pairs could in principle provide the desired rate, but our preliminary run encountered two problems. The hadron absorber destroyed the resolution. Also we had found a large counting rate of soft muons, seriously limiting the proton flux which the apparatus could tolerate. However, we discovered that most of our muon background did not originate from a single point source but rather from an effective large area. Thus the rate of background counts per cm^2 did not increase as one moved a detector closer to the target. This allowed us to move all of the detector planes as close to the target as was geometrically feasible; the resulting single counting rate in each plane (which is what limits the primary proton intensity) being proportional only to area, was no worse. The acceptance of the detector on the other hand, was *much* improved. Another important factor was the tuning of the read-out electronics and operation of the proportional wire chambers, so that these could be operated at counting rates in excess of 10 MHz. The final improvement was to use, as the hadron absorber, well machined beryllium metal, obtained free of charge from U.S. government surplus. This introduced a minimum of multiple Coulomb scattering per absorption length. Meticulous attention to the details of the region near the target (where pions are absorbed before they can decay into muons) paid off in the ratio of background, random counts to primary protons on target. All in all, the improvement over e^+e^- pairs in sensitivity was about a factor of 100! The resulting resolution was $\pm 2\%$, still very respectable. Several additional checks and background suppression ideas were installed. The apparatus was symmetrized relative to the bending plane so that $\mu^+\mu^+$, $\mu^-\mu^-$ and $\mu^+\mu^-$ pairs could simultaneously be measured with equal acceptance, which permitted us to make a direct subtraction of background (very largely accidental coincidences between left and right arms). This is because a random μ^+ is just as likely to pair with a μ^+ as with a μ^-. Since muons can pass through meters of steel unscathed it is important to verify the determination of momentum, which is based on limited sampling of the muon trajectory. A redundant solid iron magnet was installed to provide an independent determination to within an error of 15%. With this apparatus it was possible to measure $\sigma \sim 10^{-40}$ cm^2 with no background (see figs. 12 and 13).

The acceptance of the apparatus fell off at 3 GeV and was limited by kinematics to below 27 GeV. Over the region in between however, it was remarkably flat (see fig. 13). In Drell–Yan experiments the acceptance in p_T is also of extreme importance. We present the CFS acceptance curves and note that improvements are clearly possible. The primary features of CFS include the high rate capability, good background rejection and the good resolution ($\Delta m/m, \Delta p/p$).

Stepping back to look at this series of experiments in quest of high mass dileptons one discovers a long progression leading into the future. First in 1967–8 there was the BNL effort at 30 GeV. Next came the CCR work at "1500" (ISR) GeV in 1971–2, followed by the first FNAL experiment at 300 GeV in 1975–6. This yielded 50 events above 5 GeV, hints of something interesting, and a lot of experience. The fourth attempt (1977–8) was again at FNAL at an energy of 400 GeV. Here we were able to reach a compromise between resolution and luminosity. This time we recorded 100 000 events above 5 GeV and were able to distinguish the Υ, Υ' and Υ'' resonances. Furthermore the data provided a quantitative test of Drell–Yan and QCD. We obtained structure functions of gluons and of sea quarks which could be compared with neutrino data. But this was only practice. The fifth experiment will begin at

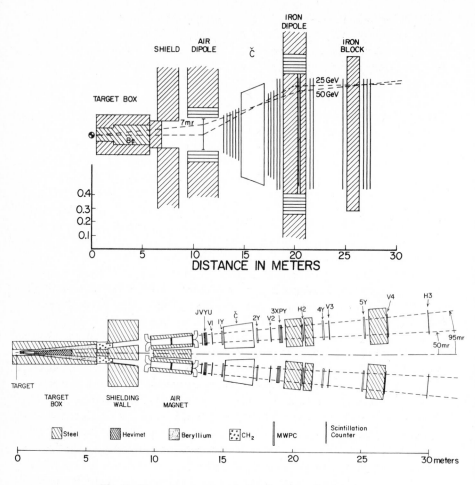

Fig. 12. Dimuon apparatus, (a) elevation view, (b) plan view.

Fermilab in the Tevatron era (1983–) at 1000 GeV with a mass resolution $\Delta m / m \approx 0.8\%$, greater acceptance and much higher rate capability.

5.11.4. Data analysis
In examining the dimuon data from E-288 we made a series of cuts on the tracks. Improving the quality of the data via "cuts" is a standard

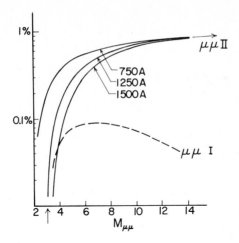

Fig. 13. Acceptance versus mass as a function of magnet current.

procedure. One must be aware of two dangers: (i) The cuts decrease the efficiency for signal. If so, the effect must be understood. (ii) The cuts may distort the distributions and possibly introduce bumps.

The first cut we applied, required that each track intercept all the MWPC and counters, project through the analyzing magnet and originate near the target ($\Delta x = 4$ cm, $\Delta z \leqslant 60$ cm). These cuts are called fiducial cuts. The second set of cuts, called 'muon' cuts, required nine out of ten hits in the MWPC, a target cut of $\Delta x < (1.5 + 50/p)$ cm, and four out of six hits in devices in back of the iron magnet. These cuts reduced the background from π and K decay to almost nothing. We tested this by examining like-sign muon pairs. These data also provided excellent checks of the cut efficiency and bias effects.

The first results were presented in June 1977. They are presented here in fig. 14 [32]. The bump at 9.5 GeV definitely peaked above the continuum background. Now the task remained to prove that it was real.

5.11.5. The agony and the ecstasy
Are the data real? Here is where the test of professionalism is most crucial. All the training, all the experience must be applied to this question.

The like-sign muon background provided us with enough information to state that the observed $\mu^+\mu^-$ pairs were not accidentals. Above 8 GeV the expected background signal was $\leqslant 0.1\%$ of our detected signal.

Fig. 14. Dilepton spectrum at an early stage of the new set up.

Reconstruction of the pairs indicated that they originated at the small target and, with the target removed, the signal vanished. Furthermore, the mass of the ψ and ψ' agreed with expected values to within 10 MeV and the resonances were narrow. Over the general spectrum we were convinced that there were no systematic errors.

Is the bump real? Can apparatus produce a spurious bump? We mixed left arm data and right arm data from different days to determine that our acceptance was smooth over the region around 9.5 GeV. We checked this by taking data at two magnet currents, with two different targets and with two different absorbers. All subsets of the data showed the bump in the same place. We also concluded that the cuts could not induce a bump. We looked at the performance of the apparatus. The integrated hit patterns in the MWPC looked reasonably uniform and furthermore were the same over several mass bins between 8 and 11 GeV, indicating that no errors were introduced there. By more data mixing, we also concluded that the software could not have introduced the resonance.

The data were real and the bump was a real effect. The next step was to get more data and measure the width of the resonance. We also needed to find a name. Why not the Υ (upsilon) since it was no longer required at 6 GeV? And so it was.

5.11.6. Measurement of the Υ width

After observing the Υ, more precise measurements were required to determine the resonance structure. To do this it was important to understand the apparatus and calibrate on the narrow J/ψ. We added upstream PWC's and mini drift chambers to improve the apparatus. The results showed three distinct peaks. Also, the width of the main peak at 9.4 GeV was completely consistent with the smearing of a very narrow peak by measurement errors and Coulomb scattering. This was confirmed by the J/ψ data and by target reconstruction data. The conclusion was of major importance: the peak was narrower than our resolution, i.e. narrower than ~ 100 MeV (and consistent with zero). Now a narrow peak, by the famous relation:

$$\Delta E \Delta t > h,$$

implies a long lifetime. But a heavy object (nine times the proton mass) has a very large number of decay channels and therefore should have a very short lifetime. Conclusion: a new law of physics is operating to forbid the decay of Υ into these channels!

5.11.7. What did we discover?

Now that we had observed and measured this new resonance we had to at least discuss what it would be. We probed the theoretical literature for candidates. It wasn't the Z^0 since a simple C-violation test of $N(E^+ > E^-) - N(E^- > E^+)$ gave null results contrary to what would be expected for the Z^0. A Higgs scalar would have produced many more μ pairs than e pairs which we ruled out from the old e^+e^- data. One speculation was that it was the J/ψ decaying into free quarks. Since these would have charge $1/3$, we would unknowingly make an error of a factor of three in momentum, also in mass! Fortunately we had enough data on ionization to rule this out.

The final conclusion was that it must be a new Quarkonium ($Q\overline{Q}$) state where Q was a new quark. The new law of physics implied earlier is simply the prohibition against decay of new into old quarks. The *three* closely spaced "narrow" particles lead to a uniquely plausible conclusion.

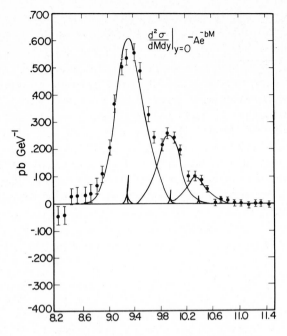

Fig. 15. Continuum-subtracted upsilon peaks.

We had found a set of states composed of the bound state of a new quark called the "b" and its antiquark. The three peak fit gave Υ'' a 4σ confidence level (see fig. 15 [33]). Table 2 shows our mass values and the currently accepted values from CESR. In 1978 results from DESY confirmed our results and $\Gamma(ee)$ implied a value of $1/3e$ for the charge of the b [34]. The mass of the b was determined to be ~ 5 GeV.

Table 2
Mass values of the upsilon resonances

Resonance	State	mass value (GeV)	
		1977 [32, 33]	current
Υ	1^3S_1	9.40 ± 0.02	9.433
Υ'	2^3S_1	10.0 ± 0.05	9.996
Υ''	3^3S_1	10.4 ± 0.12	10.323

The b quark had an amusing history dating back to a paper (apparently unknown in the West) by two Kyoto physicists, Kobayashi and Maskawa, in 1972 [35]. With the aim of incorporating *CP* violation into SU(2)×U(1), they presented several models including one that used six left handed quarks and a 3×3 matrix with generalized Cabibbo angles. Other models were generated through the 1970s, motivated by the high *y* anomaly, and using six quarks and right handed currents, but were soon discarded. In 1975 Harari [36] presented two models predicting a top (t) and bottom (b). The first did not accurately describe the ψ system but this was corrected in the second. Both replicated the results of Kobayashi–Maskawa which were still unknown. Later that year the original 1972 paper was finally discovered. Then in 1977 the $\Upsilon(b\bar{b})$ was observed.

5.11.8. What about top?

The sixth quark predicted by the Kobayashi–Maskawa model has not yet been observed. The CFS collaboration continued its search until April 1978 and covered the mass spectrum up to 19 GeV. The conclusion was that if the $t\bar{t} \to \mu^+\mu^-$ signal is as prominent as the Υ above the continuum then there is no $t\bar{t}$ in the range 5–14 GeV. However, if $\sigma(t\bar{t}) = 4\sigma(\Upsilon)$ then the mass $m(t\bar{t}) > 15$–16 GeV [37]. This limit has been extended to 36 GeV in PETRA searches.

5.12. Digression on the quark – quark force

In QCD the q − q force is derived from the exchange of a single massless gluon. QCD has been successful in explaining the bound states of heavy (non-relativistic) quarks in analogy with positronium. The $Q\bar{Q}$ system is nature's gift to theoretical physics. For $Q \gg$ few GeV the system is a non-relativistic bound state of point like objects whose binding is produced by the basic strong force. One potential for the non-relativistic Schroedinger equation is given by Eichten and Gottfried [38] as:

$$V(r) = (4/3)\alpha_s/r + r/a. \tag{14}$$

This potential is motivated by limiting arguments from QCD but is not yet derived from the quantum field theory. As data accumulates on the location, and widths and branching ratios, the interquark force will become very well defined. For these data one must go to an e^+e^- machine. "All hail CESR!"

The discovery of a new quark opened vast areas of new physics. Studies made of the production mechanisms in hadronic collisions will test QCD. Further studies include those of the decay modes and mixing angles, the lifetime, and the properties of beauty mesons and baryons.

5.13. CUSB at CESR

5.13.1. Experimental attack
Continuing the case study brings us to the Cornell Electron Storage Ring (CESR) and we find that the collaboration has grown to more than 22 people and four institutions (Columbia, Stony Brook, Max Planck Institute and Louisiana State University), originally designated CUSB.

We designed an experiment for the small, poor north area to be complementary to the large, rich south area which has over 70 collaborators and was known as CLEO. Approach was to concentrate on electrons and photons with good resolution from ~ 100 MeV up to 5 GeV. The key is a NaI detector. Sodium iodide is a scintillator of very high quantum efficiency. It also has a high Z and therefore short radiation length. It makes an ideal photon detector. Although it is much more expensive than lead glass, progress in mass production techniques made large masses of NaI thinkable. The first such use was the crystal ball at SPEAR. In contrast to the crystal ball at SPEAR, we decided to segment the NaI longitudinally as well as transversely. This would help identify high energy electrons and give us directionality with photons.

Because NaI is very expensive, we used only ten radiation lengths, and then followed these by lead glass blocks. No magnet was included. We depended on large graduate students to carry the lead glass bricks that composed the detector, down the long flight of stairs that provided the only access to the small north experimental hall at CESR. The detector, shown in fig. 16, is made of drift chamber planes inside NaI hodoscopes and lead glass calorimeters with the beam pipe passing through the center. The resolution of lead glass goes as

$$\sigma/E \simeq \left(1 + 4/\sqrt{E}\right)\%, \tag{15}$$

giving $\sigma \approx 3\%$ at 5 GeV. For NaI the resolution is much better, $\sim E^{1/4}$, giving a FWHM of 1% at 10 GeV.

The results of the first experiments give clear peaks for the Υ, Υ' and the Υ'' (see fig. 17) [39]. The resulting spectrum is plotted in fig. 18, next to the spectrum of the ψ, ψ' system. A charmonium potential and an upsilon potential have been formulated in the region affecting the data. A

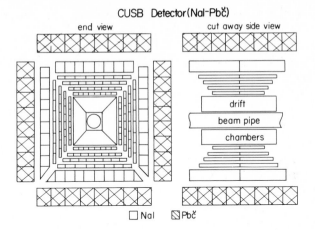

Fig. 16. CUSB detector at Cornell Electron Storage Rings.

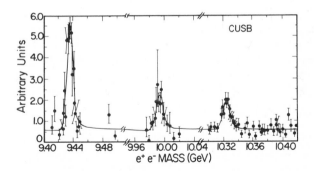

Fig. 17. Upsilon peaks as observed at CESR.

comparison of these two confirms the notion that the interquark force is flavor independent. Thus the conclusion is that the potentials are the same for $0.1 < R < 1$ fm.

The three narrow states observed in pp collisions at Fermilab are confirmed at CESR as b$\bar{\text{b}}$ bound states. These are stable, by conservation of energy, against dissociation into mesons containing b quarks. Therefore, the peaks are narrow and the allowed decay is slow:

$$T \rightarrow ggg \rightarrow hadrons. \tag{16}$$

The groups working at CESR found a fourth bump.

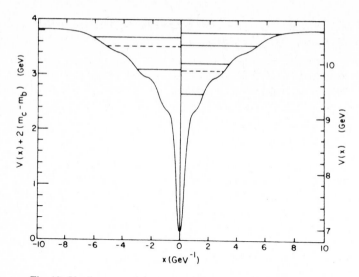

Fig. 18. Upsilon potential, compared with charmonium potential.

5.13.2. The 4S state

The three Υ states represent "hidden" bottom because the $b\bar{b}$ state does not contain a net b-quantum number. The 4S state is heavy enough to permit transitions where the d and $\bar{d}$ come out of the "sea", e.g.:

$$\Upsilon(4S) \rightarrow b\bar{d} + \bar{b}d. \tag{17}$$

This allowed transition accounts for the short lifetime of the 4S state.

Now $b\bar{d} \equiv B^0$ is a meson which exhibits b-properties (naked bottom!). The CESR data indicate that this is happening, by observing evidence for the subsequent decay chain:

$$\begin{aligned}
b\bar{d} &\rightarrow c\nu X. \\
&\quad \llcorner s\nu X' \\
&\qquad \llcorner K\nu X''
\end{aligned} \tag{18}$$

The 4^3S_1 state is a b-quark factory and will provide excellent opportunities to study "naked bottom" in future experiments:

$$e^+ + e^- \rightarrow B^0\bar{B}^0$$

One could imagine using the B particles to study b production properties and weak interactions, *CP* violation via $B\bar{B}^0$ mixing and nature of the Qq force. Perhaps even an estimate of the top mass could be determined in the B system.

CP violation in the $B^0\bar{B}^0$ system will have a strength of $\varepsilon_B \sim \tan 2\delta \gg \varepsilon_k$ ($\varepsilon_k \sim 10^{-3} = s_2 s_3 \sin\delta$). An experiment could be designed to examine the system $B\bar{B}^0$ and look for mixing by means of a charge asymmetry in the final state muon pairs,

$$\ell^- \ell^- \neq \ell^+ \ell^+,$$

since $B^\circ \to \ell^-$ and $\bar{B}^\circ \to \ell^+$. Other means of detecting *CP* violation in B's have been given by Carter and Sanda and Bander et al. [40].

All of the CUSB data so far confirms Υ as evidence for a fifth quark "b". The properties of the $b\bar{b}$ states are still being studied as well as the $b\bar{q}$ states mentioned above. The decay $\Upsilon' \to \Upsilon\pi\pi$ has been observed and the width has been measured to be $\sim 1/10[\psi' \to \psi\pi\pi]$ providing evidence for spin-1 gluons [41]. The suppression of a factor of

$$\langle r_\psi^2 \rangle^2 / \langle r_{\Upsilon'}^2 \rangle^2 \simeq 10$$

can be obtained from a gluonic radiation multipole expansion technique in QCD [42]. Our results are consistent with this prediction and indicate that the suppression is ≥ 2.

In summary, the work at CESR still goes on with the CUSB and CLEO detectors. Thus far the results have been remarkable. The narrow bound states of $b\bar{b}$ ($\Upsilon, \Upsilon', \Upsilon''$) have been observed and measured. The wide 4S resonance Υ''' has been observed. $\Gamma_{ee}(\Upsilon)$ and the branching ratio of $b \to e\nu X$ have been measured. The $\Upsilon' \to \Upsilon\pi\pi$ decay has been observed. Over the domain $0.1 \leq r < 2$ fm, QCD is consistent with the phenomenological potential determined for the Υ system measured above. The potential, when compared to the ψ, is flavor independent. Last of all the mass of the b has been determined:

$$M_b = 4.6 \pm 0.1 \text{ GeV}.$$

This concludes the "case study" of a research program which had a rich set of successes and some failures. Following the Columbia group one traces the origin of the "Drell–Yan experiment" ($h + h \to \mu^+ + \mu^-$) as a complement to lepton scattering. From this come structure functions for p, n, π and K systems, k_t of quarks, and many unique tests of QCD. Also the discovery of high p_T physics at the ISR. The group triumphed in the discovery of the b quark, but missed the J/ψ, not once, not twice but three times!

Acknowledgement

The author would like to acknowledge the essential assistance given by Patricia Mc Bride to the written version of these lectures.

References

[1]   A. Leviat et al., Phys. Lett. 22 (1966) 714.
[2]   L. Chikovani, Phys. Lett. 22 (1966) 233.
[3]   L. Braunschweig et al., Phys. Lett. 67B (1977) 243, 249.
[4]   A. Benvenuti et al., Phys. Rev. Lett. 37 (1976) 189.
[5]   D.C. Hom et al., Phys. Rev. Lett. 36 (1976) 1236.
[6]   H. Anderson and C. Lattes, Nuovo Cim. 4 (1957) 1356.
[7]   A.R. Clark et al., Phys. Rev. Lett. 26 (1971) 1667.
[8]   H. Anderson, E. Fermi, A. Martin and D. Nagle, Phys. Rev. 90 (1953) 996.
[9]   G. Homma, G. Goldhaber and L.M. Lederman, Phys. Rev. 93 (1953) 554.
[10] J. Ashkin et al., Phys. Rev. 96 (1954) 1104.
[10a] T.D. Lee and C.N. Yang, Phys. Rev. 104 (1956) 254.
[11] C.S. Wu et al., Phys. Rev. 105 (1957) 1413.
[12] R.L. Garwin, L.M. Lederman and M. Weinrich, Phys. Rev. 105 (1957) 1415.
[13] R. Cool et al., Phys. Rev. Lett. 14 (1965) 535.
[14] Taylor—D.I.S.
[15] R. Burns et al., Phys. Rev. Lett. 15 (1965) 830.
[16] R. Danby et al., Phys. Rev. Lett. 9 (1962) 529.
[17] Y. Yamaguchi, Nuovo Cim. 43 (1966) 193.
[18] J. Christenson et al., Phys. Rev. Lett. 25 (1970) 1523.
[19] J.J. Aubert et al., Phys. Rev. Lett. 33 (1974) 1404.
[20] S. Drell and T.-M. Yan, Phys. Rev. Lett. 25 (1970) 316.
[21] A.S. Ito et al., Phys. Rev. D23 (1981) 604.
[22] F. Busser et al., Phys. Lett. 46B (1973) 471.
      R.L. Cool, Proc. Int. Conf. on High Energy Physics, Batavia (Fermi National Accelerator Laboratory, 1972).
[23] S. Berman, J. Bjorken and J. Kogut, Phys. Rev. D4 (1971) 3388.
[24] A.L.S. Angelis et al., Phys. Lett. 79B (1978) 505.
[25] D. Leventhal, Columbia Univ. Thesis (1980).
[26] J.P. Boymond et al., Phys. Rev. Lett. 33 (1974) 112.
[27] J. Appel et al., Phys. Rev. Lett. 33 (1974) 722.
[28] L.M. Lederman, Phys. Rep. 26C (1976) 151.
[29] L.M. Lederman, Proc. XIX Int. Conf. on High Energy Physics, Tokyo, 1978 (Physical Society of Japan).
[30] D.C. Hom et al., Phys. Rev. Lett. 36 (1976) 1236.
[31] D.C. Hom et al., Phys. Rev. Lett. 37 (1976)1374.
[32] S. Herb et al., Phys. Rev. Lett. 39 (1977) 252.
[33] K. Ueno et al., Phys. Rev. Lett. 42 (1979) 486.
[34] C. Berger et al., Phys. Lett. 76B (1978) 243.
[35] M. Kobayashi and T. Maskawa, Prog. Theor. Phys. 49 (1973) 652.
[36] H. Harari, Stanford Conf. on Photons and Electrons (1975).
[37] C. Quigg and J. Rosner, Phys. Lett. 76B (1977) 153.

[38] S. Eichten and K. Gottfried, Phys. Lett. 66B (1977) 286.
[39] T. Bohringer et al., Phys. Rev. Lett. 44 (1980) 1111.
[40] A. Carter and A. Sanda, Phys. Rev. Lett. 45 (1980) 954.
      M. Bander et al., Phys. Rev. Lett. 43 (1979) 242.
[41] L.J. Spenser, Phys. Rev. Lett. 47 (1981) 771.
[42] K. Gottfried, Phys. Rev. Lett. 40 (1978) 598.
      T.-M. Yan, Phys. Rev. D22 (1980) 1652.

COURSE 8

RECENT RESULTS IN ELECTRON–POSITRON AND LEPTON–HADRON INTERACTIONS

K.H. MESS

Deutsches Elektronen-Synchrotron DESY, 2000 Hamburg 52, Germany

and

B.H. WIIK

II. Institut für Experimentalphysik, Universität Hamburg, 2000 Hamburg 50, Germany

M.K. Gaillard and R. Stora, eds.
Les Houches, Session XXXVII, 1981
Théories de jauge en physique des hautes énergies / Gauge theories in high energy physics
© *North-Holland Publishing Company, 1983*

Contents

1. Introduction

During the past two decades, theory and experiments have conspired to produce a cohesive and simple picture of nature. At least down to distances of 10^{-16} cm, matter is made of pointlike constituents: leptons and quarks. The leptons have electromagnetic and weak interactions and are directly observed as free particles. The quarks have colour and participate in the strong interaction. So far, only colour singlets—i.e. hadrons composed of two or three quarks—have been observed in abundance. This observation has led to the hypothesis that the quarks are confined, and do not appear as free particles.

The forces, strong, electromagnetic and weak, are predicted to arise from the exchange of gauge bosons. Based on low energy experiments, there is now a general belief that the electromagnetic and the weak interactions are unified to an electroweak interaction on a mass scale of 100 GeV. Furthermore, there are ample speculations that the strong and electroweak forces are also unified on a mass scale of 10^{15} GeV/c^2. Heroic attempts are being made to incorporate gravitation into this picture, to create a truly unified theory.

Deep inelastic lepton–hadron or electron–positron annihilation experiments have contributed much to bring this picture into focus. The existence of quasi-free pointlike constituents in the nucleon was first discovered in lepton–nucleon scattering and well defined jets of hadrons resulting from the fragmentation of quarks were observed in e^+e^- annihilation at high energies. Hadrons reflecting the existence of new quarks have been detected in e^+e^- annihilation, and both types of experiments have contributed much to the determination of quark properties. A new sequential lepton, the τ, has been found and its properties were determined in e^+e^- annihilation.

Neutrino experiments first revealed the existence of neutral weak currents and their continuation has determined much of the low q^2 properties of this new interaction.

The existence of gluons, the carrier of the strong interaction was first inferred from deep inelastic lepton–hadron experiments and the gluons were directly observed in three-jet events in e^+e^- annihilation.

Simplicity is one of the reasons why measurements of e^+e^- annihilation and deep inelastic lepton–hadron interactions have been successful. In e^+e^- annihilation the neutral timelike current—electromagnetic or weak—couples directly to the basic constituents. The current with $J^{PC} = 1^{--}$ can also produce particles with the same quantum numbers. In deep inelastic lepton–hadron interactions, the lepton interacts directly with the quarks in the nucleon via neutral and charged spacelike currents.

These lectures will start with a few remarks on detectors and beams used to study e^+e^- annihilation and deep inelastic lepton–nucleon interactions. The main part of the lectures will discuss recent results obtained from a study of these processes including a discussion on the result of recent particle searches. The picture which emerges from these data is consistent with what has become known as the standard model. However, it is important to bear in mind that the experiments so far have only investigated masses which are small compared to 100 GeV/c^2, the characteristic mass of the weak interaction. The new generation of e^+e^- and ep colliders will allow us to extend these measurements into a mass range above 100 GeV/c^2, and thus provide answers to many of the questions confronting the standard model.

2. The experiments

2.1. Experiments at electron – positron colliding rings

2.1.1. The colliders

The art of colliding electrons and positrons has developed very rapidly. In only 20 years the field has progressed from the first model machine [1] with a circumference of 4.5 m to the present LEP project [2] with a circumference of 27.4 km.

The storage ring PETRA [3] at DESY is shown in fig. 2.1 as an example of a high energy e^+e^- machine. PETRA consists of a single ring, made of eight 45° bends joined by eight straight sections, four 108 m long and four 64.8 m long. The total circumference is 2.3 km. Two bunches, of electrons and positrons, are injected and travel around the ring in opposite directions guided by a system of dipole, quadrupole and sextupole magnets. The bunches cross in the middle of four straight

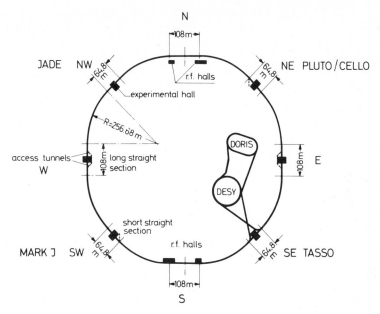

Fig. 2.1. Layout of the e^+e^- collider PETRA.

sections, and they are there focussed down to a small cross section by two pairs of quadrupoles located symmetrically with respect to the interaction point. The distance between the quadrupole pairs is a compromise between the free length needed to mount an experiment, and the peak luminosity. At PETRA, the free length has been reduced [4] from 15.0 m to 9.0 m leading to an increase in the luminosity by a factor of three, to 1.7×10^{31} cm^{-2} s^{-1}. Similar changes have also been carried out at CESR in Cornell and at PEP at Stanford with encouraging results.

An electron with energy E_0, travelling around a ring with a bending radius R radiates an energy eU_0 per turn:

$$eU_0 = 88.5 \times 10^{-6} (m/\text{GeV}^3) E_0^4 / R. \tag{2.1}$$

Thus, a 19 GeV electron in PETRA, radiates on the average 58.5 MeV per turn. This energy loss is made up by the energy gain in the RF cavities located in the long straight sections. Due to synchrotron radiation, the beam dimensions are Gaussian distributed. A bunch at the interaction point at 16 GeV is typically 0.12 mm high (σ_y), 0.5 mm wide (σ_x) and 20 mm long (σ_e) and contains on the order of 2×10^{11} electrons.

As we will discuss below, the cross section for hadron production in the continuum is very small and decreases with the square of the energy. It therefore becomes increasingly important to maximize the luminosity.

The luminosity in an e^+e^- colliding ring [5,6] is given by:

$$L = n^2 Bf / 4\pi\sigma_x^*\sigma_y^* = i^2 / 4\pi\sigma_x^*\sigma_y^* Bfe^2, \tag{2.2}$$

where n is the number of particles per bunch, B the number of bunches per beam and f the revolution frequency. σ_x^* and σ_y^* are the horizontal and vertical rms beam sizes at the interaction point, defining an effective beam cross section $F = 4\pi\sigma_x^*\sigma_y^*$. The circulating current in each beam is given by $i = enfB$.

High luminosities can only be achieved by maximizing n. However, when the bunches cross, they interact electromagnetically and the ultimate limit on the number of particles in a bunch is reached when its partner is lost due to this interaction. This beam–beam interaction is parametrized by the tune shift ΔQ with

$$\Delta Q_x = \frac{r_e}{2\pi\gamma} \frac{n\beta_x^*}{(\sigma_x^* + \sigma_y^*)\sigma_x^*}. \tag{2.3}$$

Here $r_e = 2.82 \times 10^{-13}$ cm is the classical electron radius and $\gamma = E/m_e$. It has been found at the high energy e^+e^- machines that ΔQ must be less than 0.03 to ensure stable operation. Therefore, to maximize n the value of the amplitude function β^* at the interaction point must be minimized. The ultimate limit is presumably reached when β^* is comparable to the bunch length, i.e. a few cm. However, a practical limit for existing machines with the detectors already in place is imposed by β_{max}, the largest allowed value of the amplitude function at the quadrupoles adjacent to the interaction region. Roughly speaking β_{max} should not exceed 500–1000 m. However, since β increases as $\beta(s) = \beta^* + s^2/\beta^*$ with the distance s from the interaction region, this implies that the quadrupoles must be moved towards the interaction point, i.e., the free region available to experiments must be reduced.

So far, electron–positron colliders have operated efficiently only over a rather narrow energy range centered near the peak energy. A typical luminosity curve is sketched in fig. 2.2. The rms beam size σ is given by $\sigma = (\epsilon\beta)^{1/2}$, where ϵ is the beam emittance and β the value of the amplitude function at a particular location. In electron machines ϵ is not a constant but rather determined by the synchrotron radiation and the focussing strength of the lattice and increases, for constant β-values, with

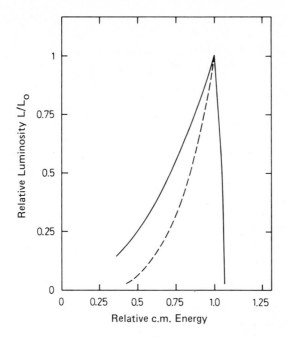

Fig. 2.2. Characteristic energy-dependence of e^+e^- luminosity. The solid line represents the luminosity achieved by optimizing the emittance at each energy. The dashed line represents the luminosity for a constant optic.

the energy squared. From eqs. (2.2) and (2.3) it then follows that the luminosity for constant β will increase as E^4 until a maximum energy E_{max}, given by the available RF power is reached. According to eq. (2.1) the energy loss per turn increases as E^4, so the RF power needed to produce the necessary energy gain per turn increases proportionally to E^8. The available RF power is increasingly used to establish the accelerating field, and the corresponding reduction in circulating current leads to a very rapid drop in luminosity above E_{max}. The luminosity at lower energies can be increased above the E^4 dependence by increasing the beam emittance and hence the beam size. The emittance can be varied by either changing the focussing strength of the lattice (variable optics) or by the use of wiggler magnets. In this case the luminosity may vary as E^2. At high energies, where the RF power is the limiting factor, the luminosity may be increased by reducing the emittance, i.e., by increasing the focussing strength of the lattice.

Table 2.1
Operational data on storage rings

Ring	Start of operation	Center of mass energies (GeV)	Max. luminosity achieved (cm^{-2}s^{-1})	Interaction regions
SPEAR, SLAC	1973	3–8	1×10^{31}	2
DORIS, DESY	1974	3–10.5 (12)	0.35×10^{31}	2
PETRA, DESY	1978	10–36.8 (45)	1.7×10^{31}	4
CESR, Cornell	1979	8–16	1×10^{31}	2
PEP, SLAC	1980	10–30	0.7×10^{31}	6

The data on e^+e^- annihilation discussed below are from experiments performed at SPEAR and PEP at SLAC, DORIS and PETRA at DESY, and CESR at Cornell. The start-up dates of these accelerators and the nominal center of mass range are listed in table 2.1.

Center of mass energies up to 12 GeV will be available at DORIS starting May 1982. The energy in PETRA will be increased in steps, 40.5 GeV will be available in the autumn of 1982 and 45 GeV in 1983.

2.1.2. The experiments
Before we discuss the detectors it might be useful to recall the basic features of e^+e^- interactions.

In lowest order, e^+e^- annihilate to a timelike current, electromagnetic or weak, which couples directly [7–10] to the basic fermions, as shown in fig. 2.3. At the small distances involved in present e^+e^- experiments, the quarks behave like free particles, and quark and lepton pairproduction have therefore similar cross sections and angular distributions. At present energies the electromagnetic current dominates and the cross section for muon pairproduction is given by:

$$\sigma(e^+e^- \rightarrow \mu^+\mu^-) = \sigma_{\mu\mu} = 4\pi\alpha^2/3s = 86.8/s \text{ nb GeV}^{-2}$$

where s is the c.m. energy squared.

The total cross section for hadron production is proportional to $\sigma_{\mu\mu}$ and the constant of proportionality R is given by:

$$R = (e^+e^- \rightarrow q\bar{q} \rightarrow \text{hadrons})/\sigma_{\mu\mu} = 3\sum_i (e_i/e)^2. \tag{2.4}$$

The sum is over all liberated quark flavours, and the factor of three accounts for colour.

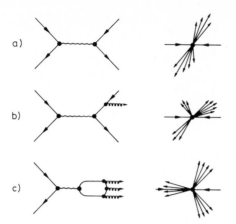

Fig. 2.3. Feynman graphs and topologies for α^2 processes in e^+e^- annihilation: (a) $e^+e^- \rightarrow q\bar{q}$, (b) $e^+e^- \rightarrow q\bar{q}g$, (c) $e^+e^- \rightarrow l^{--} \rightarrow ggg$.

As the quarks start to move apart, the confining force increases and new quark pairs are created from the vacuum. At distances on the order of 10^{-13} cm the quarks combine into colourless hadrons, resulting in two acollinear hadron jets along the direction of the original quarks (fig. 2.3a). At a center of mass energy of 36 GeV each jet on the average consists of 6–7 charged particles within a cone of opening angle 34°. Roughly 65% of the total energy is carried off by charged particles. The angular distribution of the hadron jets with respect to the beam axis is $1 + \cos^2 \theta$, reflecting the spin-1/2 nature of the quarks.

The outgoing quark may also radiate [11] a gluon [12] at a large angle, resulting in three well separated hadron jets as indicated in fig. 2.3b.

The timelike current also couples directly to vector mesons as shown in fig. 2.3c. Below the threshold for new flavour production, these resonances are narrow and they decay [13], according to QCD [14], with three gluons in the intermediate state resulting in three-jet events (fig. 2.3c). Hidden flavour states with even charge conjugation can be populated by photon emission from the vector states.

A new class of processes [15, 16], depicted in fig. 2.4a, b, occur to fourth order in α. The virtual photon cloud accompanying the electrons interacts directly: $e^+e^- \rightarrow \gamma^*\gamma^*e^+e^- \rightarrow$ hadrons e^+e^-. This spacelike interaction can be divided in two groups, depending whether the photon reveal its hadronlike or pointlike character. In the first case the photons

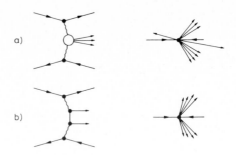

Fig. 2.4. Feynman graphs and topologies for some α^4 processes in e^+e^- interactions: (a) $e^+e^- \to$ hadrons e^+e^-, (b) $e^+e^- \to q\bar{q}e^+e^-$.

convert into vector mesons, which interact to produce a final state similar to that observed in hadron–hadron collisions, where the final-state hadrons in general emerge with low transverse momenta with respect to the beam axis. A rough estimate of the $\sigma(\gamma\gamma \to$ hadrons), using factorization yields:

$$\sigma(\gamma\gamma \to \text{hadrons}) = (\sigma_{\gamma p})^2/\sigma_{pp} \simeq 300 \text{ nb}.$$

In the pointlike piece the photon couples directly to the quarks which fragments into hadron jets. The hadrons are distributed [17] roughly as $1/p_T^4$ with respect to the beam axis.

The physics which can be investigated with an e^+e^- machine is therefore both rich and varied, and this results in conflicting demands on the detector. Some of the factors which influence the detector design can be summarized as follows.

The total cross section is low resulting in some 100 annihilation events per day for an integrated luminosity of 400 nb^{-1} at 36 GeV. The detector should therefore cover a large solid angle. All annihilation events are accepted and the trigger is used only to reduce the beam gas background to an acceptable level. Since the annihilation physics is independent of angle there is no special incentive to cover the difficult region very close to the beam.

In the continuum, the hadrons in general appear in well defined jets. The aim is to reconstruct the physics on the parton level from the particles in the jet. To this end, charged particles and photons travelling close together must be measured and identified. Good electron (muon) identification within a jet is valuable to select either new physics or weak decays of heavy flavours. For -onium spectroscopy the photon calorimeter must be able to measure low energy photons with good resolution.

Two-photon physics imposes additional constraints on the detector. The energy and the Q^2 of the virtual photon can be determined from a measurement of the angle and the energy of the scattered electrons. This requires an electron detector which extends down to small angles with respect to the beam line. However, electrons radiating nearly real photons will escape down the beam tube. In this case the c.m. energy must be determined from a measurement of the final state particles. It is then important that the detector covers extreme forward and backward angles to catch particles which travel close to the beam line.

To see how these requirements are met, we will discuss two different types of detectors, the JADE detector [18] as a representative of detectors based on a solenoidal field, and the Crystal Ball [19], a nonmagnetic detector designed to measure photons with high resolution.

The main component of the JADE detector shown in fig. 2.5 is a normal conducting solenoid which creates a longitudinal magnetic field of 0.5 T parallel to the beam axis. The coil is 3.5 m long with a diameter of 2 m. Particle tracking is done using a drift chamber with a total of 48 coordinate space points along each trajectory: The azimuthal coordinates r and ϕ are measured to 160 μm using the drift information and the axial coordinate is measured by charge division to 1.6 cm. The double track resolution is 7 mm. In addition dE/dx information is available at each coordinate point. The chamber is 2.36 m long with an inner radius of 21 cm and an outer radius of 79 cm. A lead glass detector consisting of 2688 separate blocks is mounted outside the cylindrical part of the solenoid. The endcaps are covered by 2×96 counters. The whole detector is surrounded by a segmented muon detector made of a steel concrete absorber interleaved with driftchambers. The region close to the beam is covered by counters used to determine the luminosity via a measurement of small angle Bhabha scattering or to measure the energy and direction of electrons in $e^+e^- \rightarrow e^+e^- X$.

The Crystal Ball detector [19], depicted in fig. 2.6 is a nonmagnetic detector optimized for photon detection. Tracks emerging from the interaction region are detected by a set of proportional and magnetic chambers, which measure charge over 94% of 4π and tracks over 71% of 4π defined by the last chamber. The chambers are surrounded by the main component of the detector, a 16 r.l. thick shell of NaI(Tl), which is subdivided into 672 units and stacked as a geodesic dome. The crystals are hermetically sealed in two hemispheres. These are normally in con-

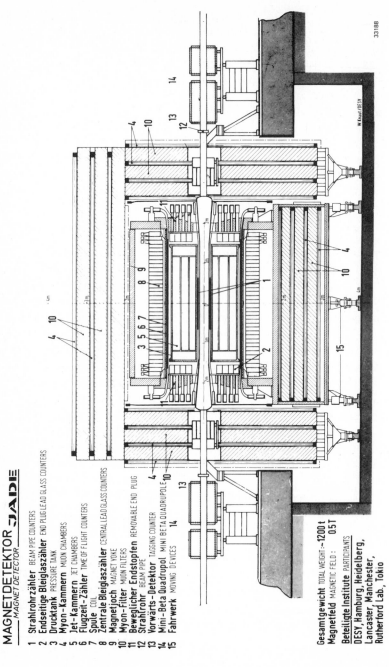

MAGNETDETEKTOR JADE
MAGNET DETECTOR

1 **Strahlrohrzähler** BEAM PIPE COUNTERS
2 **Endseitige Bleiglaszähler** END PLUG LEAD GLASS COUNTERS
3 **Drucktank** PRESSURE TANK
4 **Myon–Kammern** MUON CHAMBERS
5 **Jet–Kammern** JET CHAMBERS
6 **Flugzeit–Zähler** TIME OF FLIGHT COUNTERS
7 **Spule** COIL
8 **Zentrale Bleiglaszähler** CENTRAL LEAD GLASS COUNTERS
9 **Magnetjoch** MAGNET YOKE
10 **Myon–Filter** MUON FILTERS
11 **Beweglicher Endstopfen** REMOVABLE END PLUG
12 **Strahlrohr** BEAM PIPE
13 **Vorwärts–Detektor** TAGGING COUNTER
14 **Mini–Beta Quadrupol** MINI BETA QUADRUPOLE
15 **Fahrwerk** MOVING DEVICES

Gesamtgewicht TOTAL WEIGHT:~**1200 t**
Magnetfeld MAGNETIC FIELD: **0.5 T**
Beteiligte Institute PARTICIPANTS
DESY, Hamburg, Heidelberg,
Lancaster, Manchester,
Rutherford Lab, Tokio

Fig. 2.5. The JADE detector at PETRA.

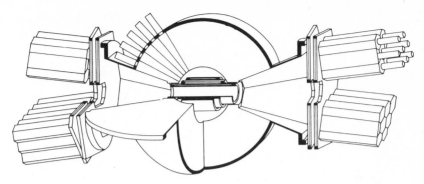

Fig. 2.6. The Crystal Ball detector at SPEAR.

tact, but might be moved apart to access the inner detector. The end cap detectors are made of planar magnetostrictive spark chambers followed by 51 cm long hexagonal NaI(Tl) crystals. Muon detectors, made of iron interleaved with proportional tubes, cover 15% of 4π and are centered at 90°.

The energy of photons and electrons is measured with a resolution of $\sigma = 0.028 E^{3/4}$ GeV and the angle of the photon is determined to about 2.0°. These properties of the Crystal Ball have led to the discovery of both the η_c and η'_c in the presence of a high background.

2.2. Beams and detectors in deep inelastic lepton – hadron interactions

2.2.1. Neutrino experiments

The basic processes which can be studied [20] in neutrino–nucleon interactions are indicated in fig. 2.7. The incoming neutrino interacts with the quarks in the target via a charged or neutral current. In the case of charged current interaction, the final state consists of a high energy muon, a hadron jet resulting from the fragmentation of the struck quark,

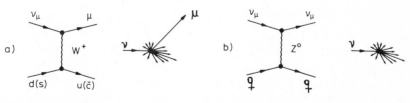

Fig. 2.7. Feynman graphs and topologies for neutrino–quark interactions: (a) $\nu_\mu q \to \mu q'$, (b) $\nu_\mu q \to \nu_\mu q'$.

and hadrons from the target fragmentation. The jet and the target fragmentation are not separated at present energies. In the case of the neutral current interaction the muon is replaced by a neutrino which escapes the detector, and a hadronic state similar to the one observed in charged current interaction.

The two processes can be distinguished by the appearance or nonappearance of a high energy muon in the final state. Every high energy neutrino detector must therefore be capable of identifying high energy muons. In the case of the charged current interaction, Q^2 and ν of the process are determined—if the energy of the incident neutrino is known —by measuring the energy and the angle of the outgoing muon. In a neutral current event these quantities must be determined from a measurement of the final state hadrons. Hence it is important to measure the momentum and the production angles of both the muon and the hadrons in the event.

The total cross section, at present energies [21], for charged current interactions are given by $\sigma(\nu_\mu N \to \mu^- X) = 0.6 \times 10^{-38} E_\nu$ cm^{-2} for an incident neutrino beam and by $\sigma(\bar{\nu}_\mu N \to \mu^+ X) = 0.28 \times 10^{-38} E_\nu$ cm^{-2} for an incident antineutrino beam. These are very small cross sections, and it requires both a high intensity neutrino beam and a large target mass to carry out detailed measurements of these processes.

The neutrino beams at high energy accelerators are tertiary beams resulting from the decay of secondary charged pions and kaons:

$$\pi^\pm \to \mu^+ (\mu^-) + \nu (\bar{\nu}),$$

$$K^\pm \to \mu^+ (\mu^-) + \nu (\bar{\nu}).$$

The two-body decay results in a neutrino spectrum which is flat up to the maximum energy, which is given by $E_\pi(1 - m_\mu^2/m_\pi^2)$, respectively $E_K(1 - m_\mu^2/m_K^2)$. If the energy of the parent pion or kaon is known then it is sufficient to measure the angle of the neutrino to determine its energy. The neutrino beams originating from these two-body decays are contaminated on the few percent level by neutrinos from the decays

$$K^0 \to e\pi\nu_e \quad \text{and} \quad K \to \mu\pi\nu_\mu.$$

The neutrino beam line at the CERN SPS is shown schematically in fig. 2.8. The primary proton beam is focussed on a production target. The secondary hadrons emerging from the target are sign selected, their momenta are defined by a focussing device, and they are passed through a long decay region. At the end of the decay region the particles enter a massive 400 m long iron shield which absorbs remaining hadrons and muons. The length of the shield is set by the requirement that all decay

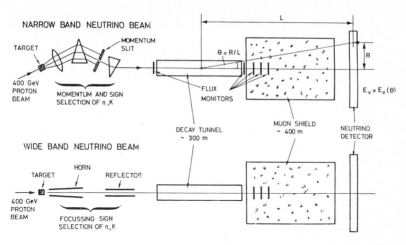

Fig. 2.8. Schematic layout of the CERN SPS neutrino beam line.

muons are absorbed, which requires roughly 0.5 m of iron per GeV of muon energy. Detectors to monitor the neutrino flux are installed in the shield.

The energy spectrum of the broad band neutrino beam resulting from the decay of hadrons without well defined momenta is shown in fig. 2.9. The flux of neutrinos and antineutrinos are given per GeV and m^2 for 10^{13} incident protons. The flux is measured with a 1 m radius detector at the position of BEBC. The spectrum has a peak around 20 GeV with a rapid, nearly exponential fall off towards larger energies. The neutrino flux near the endpoint energy is much higher than the antineutrino flux. This simply reflects the preponderance of positively charged particles at high relative momenta. The rapid fall off with energy is one of the problems associated with a precise determination of the flux. The flux is monitored by both measuring the hadron beam and by observing the muons.

The intensity of the broad band beam is high but the neutrino energy is not known. The neutrino energy can be determined by momentum-selecting the hadrons ("narrow band beam") and measuring the emission angle of the neutrino. The latter information is obtained by measuring the transverse distance from the beam center line to the event vertex. The neutrino energy as determined from a measurement by the CDHS detector of the final state particles is plotted versus the impact radius in fig. 2.10. The upper band corresponding to $K \to \mu \bar{\nu}_\mu$ decays and the lower band resulting from $\pi \to \mu \bar{\nu}_\mu$ decays are clearly seen. The knowledge of the neutrino energy typically leads to a loss in flux by a factor 100.

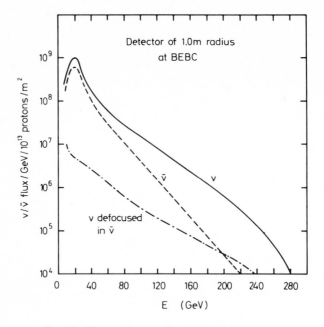

Fig. 2.9. The neutrino broad band energy spectrum.

The small cross section, the large size of the incident beam and the necessity to identify the muon and if possible measure its energy and angle makes it necessary to combine the target and the detector into a single unit. Examples of such devices are bubble chambers and calorimeter type electronic detectors.

The bubble chambers offer high granularity, and can be filled both with light and heavy fluids. The drawback of a bubble chamber is its comparatively low mass. The Big European Bubble Chamber BEBC [22] has a total volume of 32 m³ and a fiducial volume of 10–20 m³. It is surrounded by superconducting magnets producing a field of 3 T. The chamber can be filled with hydrogen, deuterium or neon. Data have also been taken with a hydrogen target (TST) surrounded by liquid neon, attempting to combine the advantages of both fluids. Downstream, the chamber is followed by electronic detectors.

The second group of detectors is finely sampling hadron calorimeters of enormous dimensions. The CDHS [23] (CERN–Dortmund–Heidelberg–Saclay) detector is the first one shown in fig. 2.11. The detector is modular, made of 19 magnetic toroids, each 3.75 m in diameter, weighing

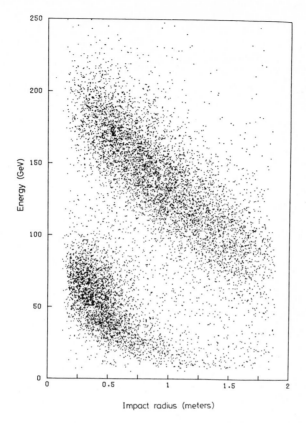

Fig. 2.10. Neutrino energy plotted versus the impact radius; the data are from CERN SPS narrow band neutrino beam.

65 tons. Each of the first seven units is subdivided into 15 five cm thick iron plates interleaved with scintillation counters. The remaining twelve units are each made of five 15 cm thick iron plates and scintillators. The iron plates are magnetized by means of two vertical coils which produce an average field of 1.65 T in the plates. Three sets of drift wires are installed between each module, to measure tracks. The detector has a fiducial good resolution mass of 400 tons and an additional 800 tons with poorer resolution.

The CHARM [24] (CERN–Hamburg–Amsterdam–Rome–Moscow) detector, shown as the second one in fig. 2.11, is a finely grained calorimeter capable of measuring the energy and direction of hadronic

Fig. 2.11. View into the CERN neutrino hall. The detector consisting of iron toroids and hexagonal driftchambers is the CDHS detector. The CHARM detector (square cross section) follows directly behind.

showers. Its primary goal is to investigate neutral current events. The detector consists of 80 sheets of 80 mm thick marble, 3×3 m^2 square. The sheets are surrounded by a picture frame magnet, used to determine the charge sign of muons produced in charged current events.

2.2.2. Charged lepton experiments

The basic process which can be measured in muon (electron)–nucleon interactions is shown in fig. 2.12. In fact it was the famous deep inelastic electron–nucleon scattering experiments [25] at SLAC which gave the first direct evidence for pointlike constituents in the nucleon.

The incoming lepton interacts with a quark in the target by the exchange of a photon or a neutral weak boson (Z^0). The cross section at present energies is dominated by one-photon exchange. The final state consists of a muon (electron) and, as in charged current interactions, of two hadron jets, one from the struck quark and the other from the target fragments. The jets are not separated at present energies. Clearly it is easier to reconstruct deep inelastic muon–hadron events than neutrino–hadron events, since the former have charged leptons both in the initial and the final state. However, muon–hadron events have larger radiative corrections, and the cross section is supressed by the photon propagator at large values of Q^2. The lepton may also interact with the quark via the charged weak current $\mu N \rightarrow \nu X$. The resulting cross section is very small at present energies and the events are more difficult to reconstruct due to the neutrino in the final state. Data on this process have not been reported.

The SLAC linear accelerator delivers a well defined high intensity beam of about 10^{14} electrons per second. The maximum energy has been increased to 33 GeV by shortening [26] the RF pulse length and increasing the peak power (SLED). It is planned to increase the energy to 50 GeV and beyond using this scheme.

The decay of kaons and pions into muons and neutrinos is the source of the high energy tertiary muon beams at high energy proton machines.

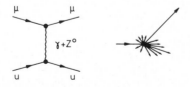

Fig. 2.12. Feynman graph and topology for μ (e)$+$q $\rightarrow \mu'$ (e$'$)$+$q.

While at FNAL both decay products are used simultaneously by bending the muons away from the neutrino beam, at CERN a special muon beam has been built. It consists of four parts. The front end accepts parent pions and kaons within an energy-dependent solid angle of 10–20 μsr in the forward direction and separates off the unused primary protons. A momentum band of $\Delta p/p(\pi, \text{K}) \leqslant \pm 10\%$ is accepted by the second part, a FODO decay channel. This 600 m long channel consists of a regular array of large aperture quadrupoles, alternatively focussing and defocussing. The channel confines with small losses both the parent pions and kaons and also the decay muons. The third part, a hadron absorber, removes most of the pions and kaons remaining at the end of the decay channel. It consists of 10 m beryllium, located at a focus within the aperture of the momentum selecting magnets. Finally, a back end consisting of two vertical bends and a 250 m FODO array selects the muons and transports them to the experiments. The effective momentum band is $\Delta p/p \simeq 5\%$. Per primary proton of 400 GeV incident on the production target, 3.8×10^{-5} positively charged muons with a momentum of 120 GeV/c are delivered to the experiment. The numbers for positively charged muons with momenta of 200 and 280 GeV/c are 1.4×10^{-5} and 1.8×10^{-6} respectively. For negative muons the fluxes are lower by a factor 2.5–5.

It is difficult to define the size of the muon beam, due to the high penetration power of muons, and special care has to be taken to avoid or at least detect halo muons. All detectors therefore start with a wall of veto counters to reject muons outside the acceptance. As a target, hydrogen, deuterium, carbon and iron have been used by the various experiments. The final state particles are measured by a magnetic spectrometer. Since the muon–nucleon cross section varies like $1/Q^4$ (with $Q^2 \approx 4EE' \sin^2 \theta/2$), it is essential to measure the energies and directions of the in- and outgoing muons with similar precision.

The detector of the EMC [27], shown in fig. 2.13, is a good example for a high resolution muon scattering detector. To achieve the desired resolution in Q^2 and ν, the spectrometer consists of sets of drift chambers ($W_{1,2}$) installed in front and ($W_{3,4,5}$) installed at the back of a large aperture dipole magnet (FSM). The tracking through the magnetic field is done with proportional chambers ($P_{1,2,3}$). The large distances between the drift chambers and their good spacial resolution result in a momentum resolution of $\Delta p/p \simeq 10^{-4} p$ GeV/c at 5 Tm.

A multicell Čerenkov counter C_2 is used for particle separation in a limited momentum range. Thus muons are identified as tracks in the drift chambers W_6 and W_7 linking with one of the tracks in $W_{4,5}$.

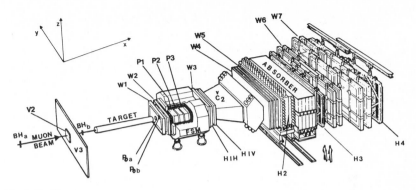

Fig. 2.13. The detector used by the European Muon Collaboration (EMC) to measure deep inelastic muon interactions.

A lead, iron, scintillator calorimeter with a depth of five interaction lengths is used to measure neutral particles. The calorimeter together with a ten interaction-length deep iron block is used to remove the hadrons. While the EMC detector like the CHIO [28] (Chicago–Harvard–Illinois–Oxford) is designed as a general purpose detector, the BCDMS [29] (Bologna–Cern–Dubna–München–Saclay) detector is designed as a high statistics experiment. It is basically a single-arm spectrometer for the muons. It consists of a 55 m long magnetized iron torus with a 40 m carbon target in the central hole. Proportional chambers and scintillator hodoscopes measure the muon tracks. To be identified as a muon, it has to transverse at least 5.28 m of iron. In low-Q^2 events the muon does not cross enough iron to be identified. The cutoff is at $Q^2 = 20$ GeV$^2/c^2$ for $E_\mu = 120$ GeV.

3. The structure of electromagnetic and weak interactions

3.1. The standard model

The standard model is based on the group $G = SU(2) \times U(1)$ for the neutral current [30], and on a V–A interaction [31] for the charged current. The group SU(2) contains two massless gauge bosons W$^\pm$ and W^0 with coupling constant g, U(1) contains the massless gauge boson B^0 with coupling constant g'. The coupling constants are related to the electric charge:

$$e^{-2} = g^{-2} + g'^{-2}.$$

$$(3.1)$$

The gauge symmetry must be spontaneously broken for the theory to be renormalizable. There is at present no consensus how this is achieved. In the standard model the symmetry is broken by introducing $I = 1/2$ scalar Higgs fields [32]. In this case we get massive $W^\pm$ bosons which mediate the charged weak current corresponding to the weak isospin raising and lowering operators $T_\mp$. The massless B^0, W^0 mix and we obtain two new bosons:

$$Z_\mu^0 = W_\mu^0 \cos\theta_W - B_\mu^0 \sin\theta_W,$$

$$A_\mu = W_\mu^0 \sin\theta_W + B_\mu^0 \cos\theta_W. \tag{3.2}$$

The A_μ remains massless and can be identified with the photon coupled to the electric charge. The Z^0, however, will acquire a mass. In the standard model with $I = 1/2$ Higgs fields the Z^0 and the $W^\pm$ mass are related:

$$\rho = m_{W^\pm}^2 / m_{Z^0}^2 \cos^2\theta_W = 1. \tag{3.3}$$

The Z_μ^0 is coupled to a neutral current J_μ^0 which is a mixture of the weak isospin current and the electromagnetic current:

$$J_\mu^0 = J_\mu^3 - 2 J_\mu^{EM} \sin^2\theta_W. \tag{3.4}$$

The mixing angle θ_W is a free parameter defined by $\tan\theta_W = g'/g$. With this definition, eq. (3.1) becomes:

$$e = g \sin\theta_W. \tag{3.5}$$

At low values of q^2 the effective strength of the charged current interactions is given by

$$G_F/\sqrt{2} = g^2/8 m_W^2. \tag{3.6}$$

The $W^\pm$ mass and the Z^0 mass can be expressed in terms of a single parameter $\sin^2\theta_W$, using eqs. (3.3), (3.5) and (3.6):

$$m_W = 37.4 \text{ GeV}/\sin\theta_W \quad \text{and} \quad m_{Z^0} = 37.4 \text{ GeV}/\sin\theta_W \cos\theta_W. \tag{3.7}$$

To complete the model the transformation properties of leptons and quarks under $SU(2) \times U(1)$ must be specified.

In the standard model all fermions are arranged in left-handed weak isospin doublets:

$$\begin{pmatrix} \nu_e \\ e \end{pmatrix}_L, \quad \begin{pmatrix} \nu_\mu \\ \mu \end{pmatrix}_L, \quad \begin{pmatrix} \nu_\tau \\ \tau \end{pmatrix}_L$$

$$\begin{pmatrix} u \\ d' \end{pmatrix}_L, \quad \begin{pmatrix} c \\ s' \end{pmatrix}_L, \quad \begin{pmatrix} ? \\ b' \end{pmatrix}_L, \tag{3.8}$$

and in right handed singlets. The fermions in the first two generations have all been found, whereas the last generation is still incomplete. Measurements from PETRA show that the mass of the top quark—if it exists—must be above 18 GeV, and the tau neutrino has not yet been observed directly.

The weak effective Lagrangian at low energies can be written as:

$$L_{\text{eff}} = \frac{G_F}{\sqrt{2}} \left(J_\mu^+ J_\mu^+ + J_\mu^0 J_\mu^0 \right). \tag{3.9}$$

The lepton charged current is given by the doublet structure:

$$\tfrac{1}{2} \nu_\ell (1 - \gamma_5) \ell, \quad \text{with } \ell = e, \mu, \tau. \tag{3.10}$$

The quark charged current can be written as:

$$J_\mu^+ = (\bar{u}, \bar{c}, \bar{t}) \tfrac{1}{2} \gamma_\mu (1 - \gamma_5) \begin{pmatrix} d' \\ s' \\ b' \end{pmatrix}. \tag{3.11}$$

It is generally assumed that the d', s' and b' quarks in the weak isospin doublets are related to the d, s and b quarks observed in the strong interactions, by the Kobayashi–Maskawa matrix [33] analogous to the Cabibbo mixing in the four quark system

$$\begin{pmatrix} d' \\ s' \\ b' \end{pmatrix} = U \begin{pmatrix} d \\ s \\ b \end{pmatrix}.$$

The mixing matrix U can be written in the form:

$$U = \begin{pmatrix} c_1 & -s_1 c_3 & -s_1 s_3 \\ s_1 c_2 & c_1 c_2 c_3 - s_2 s_3 e^{i\delta} & c_1 c_2 s_3 + s_2 c_3 e^{i\delta} \\ s_1 s_2 & c_1 s_2 c_3 + c_2 s_3 e^{i\delta} & c_1 s_2 s_3 - c_2 c_3 e^{i\delta} \end{pmatrix}, \tag{3.12}$$

where c_i and s_i are abbreviations for $\cos\theta_i$ and $\sin\theta_i$, the generalized Cabibbo angles and δ is a complex phase which permits time violation.

The neutral current was defined in eq. 3.4. The axial (g_A) and vector (g_V) coupling of the fermions to the neutral current are then given by:

$$g_A = I_L^3, \qquad g_V = I_L^3 - 2Q \sin^2 \theta_W, \tag{3.13}$$

where Q is the charge and I_L^3 the third component of weak isospin (eq. 3.8). They are explicitly listed in table 3.1 for the various fermions. A fit to the available data [34] gives $\sin^2 \theta_W = 0.224$. The vector coupling constant of the neutral current to a charged lepton is therefore rather small.

Table 3.1
Fermion neutral current coupling constants

Fermion	g_A	g_V
ν_e, ν_μ, ν_τ	1/2	1/2
e, μ, τ	$-1/2$	$-1/2 + 2\sin^2\theta_W$
u, c, t	1/2	$1/2 - \frac{4}{3}\sin^2\theta_W$
d', s', b'	$-1/2$	$-1/2 + \frac{2}{3}\sin^2\theta_W$

3.2. Test of QED

QED predictions are based on the validity of Maxwell equations and on the assumption that leptons are pointlike objects without excited states. With e^+e^- machines these assumptions can be tested [35] at very small distances in a clean environment with only small and well defined corrections due to the strong interactions.

The Feynman graphs for Bhabha scattering, lepton pair production and two-photon annihilation are shown in fig. 3.1 and the differential cross sections for these processes at 31.6 GeV c.m. energy is plotted versus scattering angle in fig. 3.2. All the cross sections scale with energy as $1/s$ with $s = (2E_{beam})^2$.

The interference of the weak and the electromagnetic currents leads to observable effects at the highest PETRA and PEP energies both in Bhabha scattering and in the forward–backward asymmetry in muon pair production. We will return to these effects after the discussion on QED limits.

The standard procedure which is used to compare data with the QED predictions can be summarized as follows: (i) The data are corrected for weak effects assuming the Glashow–Salam–Weinberg model [30] with the standard values for $\sin^2\theta_W$. These corrections are small and in most cases negligible at present energies. (ii) The measured cross section $d\sigma/d\Omega$ is corrected for radiative effects δ_R and effects due to the hadronic vacuum polarisation δ_H:

$$d\sigma/d\Omega = (1 + \delta_R + \delta_H) d\sigma_{QED}/d\Omega. \qquad (3.14)$$

(iii) The corrected cross section is compared to the QED predicted cross section and deviations are parametrized in terms of formfactors. The formfactors used for Bhabha scattering and lepton pair production can

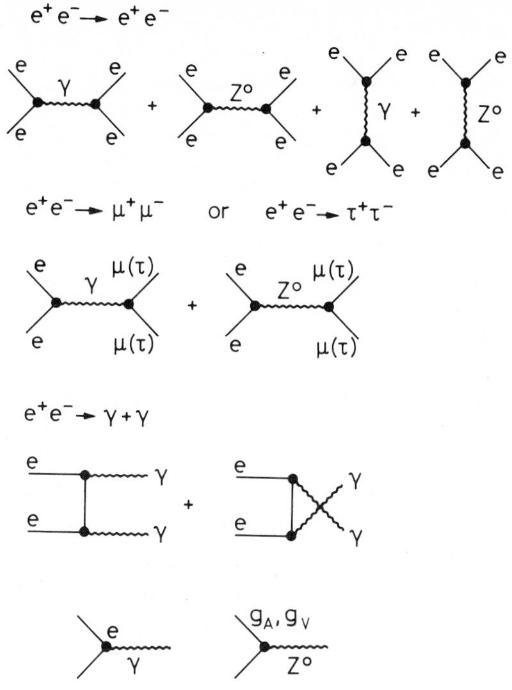

$e^+e^- \rightarrow e^+e^-$

$e^+e^- \rightarrow \mu^+\mu^-$ or $e^+e^- \rightarrow \tau^+\tau^-$

$e^+e^- \rightarrow \gamma + \gamma$

Fig. 3.1. The Feynman graphs for: (a) $e^+e^- \rightarrow e^+e^-$, (b) $e^+e^- \rightarrow \mu^+\mu^-$ ($\tau^+\tau^-$), (c) $e^+e^- \rightarrow \gamma\gamma$.

be written as:

$$F_s(q^2) = 1 \mp q^2/(q^2 - \Lambda^2_{s\pm}), \qquad F_t(q^2) = 1 \mp s/(s - \Lambda^2_{t\pm}),$$

$$(3.15)$$

where F_s and F_t are respectively the formfactors for spacelike and timelike momentum transfers squared.

The differential cross section for $e^+e^- \rightarrow e^+e^-$ as measured by the various PETRA groups and normalized to the QED cross section are plotted [35] in fig. 3.3 versus scattering angle. The data have been corrected for radiative effects and hadronic vacuum polarization, but not for electroweak effects.

The total cross sections for muon and tau pair production as measured by the PETRA groups are plotted [36] in fig. 3.4 and fig. 3.5 versus c.m. energy. The data are in good agreement with the QED predictions shown as solid lines.

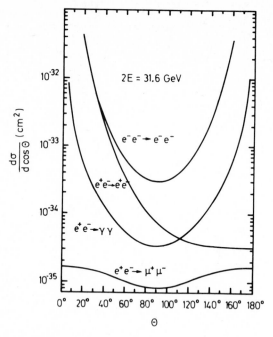

Fig. 3.2. Differential cross sections for two-body QED reactions.

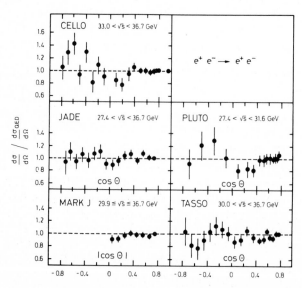

Fig. 3.3. Differential cross section for $e^+e^- \to e^+e^-$, divided by the QED prediction.

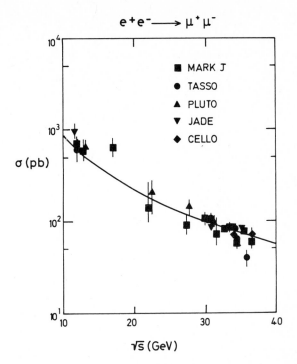

Fig. 3.4. The total cross section measured by the PETRA groups for muon pair production plotted versus c.m. energy. The solid line shows the QED prediction.

The two-photon annihilation process $e^+e^- \to \gamma\gamma$ proceeds by electron exchange in the t channel. This reaction is unique among two-body QED reactions, in that weak effects do not contribute to lowest order. The observed cross section is related to the QED cross section by:

$$d\sigma/d\Omega = (1 + \delta_R + \delta_H + \delta_\Lambda) d\sigma/d\Omega_{QED},$$

where a breakdown of QED is parametrized by $\delta_\Lambda(s, \theta)$. Two possible breakdown mechanisms have been considered, the seagull term and the exchange of a heavy electron with coupling strength e and mass m. For the seagull term:

$$\delta_\Lambda(s, \theta) = \pm \frac{s^2}{2\Lambda_{\pm}^4} \frac{\sin^4\theta}{1 + \cos^2\theta}, \tag{3.16}$$

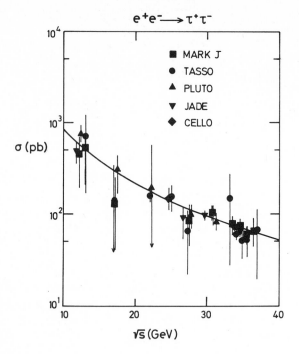

Fig. 3.5. The total cross section measured by the PETRA groups for tau pair production plotted versus c.m. energy. The solid line shows the QED prediction.

and for the exchange of a heavy electron

$$\delta_\Lambda(s,\theta) = \frac{s^2}{2\Lambda^4}\sin^2\theta. \qquad (3.17)$$

The measured cross sections normalized to the QED prediction are plotted in fig. 3.6. The data agree well with the QED prediction. Two-body QED reactions have been measured [37] by the MAC and by the MARK II Collaborations at PEP and the data obtained by these groups are also in agreement with QED.

The lower limits on the formfactor Λ extracted from a comparison between the data and the QED predictions are listed in table 3.2. Equation (3.16) was used in the case of $e^+e^- \to \gamma\gamma$. These data show that QED is valid down to distances of 10^{-16} cm and exclude a heavy electron with mass less than about 50 GeV and with a coupling strength e.

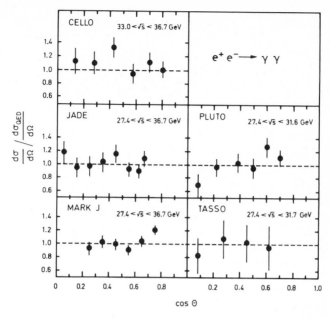

Fig. 3.6. The cross section for $e^+e^- \to \gamma\gamma$ measured by the PETRA groups, normalized to the QED prediction.

Table 3.2
QED parameters: 95% confidence lower limits in GeV

Experiment	$e^+e^- \to e^+e^-$		$\mu^+\mu^-$		$\tau^+\tau^-$		$\gamma\gamma$	
	Λ_+	Λ_-	Λ_+	Λ_-	Λ_+	Λ_-	Λ_+	Λ_-
CELLO	83	155			139	120	43	48
JADE	112	106	142	126	111	93	47	44
MARK J	128	161	194	153	126	116	55	38
PLUTO	80	234	107	101	79	63	46	–
TASSO	140	296	127	136	104	189	34	42
MARK II							50	41

3.3. The neutral weak current in purely leptonic interactions

Neutrino–electron scattering gives information on the neutral weak interaction in a purely leptonic system. This current and its interference with the electromagnetic current have also been studied in $e^+e^- \to e^+e^-$,

$e^+e^- \to \mu^+\mu^-$ and $e^+e^- \to \tau^+\tau^-$. The experimental results are discussed in this section.

3.3.1. Neutrino–electron scattering

In principle, four neutrino–electron scattering processes are possible:

$$\nu_e e^- \to \nu_e e^-, \tag{3.18a}$$

$$\bar{\nu}_e e^- \to \bar{\nu}_e e^-, \tag{3.18b}$$

$$\nu_\mu e^- \to \nu_\mu e^-, \tag{3.18c}$$

$$\bar{\nu}_\mu e^- \to \bar{\nu}_\mu e^-. \tag{3.18d}$$

The processes (3.18a) and (3.18b) can occur through charged and neutral current interactions. The interference between these interactions gives information on possible scalar (S), pseudoscalar (P) and tensor (T) contributions to the neutral current amplitudes in addition to the vector (V) and axialvector (A) part. The cross sections [38] can be related to the V, A, S, P, T coupling constants:

$$\frac{d\sigma}{dy}(\nu_i e \to \nu_i e) = \frac{G^2 m_e E_\nu}{2\pi} \left[A_i + 2B_i(1-y) + C_i(1-y)^2 \right], \tag{3.19}$$

where $i = \mu$ or e, and $y = \nu/\nu_{max}$.

$$A_\mu = (g_V^e + g_A^e)^2 + [(C_S + C_P)/4]^2 + [(C_S - C_P - 4C_T)/4]^2,$$

$$B_\mu = B_e = C_T^2 - [(C_S + C_P)/4]^2 - [(C_S - C_P)/4]^2,$$

$$C_\mu = C_e = (g_V^e - g_A^e)^2 + [(C_S + C_P)/4]^2 + [(C_S - C_P + 4C_T)/4]^2,$$

$$A_e = \underset{NC}{A_\mu} + \underset{CC}{4} + \underset{interference}{4(g_V^e + g_A^e)^2}. \tag{3.20}$$

Neglecting S, P and T terms the total cross sections are:

$$\sigma(\nu_e e^-) = \frac{G^2 m_e E_\nu}{2\pi} \left\{ (g_V^e + g_A^e + 2)^2 + \tfrac{1}{3}(g_V^e - g_A^e)^2 \right.$$

$$\left. - \frac{m_e}{2E_\nu} \left[(g_V^e + 1)^2 - (g_A^e + 1)^2 \right] \right\},$$

$$\sigma(\bar{\nu}_e e^-) = \frac{G^2 m_e E_\nu}{2\pi} \left\{ (g_V^e - g_A^e)^2 + \tfrac{1}{3}(g_V^e + g_A^e + 2)^2 \right.$$

$$\left. - \frac{m_e}{2E_\nu} \left[(g_V^e + 1)^2 - (g_A^e + 1)^2 \right] \right\},$$

$$\sigma(\nu_\mu e^-) = \frac{G^2 m_e E_\nu}{2\pi} \left\{ (g_V^e + g_A^e)^2 + \tfrac{1}{3}(g_V^e - g_A^e)^2 - \frac{m_e}{2E_\nu} \left[g_V^{e2} - g_A^{e2} \right] \right\},$$

$$\sigma(\bar{\nu}_\mu e^-) = \frac{G^2 m_e E_\nu}{2\pi} \left\{ (g_V^e - g_A^e)^2 + \tfrac{1}{3}(g_V^e + g_A^e)^2 - \frac{m_e}{2E_\nu}\left[g_V^{e2} - g_A^{e2} \right] \right\}.$$

(3.21)

These equations describe four ellipses in a $g_V^e - g_A^e$ plane which intersect in two points. Due to the small mass of the electron, the cross sections are very small and high neutrino fluxes are needed to carry out the experiments.

There are no data on (3.18a) since sufficiently high ν_e fluxes are not available. The reaction (3.18b) has been measured [39] at a 1800 MW fission reactor with a flux of $2.2 \times 10^{13}\bar{\nu}_e$ cm^{-2} s^{-1}. The main backgrounds are caused by inverse β decay ($\bar{\nu}_e p \rightarrow e^+ n$) and reactions induced by neutrons and photons. The inverse β decay was identified and rejected by the e^+ annihilation and neutron capture. The latter processes were determined by varying the shielding and found to contribute less than 10%. A signal of 5.9 ± 1.4 events per day with $1.5 < E_e < 3.0$ MeV remained. The statistics are too poor to allow a definite conclusion on the

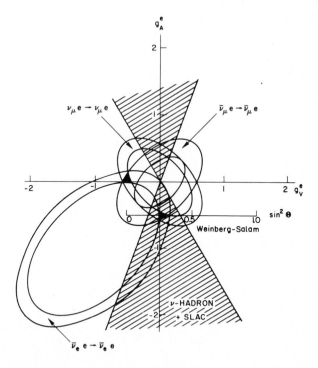

Fig. 3.7. Regions in the g_A^e, g_V^e plane allowed by the data for $\bar{\nu}_e e \rightarrow \bar{\nu}_e e$, $\nu_\mu e \rightarrow \nu_\mu e$ and $\bar{\nu}_e e \rightarrow \bar{\nu}_e e$. The ellipsoidal bands are the 60% confidence limits.

interference term, but a destructive interference is slightly preferred [38]. The data confine the allowed g_A^e–g_V^e values to an ellipsoidal band as shown in fig. 3.7.

The reactions (3.18c) and (3.18d) have been studied in high energy neutrino and antineutrino beams. The observation of $\bar{\nu}_\mu e^- \rightarrow \bar{\nu}_\mu e^-$ was in fact the first evidence for a neutral weak current. The discovery was made in the heavy liquid bubble chamber Gargamelle [40] at the CERN PS. Subsequently this reaction (3.18d) and the reaction $\nu_\mu e^- \rightarrow \nu_\mu e^-$ were investigated with bubble chambers [41–46], visual spark chambers [47] and electronic detectors [48,49].

The world-averaged cross sections for neutrino–lepton scattering are summarized in table 3.3.

The new results of the CHARM experiment [49] on the antineutrino cross section are not included in this average. Figure 3.8 shows the rate of observed events with an electron shower and no visible hadron in the final state as a function of $E^2\theta^2$ (square of the momentum transfer). The $(72 \pm 16)\bar{\nu}_\mu e^-$ events of this experiment with $q^2 = E^2\theta^2 < 0.12$ GeV^2 correspond to

$$\sigma/E = (1.7 \pm 0.33) \times 10^{-42} \text{ cm}^2/\text{GeV},$$

or in the GWS model:

$$\sin^2 \theta_W = 0.29 \pm 0.05.$$

The combined neutrino–electron scattering data (fig. 3.6) restrict the possible g_A^e, g_V^e values to the overlap regions of three elliptical bands. Two possible solutions are left. The axial-dominant solution has been fitted to

$$g_A^e = -0.52 \pm 0.06, \qquad g_V^e = 0.06 \pm 0.08.$$

This solution is in excellent agreement with the Glashow–Weinberg–Salam model which predicts, for $\sin^2 \theta_W = 0.23$, the values

Table 3.3
$\nu_\mu e^-$ and $\bar{\nu}_\mu e^-$ cross sections

Process	$\sigma/E(10^{-42} \text{cm}^2/\text{GeV})$	$\sin^2 \theta_W$
$\nu_\mu e^- \rightarrow \nu_\mu e^-$	1.5 ± 0.3	$0.24^{+0.06}_{-0.04}$
$\bar{\nu}_\mu e^- \rightarrow \bar{\nu}_\mu e^-$	1.3 ± 0.6	$0.23^{+0.09}_{-0.23}$

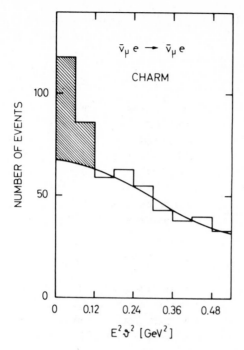

Fig. 3.8. Number of candidate events for $\bar{\nu}_\mu e \to \bar{\nu}_\mu e$ observed by the CHARM collaboration plotted versus $E^2\theta^2$.

$g_A^e = -0.5$ and $g_V^e = 0.04$. The second possible solution gives:

$$g_A^e = 0.06 \pm 0.08, \qquad g_V^e = -0.52 \pm 0.06.$$

Data either on neutrino–quark and electron–quark scattering or from the e^+e^- storage rings are needed to resolve the ambiguity. The data from e^+e^- reactions are now sufficiently precise to observe weak electromagnetic interference effects such that a unique solution can be found using only leptonic data.

3.3.2. Electroweak effects in e^+e^- annihilation

Including the neutral weak current will change [50] the normalized QED cross section for muon and tau pair production by ΔR and lead to a forward–backward charge asymmetry A in the final state.

At present PETRA and PEP energies the change in the normalized cross section, $\Delta R = [\sigma(\gamma + Z^0) - \sigma(\gamma)]/\sigma(\gamma)$, is given by:

$$\Delta R = + \left(\frac{G_F}{2\sqrt{2}\,\pi\alpha}\right) \frac{2sg_V^2}{(s/m_Z^2 - 1)} + \left(\frac{G_F}{2\sqrt{2}\,\pi\alpha}\right)^2 \frac{s^2\left(g_A^2 + g_V^2\right)^2}{\left(s/m_Z^2 - 1\right)^2}.$$

(3.22)

The forward–backward charge asymmetry is defined as:

$$A = \frac{(d\sigma/d\Omega)(\theta < 90°) - (d\sigma/d\Omega)(\theta > 90°)}{(d\sigma/d\Omega)(\theta < 90°) + (d\sigma/d\Omega)(\theta > 90°)}.$$

A negative asymmetry results from an excess of negative muons along the initial e^+ direction. The asymmetry integrated over 4π can be written at present energies as:

$$A_{\mu\mu} = \tfrac{3}{2}\left(G_F/2\sqrt{2}\,\pi\alpha\right)sg_A^2/\left(s/m_Z^2 - 1\right).$$

(3.23)

These quantities are plotted in fig. 3.9 as a function of $s = (2E)^2$ for $m_Z = 93$ GeV and $\sin^2\theta_W = 0.23$. Note that the Z^0 mass is increased from 89 GeV to 93 GeV by radiative corrections. At $\sqrt{s} = 35$ GeV the model predicts an asymmetry $A_{\mu\mu} = -9.6\%$ of which some 1.5% is due to the finite mass of the Z^0. ΔR is about 1.5% at 35 GeV and this results from the finite Z^0 mass. At 45 GeV, the highest PETRA energy with conventional cavities, it should be possible to determine the Z^0 mass by combining the total cross section and the asymmetry data. Mass effects can easily be measured at 60 GeV, an energy possible with superconducting cavities in PETRA.

The observed asymmetry must be corrected for radiative effects [51, 52]. Some of the higher order QED diagrams produce muon pairs with even C and they interfere with the diagrams which lead to muon pairs with odd C to produce a positive charge asymmetry. The size of this asymmetry depends on the c.m. energy and on the experimental cuts which are used to select the data, i.e. minimum muon energy E_μ, the maximum value of the acollinearity angle ξ, and the acceptance in production angle θ. For typical conditions like $\xi \leqslant 20°$, $E_\mu \geqslant 0.5\,E_{\text{beam}}$ and $|\cos\theta| < 0.8$ this asymmetry is $+1.5\%$ at $\sqrt{s} = 35$ GeV. It is thus smaller than the asymmetry caused by the electroweak interference term and of opposite sign. All the data presented below have been radiatively corrected.

The angular distribution of the muons measured [53] by the JADE group at high energies is plotted in fig. 3.10. The dotted line indicates the

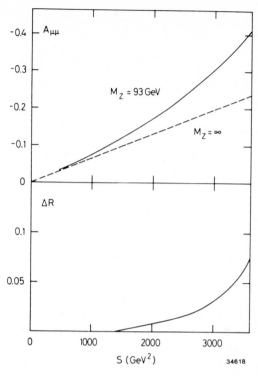

Fig. 3.9. Muon pair production in the Salam–Weinberg model, evaluated for $\sin^2 \theta_W = 0.23$. In the upper figure, the forward–backward asymmetry is plotted versus energy for $M_Z = 93$ GeV and $M_Z = \infty$. The lower figure shows the change in the cross section, normalized to the QED cross section, $\Delta R = [\sigma(\gamma + Z^0) - \sigma(\gamma)]/\sigma(\gamma)$.

$1 + \cos^2 \theta$ distribution for a pure electromagnetic current, the solid line shows the expected angular distribution including weak effects.

The value of the asymmetry extracted from these data is shown in table 3.4, together with values [36] obtained by other PETRA groups. The value quoted by TASSO was obtained by extrapolation to $|\cos \theta| \leqslant 1$. For the other experiments the asymmetry integrated over the detector acceptance, typically $|\cos \theta| < 0.8$, is given.

The systematic uncertainties are quite small and the data from the various groups can be combined. The resulting asymmetry [36] is $A_{\mu\mu} = -7.7 \pm 2.4\%$ with a χ^2 of 3.6 for three degrees of freedom. The result is in good agreement with the -7.8% predicted by the standard model, and it

Table 3.4
Charge asymmetry in muon and tau pair production

Asymmetry	JADE	MARK J	PLUTO	TASSO
$A_{\mu\mu}$ measured	-11 ± 4	-3 ± 4	7 ± 10	-11.3 ± 5
$A_{\mu\mu}$ predicted	-7.8	-7.1	-5.8	-8.7
$A_{\tau\tau}$ measured	$-$	-6 ± 11	$-$	0 ± 11
$A_{\tau\tau}$ predicted	$-$	-5	$-$	-7

yields:

$$|g_A^e| = 0.5^{+0.07}_{-0.09}.$$

This is the first evidence for a weak neutral current contribution in e^+e^- annihilation.

At PEP the muon asymmetry has been measured [37] by the MAC and the MARK II Collaboration at $\sqrt{s} = 29$ GeV. MAC finds $A_{\mu\mu} = -0.9\pm 5.2\pm1.5\%$ and MARK II $A_{\mu\mu} = -4.0\pm3.5\%$.

The Bhabha data can now be used to set limits on $\sin^2\theta_W$. The difference between the observed cross section for Bhabha scattering and the QED prediction as measured by the MARK J Collaboration for center-of-mass energies between 14 GeV and 37 GeV is plotted versus

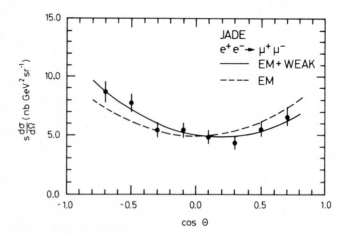

Fig. 3.10. The angular distribution in $e^+e^- \to \mu^+\mu^-$ as measured by the JADE collaboration at c.m. energies of 35 GeV.

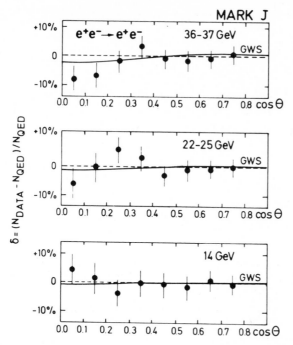

Fig. 3.11. The difference between the cross section for $e^+e^- \to e^+e^-$ as measured by the MARK J collaboration and the QED prediction, normalized to the QED prediction and plotted versus scattering angle.

$\cos \theta$ in fig. 3.11 Only scattering angles between $0°$ and $90°$ can be measured, since the charge sign of the outgoing lepton is not determined. The solid lines indicate the predictions based on the GWS model, with $\sin^2 \theta_W = 0.23$.

The ratio of the experimental cross section to the QED cross section at a center of mass energy of 35 GeV is plotted versus $\cos \theta$ in Fig. 3.12. The data, obtained by the TASSO Collaboration are also compared with the standard theory for various values of $\sin^2 \theta_W$. Although the present data are not yet sufficient to give a precise value of $\sin^2 \theta_W$, they do limit the value of $\sin^2 \theta_W$. Using all the available information, the various groups have extracted the values for $\sin^2 \theta_W$. These values are listed in table 3.5.

The 95% C.L. boundaries in a g_V, g_A plane as measured by the PETRA groups are shown in fig. 3.13. They describe two well separated regions centered at $g_A = \pm 1/2$. If these data are combined with the

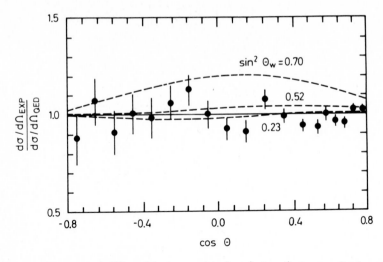

Fig. 3.12. The measured differential cross section for $e^+e^- \to e^+e^-$, normalized to the QED cross section, plotted versus $\cos\theta$. Predictions by the GWS model for various values of $\sin^2\theta_W$ are shown by the dashed lines.

elastic neutrino–electron scattering data discussed above and replotted in fig. 3.13, then we find a unique solution, consistent with the assignment made in the GWS model, i.e. the axial coupling of the charged lepton is $g_A = -1/2$.

It is possible to construct models [54,55] which reproduce the low energy data but which have a much richer spectrum of vector bosons. In

Table 3.5
Values of $\sin^2\theta_W$ for purely leptonic reactions

Group	$\sin^2\theta_W$	input
CELLO	$0.22^{+0.15}_{-0.10}$	$e^+e^-, \tau^+\tau^-, A_{\tau\tau}$
JADE	0.25 ± 0.15	$e^+e^-, \mu^+\mu^-, A_{\mu\mu}$
MARK J	0.25 ± 0.11	$e^+e^-, \mu^+\mu^-, \tau^+\tau^-, A_{\mu\mu}$
PLUTO	0.23 ± 0.17	$e^+e^-, \mu^+\mu^-$
TASSO	0.25 ± 0.10	$e^+e^-, \mu^+\mu^-, A_{\mu\mu}$
MAC	0.24 ± 0.16	e^+e^-
MARK II	$0.36^{+0.09}_{-0.21}$	$e^+e^-, \mu^+\mu^-$

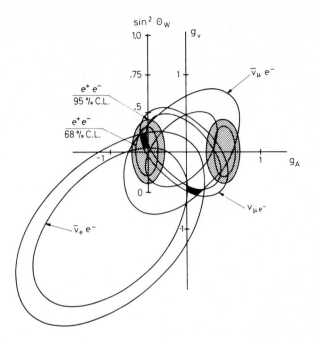

Fig. 3.13. The 95% confidence limits for g_A and g_V as determined by combined PETRA data and compared to the 68% confidence limits extracted from the elastic neutrino–electron scattering data.

particular it has been pointed out that the low energy data can be reproduced in a model with an effective Lagrangian:

$$\hat{L}_{\text{eff}} = L_{\text{eff}} + C\frac{4G}{\sqrt{2}}(J_{\text{EM}})^2 \tag{3.24}$$

In this class of models, the vector coupling is modified to:

$$g_V^2 = \left(1/2 - 2\sin^2\theta_W\right)^2 + 4C, \tag{3.25}$$

and g_A and g_A^2 remain unchanged.

The various groups working at PETRA have derived upper limits [36] on the value of C, using Bhabha scattering and muon pair production data as an input. The results are shown in table 3.6.

There are various ways to realize such models. For example, [56] $SU(2) \times U(1) \times U'(1)$ will have only one charged but two neutral vector

Table 3.6
95% confidence upper limits on C

Group	Limit
CELLO	< 0.032
JADE	< 0.039
MARK J	< 0.027
PLUTO	< 0.06
TASSO	< 0.03

bosons. In this case

$$C = \cos^4\theta_W\left(m_Z^2/m_1^2 - 1\right)\left(1 - m_Z^2/m_2^2\right). \tag{3.26}$$

In this equation m_Z is the mass of the Z^0 in the standard $SU(2)\times U(1)$ model and m_1 and m_2 are the masses of two neutral bosons in the extended model.

It is also possible to construct a model with $SU(2)\times SU'(2)\times U(1)$. Such a model [57] will have two charged and two neutral vector bosons, and in this case the $\cos^4\theta_W$ is replaced by $\sin^4\theta_W$. The limit on C can therefore be translated into limits on m_1 and m_2 using the expressions given above. The results are plotted in fig. 3.14 for $C \leqslant 0.027$ (MARK J).

3.4. The neutral weak current in mixed lepton – quark interactions

3.4.1. Neutral current neutrino – quark interactions

Neutrino–hadron interactions, inclusive, semiinclusive and exclusive, determine [34, 58, 59] the neutral-current couplings to the light quarks u and d, and the first results on the neutral-current couplings to s and c quarks have also been extracted from these data. As will be explained in section 9, the cross section for charged and neutral current neutrino interactions with an *isoscalar* target can be written as:

$$\frac{d\sigma^{CC}}{dy}(\nu N) = \frac{G^2 M_N E_\nu}{\pi}\left[Q + \overline{Q}(1-y)^2\right],$$

$$\frac{d\sigma^{CC}}{dy}(\bar\nu N) = \frac{G^2 M_N E_\nu}{\pi}\left[\overline{Q} + Q(1-y)^2\right],$$

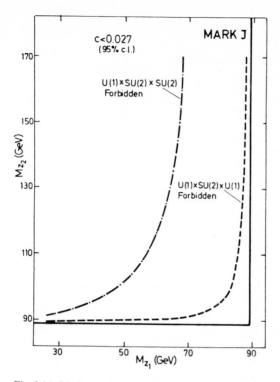

Fig. 3.14. Limits on the mass of neutral vector bosons.

$$\frac{d\sigma^{NC}}{dy}(\nu N) = \frac{G^2 M_N E_\nu}{\pi} \left\{ \left[u_L^2 + d_L^2 \right] \left[Q + \overline{Q}(1-y)^2 \right] \right.$$

$$\left. + \left[u_R^2 + d_R^2 \right] \left[\overline{Q} + Q(1-y)^2 \right] \right\},$$

$$\frac{d\sigma^{NC}}{dy}(\bar{\nu} N) = \frac{G^2 M_N E_\nu}{\pi} \left\{ \left[u_L^2 + d_L^2 \right] \left[\overline{Q} + Q(1-y)^2 \right] \right.$$

$$\left. + \left[u_R^2 + d_R^2 \right] \left[Q + \overline{Q}(1-y)^2 \right] \right\}. \quad (3.27)$$

The contributions from the strange and the charmed sea are neglected, $y = E_{had}/E_\nu$ is the inelasticity. Q and $\overline{Q}$ are the total quark and anti-quark content, respectively, u_L, d_L, u_R and d_R are the chiral coupling constants of the neutral current to u and d quarks, respectively, assuming vector (V) and/or axialvector (A) couplings only. The chiral coupling constants are related to the vector- and axial vector couplings by the simple relations:

$$g_A^u = u_L - u_R, \qquad g_V^u = u_L + u_R,$$
$$g_A^d = d_L - d_R, \qquad g_V^d = d_L + d_R. \tag{3.28}$$

The chiral coupling in the GWS model is related to the weak mixing angle θ_W as follows:

$$u_L^2 = \tfrac{1}{4} - \tfrac{2}{3} \sin^2 \theta_W + \tfrac{4}{9} \sin^4 \theta_W,$$
$$u_R^2 = \tfrac{4}{9} \sin^4 \theta_W,$$
$$d_L^2 = \tfrac{1}{4} - \tfrac{1}{3} \sin^2 \theta_W + \tfrac{1}{9} \sin^4 \theta_W,$$
$$d_R^2 = \tfrac{1}{9} \sin^4 \theta_W. \tag{3.29}$$

It is sometimes convenient to use Sakurai's notation [60]:

$$u_L = \tfrac{1}{4}(\alpha + \beta + \gamma + \delta),$$
$$u_R = \tfrac{1}{4}(\alpha - \beta + \gamma - \delta),$$
$$d_L = \tfrac{1}{4}(-\alpha - \beta + \gamma + \delta),$$
$$d_R = \tfrac{1}{4}(-\alpha + \beta + \gamma - \delta). \tag{3.30}$$

If we assume that the neutral weak current does not change isospin by more than one unit, in analogy to the electromagnetic and the charged weak currents, then α, β, γ and δ have a very simple interpretation. In this case α represents the isovector vector, β the isovector axialvector, γ the isoscalar vector and δ the isoscalar axialvector fraction of the hadronic neutral current.

By adding or subtracting the NC cross sections [eq. (3.27)] and normalizing to the CC cross sections one arrives at:

$$R_+ = \frac{\sigma_{NC}(\nu N) + \sigma_{NC}(\bar{\nu} N)}{\sigma_{CC}(\nu N) + \sigma_{CC}(\bar{\nu} N)} = u_L^2 + d_L^2 + u_R^2 + d_R^2 + \text{corrections},$$

$$\tag{3.31}$$

$$R_- = \frac{\sigma_{NC}(\nu N) - \sigma_{NC}(\bar{\nu} N)}{\sigma_{CC}(\nu N) - \sigma_{CC}(\bar{\nu} N)} = u_L^2 + d_L^2 - u_R^2 - d_R^2 + \text{corrections}.$$

$$\tag{3.32}$$

Equation (3.31) describes total strength of the neutral current coupling, whereas eq. (3.32) measures the VA interference.

If both right handed couplings, u_R and d_R, are zero, the interaction is pure $V-A$. Accordingly, it is pure $V+A$ if the left handed couplings vanish. Pure V or A couplings require $u_L = u_R$ and $d_L = d_R$.

By defining:

$$R_\nu = \frac{\sigma_{NC}(\nu N)}{\sigma_{CC}(\nu N)}, \qquad R_{\bar\nu} = \frac{\sigma_{NC}(\bar\nu N)}{\sigma_{CC}(\bar\nu N)}, \qquad r = \frac{\sigma_{CC}(\bar\nu N)}{\sigma_{CC}(\nu N)},$$

eqs. (3.31) and (3.32) can be combined to:

$$u_L^2 + d_L^2 = \frac{R_\nu - r^2 R_{\bar\nu}}{1 - r^2} + \text{corrections}, \tag{3.33}$$

$$u_R^2 + d_R^2 = \frac{r(R_{\bar\nu} - R_\nu)}{1 - r^2} + \text{corrections}. \tag{3.34}$$

In the context of the GWS model, eq. (3.32) reduces to the well known Paschos–Wolfenstein relation [61]:

$$\frac{R_\nu - R_{\bar\nu}}{1 - r} = \frac{\sigma_{NC}(\nu N) - \sigma_{NC}(\bar\nu N)}{\sigma_{CC}(\nu N) - \sigma_{CC}(\bar\nu N)} = \tfrac{1}{2}\left(1 - 2\sin^2\theta_W\right) + \text{corrections}. \tag{3.35}$$

Recent data from CDHS [62], CHARM [63] and CITFRR [64] are summarized in table 3.7. Similar results have been reported from earlier experiments. Clearly, the neutral current interactions are predominantly $V-A$ but with an admixture of $V+A$, since $u_R = d_R = 0$ is ruled out.

To obtain more information on u_L, u_R, d_L and d_R one has to measure deep inelastic scattering on protons and neutrons or semi-inclusive

Table 3.7
Recent data on hadronic weak neutral current interactions

Quantity	CDHS	CHARM	CITFRR
R_ν	0.307 ± 0.008	0.320 ± 0.010	–
$R_{\bar\nu}$	0.373 ± 0.025	0.377 ± 0.020	–
r	–	0.498 ± 0.019	–
R_+	–	–	0.330 ± 0.017
R_-	–	–	0.254 ± 0.037
$\sin^2\theta_W$	0.228 ± 0.018	0.230 ± 0.023	0.243 ± 0.015
$\|u_L\|^2 + \|d_L\|^2$	0.300 ± 0.015	0.305 ± 0.013	0.292 ± 0.020
$\|u_R\|^2 + \|d_R\|^2$	0.024 ± 0.008	0.036 ± 0.013	0.038 ± 0.020

reactions. Ignoring the sea, the NC/CC cross section ratios on protons and neutrons can be written as

$$R_p = \sigma^{NC}(\nu p)/\sigma^{CC}(\nu p) = 2u_L^2 + d_L^2 + \text{corrections}, \qquad (3.36)$$

$$R_n = \sigma^{NC}(\bar{\nu}N)/\sigma^{CC}(\nu N) = 2d_L^2 + u_L^2 + \text{corrections}. \qquad (3.37)$$

This determines the dominant left handed couplings up to a sign. The IMSTT (Illinois–Maryland–Stony Brook–Tohoku–Tufts) collaboration obtains the following values [65] from a measurement of neutrino-deuteron interactions in a bubble chamber:

$$R_p = 0.47 \pm 0.05, \qquad R_n = 0.22 \pm 0.02, \qquad R_\nu = 0.30 \pm 0.02,$$

and derives

$$u_L^2 = 0.19 \pm 0.07, \qquad d_L^2 = 0.11 \pm 0.03,$$

in reasonable agreement with previous results [66–70] and with the preliminary result of the BEBC–TST collaboration [71]:

$$R_p = 0.530 \pm 0.045 \pm 0.030.$$

The right handed couplings can be obtained from semi-inclusive π^+, π^- production on isoscalar targets (the left handed couplings too but the above described method gives more precise results).

The charged current interaction $\nu_\mu + d \rightarrow \mu^- + u$ on a valence d quark produces a u quark which subsequently fragments into charged pions with the probability D_u^+ and D_u^-. The number of observed pions coming from this primary process (i.e. with a high fractional energy $z = E_\pi / E_{had}$) is therefore proportional to D_u. The ratios of energetic positively and negatively charged pions resulting from ν–d or $\bar{\nu}$–u interactions are related by

$$\frac{N_{\pi^+}}{N_{\pi^-}}\bigg|_{\nu \rightarrow \mu} = \frac{N_{\pi^-}}{N_{\pi^+}}\bigg|_{\bar{\nu} \rightarrow \mu} = \frac{D_u^{\pi^+}}{D_u^{\pi^-}} = \frac{D_d^{\pi^-}}{D_d^{\pi^+}}. \qquad (3.38)$$

For the neutral current reactions [72–74], both quark helicities are involved, the right handed being only $1/3$ as effective since $\int_0^1 (1 - y)^2 dy = 1/3$. In an isoscalar target the π^+/π^- ratios are:

$$\frac{N_{\pi^+}}{N_{\pi^-}}\bigg|_{\nu \rightarrow \nu} = \frac{\left[u_L^2 + \frac{1}{3}u_R^2\right]D_u^{\pi^+} + \left[d_L^2 + \frac{1}{3}d_R^2\right]D_u^{\pi^-}}{\left[u_L^2 + \frac{1}{3}u_R^2\right]D_u^{\pi^-} + \left[d_L^2 + \frac{1}{3}d_R^2\right]D_u^{\pi^+}},$$

$$\frac{N_{\pi^+}}{N_{\pi^-}}\bigg|_{\bar{\nu} \rightarrow \bar{\nu}} = \frac{\left[u_R^2 + \frac{1}{3}u_L^2\right]D_u^{\pi^+} + \left[d_R^2 + \frac{1}{3}d_L^2\right]D_u^{\pi^-}}{\left[u_R^2 + \frac{1}{3}u_L^2\right]D_u^{\pi^-} + \left[d_R^2 + \frac{1}{3}d_L^2\right]D_u^{\pi^+}}. \qquad (3.39)$$

Experimentally [75] these ratios are

$$\frac{N_{\pi^+}}{N_{\pi^-}}\bigg|_{\nu \to \nu} = 0.77 \pm 0.14, \qquad \frac{N_{\pi^+}}{N_{\pi^-}}\bigg|_{\bar{\nu} \to \bar{\nu}} = 1.64 \pm 0.37,$$

yielding [74]

$$u_R^2 = 0.03 \pm 0.015, \qquad u_R^2 = 0.00 \pm 0.015.$$

So far, the experiments allow a unique determination of the squares of the coupling constants, in particular d_R^2 is very small. Ignoring d_R^2 and defining u_L to be positive, the signs of the two products $u_L \times d_L$ and $u_L \times u_R$ have still to be determined. These products can be written in terms of the model independent notation [58].

$$u_L d_L = \tfrac{1}{8}\left[-(\alpha + \beta)^2 + (\gamma + \delta)^2 \right],$$

$$u_L u_R = \tfrac{1}{8}\left[(\alpha + \gamma)^2 - (\beta + \delta)^2 \right]. \tag{3.40}$$

$u_L d_L < 0$ would be indicated by a strong isospin changing component in the hadronic neutral current; α and β are the isovector ($|\Delta I| = 1$) parts, likewise $u_L u_R < 0$ means a dominantly axial vector current (β and δ dominant).

The GGM collaboration [76] observed a strong enhancement at the Δ in the reaction

$$\nu p \to \nu + p + \pi^0,$$

which indicates a large isovector ($|\Delta I| = 1$) component in the neutral current, i.e., $(\alpha + \beta)$ is dominant and $u_L d_L < 0$.

The neutrino disintegration of the deuteron [77],

$$\bar{\nu}_e + D \to \bar{\nu}_e + n + p,$$

is a pure isovector–axialvector (Gamov–Teller) transition near threshold. The $^3S_1(I = 0)$ D state goes into the $^1S_0(I = 1)$ n–p state. This means that the transition is sensitive only to β.

Experimentally [77] one finds $|\beta| = 0.9 \pm 0.1$. Using the above data, the sign ambiguity of $u_L u_R$ can be transformed into two predictions for $|\beta|$:

$$|\beta|_{\text{predicted}} = \begin{cases} 0.92 \pm 0.14, & |\beta| \text{ large} = \text{axial vector dominant,} \\ 0.58 \pm 0.14, & |\beta| \text{ small} = \text{vector dominant.} \end{cases}$$

Clearly the solution $u_L u_R < 0$ is preferred.

Kim et al. [34] have made a fit to all data available in 1980, arriving at the following values for the chiral coupling constants:

$$u_L = 0.340 \pm 0.033 \quad (0.347),$$

$$d_L = -0.424 \pm 0.026 \quad (-0.423),$$

$$u_R = -0.179 \pm 0.019 \quad (-0.153),$$

$$d_R = -0.017 \pm 0.058 \quad (0.077).$$

The numbers in brackets are the predictions of the GWS model for $\sin^2 \theta_W = 0.23$.

Within the standard model the neutral current coupling to the quarks depends only on the charge and is independent [78] of the generation. Two experiments have reported measurements of the couplings to quarks belonging to higher generations.

With the assumption that the weak currents contain only V and A parts, the differential cross section for deep inelastic neutrino and antineutrino scattering on an isoscalar target can be written [79] as:

$$(d\sigma/dy)_{\nu \to \mu^-} = B\left[(1-\alpha^\nu)+\alpha^\nu(1-y)^2\right],$$

$$(d\sigma/dy)_{\bar{\nu} \to \mu^+} = B\left[(\alpha^{\bar{\nu}}+(1-\alpha^{\bar{\nu}})(1-y)^2\right],$$

$$(d\sigma/dy)_{\overset{(-)}{\nu} \to \overset{(-)}{\nu}} = B\Big\{\left(u^2_{L(R)}+d^2_{L(R)}\right)\left[(1-\alpha^{\bar{\nu}})+\alpha^\nu(1-y)^2\right]$$

$$+ \left(u^2_{R(L)}+d^2_{R(L)}\right)\left[\alpha^\nu+(1-\alpha^{\bar{\nu}})(1-y)^2\right]$$

$$+ \left(s^2_R + s^2_L\right)(\alpha^{\bar{\nu}}-\alpha^\nu)\left[1+(1-y)^2\right]\Big\}. \quad (3.41)$$

By fitting these expressions to their measured y distributions for neutrino and antineutrino induced charged and neutral current events, the CHARM [80] collaboration obtains for the coupling of the strange quark

$$s^2_L + s^2_R = (1.39 \pm 0.43)\left(d^2_L + d^2_R\right),$$

in agreement with the expected universality.

The CDHS collaboration [81] recently reported a value for the corresponding coupling of the charmed quark. The dimuon mass spectrum of their neutrino-induced dimuons shows a peak at the J/ψ mass, which is interpreted as the $c\bar{c}$ bound state produced via Z^0 and gluon fusion. Comparing the observed cross section for this channel, $(4.2 \pm 1.5) \times 10^{-41}$ cm^2/nucleon, with that of J/ψ production by an incident muon, as measured by the European Muon Collaboration (EMC) and applying the gluon fusion model to both the photon–gluon and the Z^0–gluon vertex, a connection between the $u\bar{u}Z^0$ and the $c\bar{c}Z^0$ coupling can be derived. In this model dependent way the authors conclude

$$c_L^2 + c_R^2 = (1.6 \pm 0.5)(u_L^2 + u_R^2),$$

consistent with universality.

In summary, the neutral current coupling constants to the up, down, strange and charm quarks have been (at least partially) determined. The standard model with $\sin^2 \theta_W \approx 0.23$ agrees extraordinarily well with all the available data.

In general, in SU(2)×U(1) models the relative coupling strength $\rho = m_W^2 / m_Z^2 \cos^2 \theta_W$ [eq. (3.3)] is related to the Higgs fields and is equal to one if all Higgs bosons are in doublets.

In a general model the constants u_R, d_R, u_L, d_L etc. have to be replaced by [34]:

$$u_R \rightarrow \rho u_R + \rho T_{3R}(u), \qquad d_R \rightarrow \rho d_R + \rho T_{3R}(d),$$

$$u_L \rightarrow \rho u_L, \qquad d_L \rightarrow \rho d_L.$$

where $T_{3R}(i)$ is the third component of the weak isospin for the right handed component of the fermion i.

Kim et al. [34] have shown that the data are consistent with all right handed fermions being in singlets [$T_{3R}(i) = 0$]. They find, in a general fit to all available data:

$$\rho = 1.018 \pm 0.045, \qquad \sin^2 \theta_W = 0.249 \pm 0.031,$$

$$T_{3R}(u) = -0.010 \pm 0.040, \qquad T_{3R}(d) = -0.101 \pm 0.058,$$

$$T_{3R}(e) = 0.039 \pm 0.047.$$

Assuming that the third components of the weak isospin are all zero, they find, fitting polarized electron–deuteron and deep inelastic scattering data:

$$\rho = 0.992 \pm 0.017(\pm 0.013), \qquad \sin^2 \theta_W = 0.224 \pm 0.015(\pm 0.015),$$

where the numbers in parentheses are the theoretical uncertainties. With the same assumptions the CHARM collaboration [63] obtained, from the total cross section for neutral current and charged current (anti)neutrino–nucleon reactions, for the left and right handed coupling con-

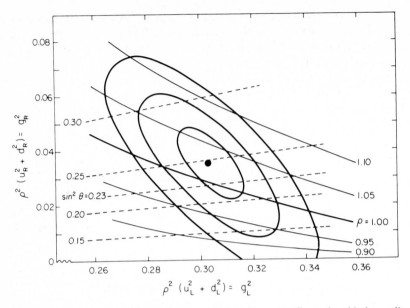

Fig. 3.15. Best fit and confidence limits of 39%, 87%, and 99% on the chiral coupling constants as determined by the CHARM experiment. The solid curves are lines of constant ρ, and the dashed curves are lines of constant $\sin^2 \theta_{\rm W}$.

stants:

$$g_L^2 = \rho^2 \left(u_L^2 + d_L^2 \right) = 0.305 \pm 0.013,$$
$$g_R^2 = \rho^2 \left(u_R^2 + d_R^2 \right) = 0.036 \pm 0.013.$$

Without constraining ρ, a nonzero value for the right handed coupling was measured (more than 90% C.L.).

The result can also be expressed in terms of the parameters ρ and $\sin^2 \theta_{\rm W}$:

$$\rho = 1.027 \pm 0.023, \qquad \sin^2 \theta_{\rm W} = 0.247 + 0.038.$$

Figure 3.15 shows this result with the 1, 2 and 3σ contours. The measurements favour the assumption that right handed fermions are grouped in singlets, and Higgs bosons are in doublets, as in the standard model.

3.4.2. Electroweak interference effects in neutral current quark interactions
Electroweak interference effects have been studied over a wide range in

q^2, from parity violations in atomic levels, via deep inelastic electron–deuterium scattering to electron–positron annihilation into hadrons. The available data are reviewed briefly below in order of increasing values of q^2.

The neutral weak current interaction between the electrons in the atom and the quarks in the nuclei induces [82] a small opposite-parity part in the wavefunctions of the atomic levels. This parity violation is reflected in the radiative transitions; for example a dominant M1 transition will have a small E1 admixture. This parity violation is expected to be of the order of 10^{-4} (q^2/m_p^2). In atomic transitions $q^2 = (1/R_{atom})^2 = 10^{-5}$ MeV^2, resulting in a signal on the order of 10^{-15}. It was realized [83] that the effect scales as Z^3 resulting in a large gain for transitions in heavy nuclei. However, in order to extract a quantitative value, the electron density at the nuclei must be computed, and this introduces rather large uncertainties for heavy nuclei.

The parity violation in atomic transitions results from a linear combination of $V_q A_e$ and $A_q V_e$. V and A are the axial and vector coupling constants of quarks and electrons as denoted by the subscript. The dominant contribution is from $V_q A_e$ since the nucleons contribute coherently. The nucleon spin contributes to A_q, but the contribution from individual nucleons tends to cancel. The matrix element of the V_q is in general expressed through the weak charge

$$Q_W = -4V_u A_e (2Z+N) + V_d A_e (Z+2eN).$$

The experimental results are quoted using different quantities as the weak charge Q_W, R, the ratio of the matrix elements for $E1$ to $M1$, or ϕ, the amount of rotation due to parity violation. The results are listed in table 3.8.

Table 3.8
Results on parity violations in atomic physics

Element	λ	Quantity measured	Result	Prediction	Groups	Ref.
Bi	6476	R	$(-20.2 \pm 2.7) \times 10^{-8}$	$-(11-17) \times 10^{-8}$	Novosibirsk	[84]
Bi	6476	R	$(2.7 \pm 4.7) \times 10^{-8}$	$-(11-17) \times 10^{-8}$	Oxford	[85]
			$(-10.7 \pm 1.5) \times 10^{-8}$			[86]
Bi	6476	$\Delta\phi_{PNC}$	$(-0.22 \pm 1.0) \times 10^{-8}$	10^{-7}	Moscow	[87]
Bi	8757	R	$(-0.7 \pm 2.1) \times 10^{-8}$	$-(8-13) \times 10^{-8}$	Washington	[88]
			$(-10.4 \pm 1.7) \times 10^{-8}$			[89]
Bi	2927	Q_W	-155 ± 63	-116.5	Berkeley	[90]

The range in predicted values for the bismuth data reflects the spread in the published calculations [91]. Thallium has a simpler electronic configuration and the estimates might therefore be more reliable. The data are now in reasonable agreement with the standard model predictions. Note, however, that there are still discrepancies by a factor of two between different groups. These are very difficult experiments and a complete discussion can be found in reference [92].

The observation of a parity violation in deep inelastic electron (muon)–hadron interactions is a unique signature for the interference between the electromagnetic and the neutral weak current (fig. 3.1). In the SLAC experiment [93], parity violation was observed by measuring the cross section for inelastic electron–deuterium scattering using right and left handed electrons.

In the experiment, a 40% polarized electron beam with energies between 16 GeV and 22 GeV was passed through a deuterium target and the scattered electrons were observed at 4°, using a single-arm spectrometer. The quantity $A = [d\sigma_+ - d\sigma_-]/[d\sigma_+ + d\sigma_-]$, where $d\sigma_+$ $(d\sigma_-)$ stands for the double differential cross section $d^2\sigma/d\Omega dE'$ for right (left)

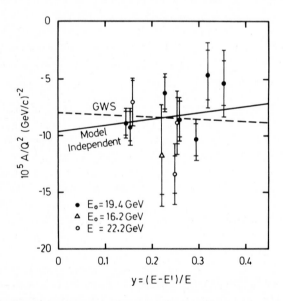

Fig. 3.16. The asymmetry A, observed in polarized electron–deuteron scattering, plotted versus y. The dashed line shows the best fit to the data in the GWS model, resulting in $\sin^2\theta_{WS} = 0.224 \pm 0.02$. The solid line is a model-independent fit.

handed electrons, is then a measure of the parity violation. The experimental data are shown in fig. 3.16.

The quantity A has been evaluated by Cahn and Gilman [94] in the parton model, neglecting the sea contribution. They find:

$$A/q^2 = -\frac{6G_F}{5\pi\alpha\sqrt{2}}\left[(V_uA_e - \tfrac{1}{2}V_dA_e) + F(y)(A_uV_e - \tfrac{1}{2}A_dV_e)\right].$$

(3.42)

V and A are the vector and axial vector coupling constants for u and d quarks and for electrons as indicated by the subscript. $F(y)$ is defined by

$$F(y) = \frac{1 - (1 - y)^2}{1 + (1 - y)^2}, \quad \text{with } y = v/v_{max}.$$

A fit to the data [93] yields:

$$A/q^2 = [(-9.7 \pm 2.6) + (4.9 \pm 8.1)F(y)] \times 10^{-5}.$$

Expressed by the coupling constants this results in:

$$V_uA_e - \tfrac{1}{2}V_dA_e = -0.23 \pm 0.06,$$

$$A_uV_e - \tfrac{1}{2}A_dV_e = 0.11 \pm 0.19.$$

In the standard model V and A are expressed by a single parameter $\sin^2\theta_W$ as listed in table 3.1. A single parameter fit to the data yields

$$\sin^2\theta_W = 0.224 \pm 0.12 \pm 0.008,$$

in good agreement with the values found in neutrino-induced reactions.

Including the weak neutral current [50] in $e^+e^- \to q\bar{q} \to$ hadrons will lead to a forward–backward asymmetry and to a change in the total cross section. The size of the effects can be estimated within the GWS model using eqs. (3.22) and (3.23) suitably modified with the coupling constants listed in table 3.1. At PETRA/PEP energies, the expected asymmetry in $e^+e^- \to s\bar{s}$ ($d\bar{d}, b\bar{b}$) would be on the order of 25%. This is a sizable effect, however, it is at present not possible to identify the quark flavour from the hadrons in a jet. Therefore only the total cross section has been used to test the theory.

As an example, the normalized total cross section for hadron production as measured [95] by MARK J is plotted in fig. 3.17 versus the c.m. energy. The data are compared to the theoretical expectations for various values of $\sin^2\theta_W$. The prediction also includes the first order QCD correction $(1 + \alpha_s/\pi)$. A similar analysis has also been done by the

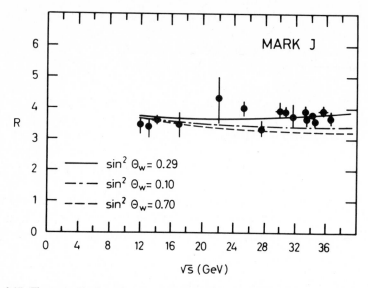

Fig. 3.17. The normalized total cross section for $e^+e^- \to$ hadrons plotted versus the c.m. energy. The data obtained by MARK J are compared to the expected weak effects for various values of $\sin^2 \theta_W$.

JADE collaboration [96]. Including the leptonic data the groups find:

$$\text{MARK J:} \quad \sin^2 \theta_W = 0.27^{+0.06}_{-0.04},$$

$$\text{JADE:} \quad \sin^2 \theta_W = 0.22 \pm 0.08.$$

The error quoted by the JADE group includes a statistical error of 3%, an error of 3% resulting from the assumed uncertainty of 20% in the value of α_s, and a 7% systematic uncertainty including the normalization error.

3.5. Limits on flavour changing neutral currents

The $K^0\overline{K}^0$ mass mixing puts stringent limits on the amount of neutral current induced $s \leftrightarrow d$ transitions. In the standard model with the GIM mechanism, the neutral current is diagonal [97] in all flavours, forbidding decays of the type shown in fig. 3.18.

The MARK II collaboration at SPEAR obtained limits on the amount of $c \leftrightarrow u$ transitions from a measurement of $D^0-\overline{D}^0$ mixing. From a

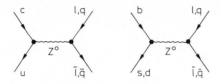

Fig. 3.18. Possible neutral current induced flavour changes.

sample of initial $D^0\overline{D}^0$ events Goldhaber et al. [98] find

$$\frac{N(K^+K^+)-N(K^-K^-)}{N(K^+K^-)} < 18\%, \quad (90\% \text{ upper C.L.}).$$

The same groups [99] also studied the process $e^+e^- \to D^{*+}D^{*-} \to (\pi^+D^0)(\pi^-\overline{D}^0)$. The amount of D^0–$\overline{D}^0$ mixing can now be determined from the number of wrong charge kaons. They did not observe a signal and from the data they set the limit

$$\frac{N(\pi^-K^-\pi^+)+N(\pi^+K^+\pi^-)}{N(\pi^-K^+\pi^-)+N(\pi^+K^-\pi^+)} < 16\%.$$

Combining the two experiments yields a theoretical upper limit on the coupling strength of $10^{-7}\,G_F$ for the strength of the flavour changing neutral current:

$$F_{FC}/\sqrt{2} \lesssim 10^{-7}G_F/\sqrt{2}\,.$$

Recently the CLEO group [100] at CESR has set a limit on the transition $b \leftrightarrow d$ (s). As a signature they used $e^+e^- \to B\overline{B} \to \ell^+\ell^- +$ hadron. Such final states will of course also arise from pair production of charm or beauty followed by two semileptonic decays $q \to \ell\bar{\nu}_e +$ hadrons. The number of observed and expected dilepton events are listed in table 3.9. The number of observed events is clearly consistent with the predicted background, and they quote as an upper 90% C.L.:

$$B(b \to \ell^+\ell^- + X) \leqslant 7.4 \times 10^{-3}.$$

Table 3.9
Number of mixed dilepton + hadron events

Final state	Observed	Expected background
ee + hadrons	5	3.04
eμ + hadrons	5	6.04
μμ + hadrons	0	3.04

3.6. The structure of the charged weak current

The standard $V - A$ picture of charged weak currents [31] has been tested extensively [101] both in purely leptonic reactions and in reactions involving hadrons. Of the leptonic reactions, muon decay: $\mu^- \rightarrow e^- \bar{\nu}_e \nu_\mu$ has given the most precise information [102]. Both the shape of the spectrum and polarization of the electron are in good agreement with the $V - A$ prediction. A right handed $W^\pm$ must have a mass greater than 250 GeV if the corresponding neutrino is massless. The data on leptonic tau decay are also in agreement with a $V - A$ structure.

The charged weak current involving hadrons is defined in the standard picture by eq. (3.11). The matrix [33] U describing the mass mixing is defined by eq. (3.12). For a four quark system this reduces to the GIM model [78] with one parameter, the Cabibbo angle. A recent analysis [103] of Ke_3 data and baryonic decay data gave $\sin \theta_C = 0.219 \pm 0.011$. A large sample of hyperon semileptonic decays [104] gave $\sin \theta_C = 0.228 \pm 0.012$. Data on the couplings of the charm quark to lighter quarks are consistent with this value as discussed in sections 5 and 9. Data [100] from Cornell (section 5) show that the b quark preferentially mixes with the s quark, i.e., the decay $b \rightarrow W^- c$ dominates $b \rightarrow W^- u$. Sakurai [101] has recently discussed in detail how the various elements U_{ik} are determined from the available data, and finds:

$$U = \begin{pmatrix} 0.974 & 0.219 & 0.059 \\ -0.213 & 0.845 & -0.488 \\ -0.057 & 0.489 & -0.870 \end{pmatrix}.$$

Also the data (section 4) on hadronic tau decays, although of limited statistics, support the standard picture.

Most of the tests have been carried out at low values of Q^2. Recently two high energy neutrino experiments have shown that the $V - A$ structure is also dominant at high values of Q^2. We will now discuss these experiments in more detail.

3.6.1. Muon polarization

The differential cross section [105] $d\sigma/dy$ of the reaction $\bar{\nu}_\mu N \rightarrow \mu^+ X$, where $y = E_X/E_\nu$, is proportional to

$$d\sigma/dy \sim 2(g_V - g_A)^2 + 2(g_V + g_A)^2 (1 - y)^2 + \left(|g_S|^2 + |g_P|^2 \right) y^2$$
$$+ 32 |g_T|^2 (1 - y/2)^2 + 8R_e \left[g_T (g_S^* + g_P^*) \right] y (1 - y/2).$$
$$(3.43)$$

The g_i (i = V, A, S, P, T) denote the various coupling constants. An increase of $d\sigma/dy$ for $y \to 1$ is clear evidence for a S or/and P term. However, a y distribution decreasing with y, although expected, is *not* a proof [106] for a pure V and A coupling. Such a behaviour may also result from a mixture of S, P and T interactions.

In a recent experiment [107], performed by combining the CDHS [23] and CHARM [24] detectors in the CERN–SPS neutrino beam, the helicity of the μ^+ in the final state was measured. The incident antineutrinos are produced in π or K decay and have positive helicity, while S, P or T interactions result in a negative helicity of the μ^+. The CDHS detector was used as an active target, while the CHARM detector was used as a polarimeter for the fraction (5%) of produced μ^+, which stopped in the CHARM detector and decayed via $\mu^+ \to e^+ \bar{\nu}_\mu \nu_e$. Due to the V–A structure of this decay, the high energy positrons are emitted preferentially in the direction of the muon spin. A uniform magnetic field of 5.8×10^{-3} T caused the muon spin to process.

The positron yield detected in the scintillator plane forward (in μ flight direction) or backward relative to the stopping plane is thus periodically modulated with the Larmor frequency ω.

$$N_{B,F}(t) = N_0 e^{-t/\tau}\{1 + R_0 \cos(\omega t + \phi)\}. \tag{3.44}$$

The asymmetry

$$R(t) = \frac{N_B - N_F}{N_B + N_F} = R_0 \cos(\omega t + \phi), \tag{3.45}$$

is a periodic function. The phase ϕ describes the sign of the polarization ($\phi = 0$ for negative polarization or S, P or T contribution). R_0 is proportional to the magnitude of the polarization. The data (fig. 3.19) show a sinusoidal oscillation pattern ($\phi = -3.1 \pm 0.2$) with no evidence for spin flip contributions. By evaluating the analyzing power with a Monte Carlo calculation, the polarization was determined to be $P = +(1.09 \pm 0.22)$. The results put an upper limit of

$$\sigma_{SPT}/\sigma_{tot} < 0.18, \quad 95\% \text{ C.L.},$$

on S, P, or T contributions.

3.6.2. Inverse muon decay

The inverse muon decay $\nu_\mu e^- \to \mu^- \nu_e$ requires a threshold energy of $E_\nu \approx 11$ GeV. It had therefore not been observed until recently by the Gargamelle [108] and CHARM [109] experiments in the CERN–SPS neutrino beam. Motivated by the results of the polarization experiment,

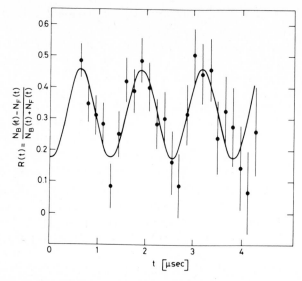

Fig. 3.19. Oscillation pattern of the forward–backward asymmetry of the muon decay.

one can assume the absence of S, P and T contributions in this reaction. Then the differential cross section can be expressed as

$$d\sigma/dy \sim (1+P)(1-\lambda)y^2 + (1-P)(1-\lambda), \tag{3.46}$$

where $\lambda = -2R_e(g_V^* g_A)/(|g_V|^2 + |g_A|^2)$ is the relative V − A contribution and

$$P = \frac{N(\nu_R) - N(\nu_L)}{N(\nu_R) + N(\nu_L)} \tag{3.47}$$

is the polarization of the incident neutrino beam.

In events of the type $\nu_\mu e^- \rightarrow \mu^- \nu_e$, the muon is produced in a very narrow cone, $\theta_\mu = [2m_e(1-y)/E_\mu]^{1/2}$ of less than 10 mrad. No recoil is visible at the vertex. These criteria have been used to discriminate against the four orders of magnitude more abundant events of the type $\nu_\mu N \rightarrow \mu^- X$. Additionally, the remaining background was experimentally determined using the data from an antineutrino exposure, where the inverse muon decay cannot occur. Fig. 3.20 shows the observed Q^2 distribution of the 171 ± 29 events remaining after subtraction of the background (CHARM). The dashed curve shows the shape of the Q^2 distributions for the 175 ± 5 events predicted by the V − A theory. A V + A interaction would result in 65 ± 2 events. Figure 3.21 shows the results of both the

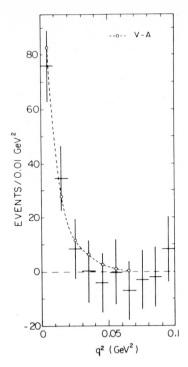

Fig. 3.20. Observed candidates for the inverse μ decay as a function of Q^2 (CHARM).

CHARM and GGM experiments in terms of upper limits (90% confidence) in the $P - \lambda$ plane.

Both experiments are consistent with a $V - A$ structure of the charged weak current.

3.7. Summary

The standard picture of the weak interaction outlined in section 3.1 has met with great success.

A wide range of neutral current phenomena ranging from atomic physics to purely leptonic reactions and mixed lepton–hadron interactions are described by the standard theory, using only one free parameter $\sin^2 \theta_W$ with [34] $\sin^2 \theta_W = 0.233 \pm 0.009 \pm 0.005$. The best value [34] of the parameter ρ is $1.002 \pm 0.015 \pm 0.011$, in agreement with the value of 1, predicted if the symmetry breaking is caused by scalar Higgs fields with the Higgs particle in an isospin doublet. The standard model has only one Higgs doublet, but the predictions are independent on the number of

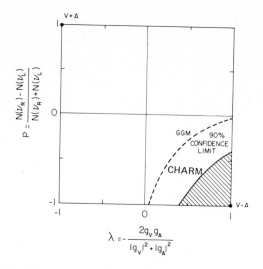

Fig. 3.21. The 90% confidence areas for the relative $V-A$ contribution and the polarization of the incident neutrino beam. Results from GGM and CHARM.

such doublets. The value of $\sin^2 \theta_W$ is determined in models which attempt to unify [110, 111] the electroweak and the strong interaction. The simplest of such models, SU(5) [111] predicts [112] $\sin^2 \theta_W = 0.2109$ for purely leptonic reactions at $q^2 = 0$, consistent with the experimental value.

All the experimental data on charged currents are consistent with a $V-A$ structure of the interaction and mixing angles as given by the GIM mechanism [78] for a four quark system or by the Kobayashi–Maskawa [33] matrix for a six quark system.

The great success of the standard picture however, should not lead to the conclusion that weak interactions are solved. Anybody can write his own list of unsolved questions, some of which will hopefully be answered by the next generation of high energy accelerators.

4. The tau lepton

The first evidence [113] for the tau lepton, presented by Perl and collaborators, was based on the observation of events of the type $e^+ e^- \rightarrow$

$e^{\mp}\mu^{\pm}$ + nothing. By now the tau is firmly established and a wealth of information [114] on its properties has become available.

4.1. The mass

The mass of a heavy lepton can be determined with rather high precision from a measurement of the sharply rising production cross section near threshold. Such measurement, first made by the DASP collaboration [115], has been refined and the DELCO collaboration [116] finds $m_{\tau} = 1782^{+3}_{-4}$ MeV/c^2. The mass of the tau neutrino is rather poorly known. From a measurement of the electron spectrum in $\tau \to \nu_{\tau}\bar{\nu}_e e$ the DELCO collaboration [116] establishes a 95% upper confidence limit of the $m_{\tau} < 250$ MeV/c^2.

4.2. The space – time structure of the current

The DELCO collaboration has determined [117] the Michel parameter $\rho = 0.72 \pm 0.10$ from a measurement of the electron momentum spectrum in $\tau \to \nu_{\tau}\bar{\nu}e$. A V − A interaction yields $\rho = 0.75$ in agreement with the data, whereas $\rho = 0$, corresponding to a V + A current, is excluded. Pure V or A, yielding $\rho = 0.375$, are strongly disfavoured.

4.3. Decay modes

A selected set of decay modes is listed in table 4.1, together with the theoretical predictions [118]. These decay modes can be predicted with little ambiguity if the τ is a sequential lepton which couples to the weak current with the normal strength G_F.

Table 4.1
Selected tau decay modes

Decay mode	Theoretical B (%)	Experimental B (%)
$\nu_{\tau} + e^- \bar{\nu}_e$	17.6	17.5 ± 1.2
$\nu_{\tau} + \mu^- + \bar{\nu}_{\mu}$	17.2	17.1 ± 1.2
$\nu_{\tau}\pi$	10.5	$11.7 \pm 0.4 \pm 1.8$
$\nu_{\tau}K$	0.66	$1.2 \pm 0.4 \pm 0.2$
$\nu_{\tau}\rho$	21.5	$21.6 \pm 1.8 \pm 3.2$
$\nu K^*(890)$	1.46	1.7 ± 0.7
νA_1	8.7	10.4 ± 0.03

The leptonic decay widths,

$$\Gamma_e(\tau \to \nu_\tau \bar{\nu}_e e) = 1.028 \Gamma_\mu(\tau \to \nu_\tau \bar{\nu}_\mu \mu) = G_F^2 m_\tau^5/192\pi^3, \qquad (4.1)$$

can be computed unambiguously. The decay $\tau \to \nu_\tau \pi$ tests the axial vector current and is directly related to $\pi \to \bar{\nu}_\mu \mu$. The decay $\tau \to \nu_\tau \rho$ tests the vector current and is related to $e^+e^- \to \rho^0$ by CVC. Recent data on $\tau \to \nu_\tau \pi^\pm$ [119] and on $\tau \to \nu_\tau \rho$ [120] are listed in table 4.1. They are in good agreement with the predictions.

The expected suppression of strangeness changing decays can be checked by a measurement of the ratio $B(\tau \to \nu_\tau K^*(890))/B(\tau \to \nu_\tau \rho)$. The standard picture of charged weak currents predicts that this ratio is given by $0.93 \tan^2 \theta_C$, where 0.93 results from a small phase space correction, and θ_C is the Cabibbo angle.

Experimentally, the MARK II group determines [121] this ratio to 0.085 ± 0.038, in agreement with the theoretical value of 0.05.

The decay $\tau \to \nu_\tau K$ is the Cabibbo suppressed partner to the decay $\tau \to \nu_\tau \pi$, and the relative rates are given by:

$$\frac{\Gamma(\tau \to \nu_\tau K)}{\Gamma(\tau \to \nu_\tau \pi)} = \tan^2 \theta_C \left(\frac{1 - (m_K/m_\tau)^2}{1 - (m_\pi/m_\tau)^2} \right). \qquad (4.2)$$

The MARK II group recently published [122] data on $\tau \to \nu_\tau K$. The process is identified by selecting events of the type $e^+e^- \to \tau^+\tau^- \to (\ell \bar{\nu}_\ell \nu_\tau)(\bar{\nu}_\tau K^+)$. The leptons are positively identified and the kaons separated from the pions by time of flight. From a sample of 47 000 $\tau^+\tau^-$ pairs, 15 events satisfied the selection criteria with a pion contamination of 3 ± 0.6 events and a background of 2.1 ± 0.9 events from $\tau \to \nu K^*(890)$. They find $B(\tau \to \nu_\tau K) = (1.2 \pm 0.4 \pm 0.2)\%$ in reasonable agreement with the predicted value of 0.0066. The first error is statistical and the second systematic. Tau decays involving strange particles are thus suppressed relative to non strange τ decay by the usual Cabibbo angle.

4.4. The lifetime

In the standard model, the tau life time is given by:

$$\tau_\tau = B_e(m_\mu/m_\tau)^5, \qquad \tau_\mu = (2.8 \pm 0.2) \times 10^{-13} \text{ s}. \qquad (4.3)$$

This lifetime is quite large and, on the average, 15 GeV tau's travel 0.7 mm before decaying. Hence the lifetime can be determined from a measurement of the vertex distribution of tau decays.

TASSO has placed an upper limit on the tau lifetime, $\tau_\tau < 5.7 \times 10^{-13}$ s, from their measurements of the vertex distribution. The MARK II collaboration now reports [123] a measurement of the tau lifetime. They measure the vertex distribution by selecting topologies consisting of three charged particles recoiling against a single charged particle or three charged particles. This topology separates tau pair production from hadron production, and the intercept of the momentum vectors of the charged particles defines the vertex. The observed decay length distribution is plotted in fig. 4.1 for events with vertex uncertainties of 8 mm and 4 mm. The beam center is indicated by the dashed line. A careful analysis shows that the center of the measured distribution is shifted by (1.07 ± 0.37) mm with respect to the beam center. Various checks show that this shift is real and not due to systematic uncertainties. The resulting value of the lifetime, $\tau_\tau = (4.9 \pm 1.8) \times 10^{-13}$ s is in agreement with the theory. Thus the weak coupling strength of the tau is consistent with G_F, the weak coupling strength observed for the electron and the muon.

So far, the tau was treated as a new sequential lepton arranged in a weak doublet with its own neutrino like the electron and the muon. Indeed all the data [114] are consistent with this assumption. If the τ^- has the same lepton number as the e^+ or the μ^+, then the branching ratios for leptonic decays of the tau into electrons or muons would differ by a factor of two. This is excluded experimentally. It has been found that the $\tau - \nu_\mu$ coupling strength is less than 0.025 of the $\mu - \nu_\mu$ coupling

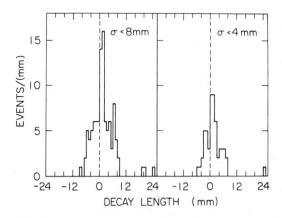

Fig. 4.1. The distribution of τ flight distances at $\sqrt{s} = 29$ GeV for events with vertex uncertainties less than 8 mm (left) and less than 4 mm (right).

strength, excluding that the τ^- has the lepton number of the μ^-. It remains to be shown that the τ^- has not the e^- lepton number. Furthermore the tau neutrino has not been observed directly but inferred from the observed threebody decay $\tau \to \nu_\tau \ell \nu_e$.

5. Production of new flavours in e^+e^- annihilation

A new quark flavour q has very striking signatures in e^+e^- annihilation. Below threshold the production of bound $q\bar{q}$ 1^{--} states shows up in the total cross section as narrow peaks. Above threshold the bound $q\bar{q}$ states become wider rapidly and the normalized total cross section increases by $\Delta R = 3\Sigma e_q^2$. In this chapter we will review the data on charm and beauty production in e^+e^- annihilation. The data will be compared to the QCD predictions in section 7.

A $q\bar{q}$ system of heavy quarks bound in a steeply rising potential, will lead to the level scheme shown in fig. 5.1. The levels [124] are labeled by J^{PC} with $P = (-1)^{L+1}$ and $C = (-1)^{L+S}$. For each value of angular momentum L there are two bands of radial excitations with opposite charge conjugation, depending whether the total spin S is 0 or 1. The spectroscopic notation $n^{2S+1}L$, where $n-1$ is the number of radial nodes, is used to label the levels. The P levels will split into one state $^1P_1$

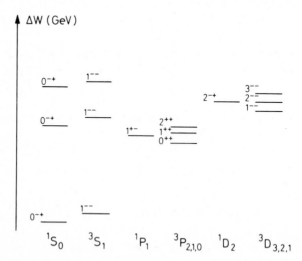

Fig. 5.1. The level scheme of two heavy quarks bound in a steeply rising potential.

with odd, and three states $^3P_{2,1,0}$ with even charge conjugation. In a pure coulombic potential the first set of P levels will be degenerated with the 2^3S_1 level. The addition of a confining potential pushes the mass of the 1P levels below the mass of the 2^3S_1 level.

The D levels will split into one state $^2D_1$ with even, and three states $^3D_{3,2,1}$ with odd charge conjugation. The 1^3D_1 state has the quantum number of a photon. The wavefunction of this state may acquire a finite value at the origin by mixing with the nearby 2^3S_1 state and can be produced directly in e^+e^- collisions.

The number and quantum numbers of the predicted levels reflect the spin-1/2 nature of the quark. The level spacing and the level widths on the other hand are strongly model dependent and can be used to test the theory.

QCD makes clear first order predictions for various decay modes. The $^3S_1$ state decays to lowest order via a three gluon intermediate state, since one gluon is forbidden by colour and two gluons by charge conjugation. The resulting hadronic width [125] is given by:

$$\Gamma\left(^3S_1 \to ggg \to \text{hadrons}\right) = \tfrac{160}{81}(\pi^2 - 9)\alpha_s^3|^3S_1(0)|^2/M^2. \qquad (5.1)$$

M is the mass of the quark and $^3S_1(0)$ denotes the wavefunction at the origin. The strong interaction constant α_s is given by

$$\alpha_s(Q^2) = \frac{12\pi}{(33 - 2N_f)\ln(Q^2/\Lambda^2)}, \qquad (5.2)$$

where N_f is the number of flavours and Λ the characteristic strong interaction mass. The width [126] for the decay into a pair of leptons is given by:

$$\Gamma\left(^3S_1 \to e^+e^-\right) = 16\pi\alpha^2 e_q^2|^3S_1(0)|^2/M^2. \qquad (5.3)$$

The pseudoscalar $^1S_0$ states which are expected to lie below the vector states, can decay into ordinary hadrons by a two gluon intermediate state. The width [125] is given by:

$$\Gamma\left(^1S_0 \to gg \to \text{hadrons}\right) = \tfrac{32}{3}\pi\alpha_s^2|^1S_0(0)|^2/M^2. \qquad (5.4)$$

The pseudoscalar states are therefore much wider than the corresponding vector states. Indeed, if the two states have similar wavefunctions at the origin as expected, then:

$$\Gamma\left({}^1S_0 \to \text{hadrons}\right) = \frac{27}{5} \frac{\pi}{(\pi^2 - 1)\alpha_s} \Gamma({}^3S_1 \to \text{hadrons})$$

$$\approx 100\,\Gamma({}^3S_1 \to \text{hadrons}). \tag{5.5}$$

However, the characteristic distances involved in $c\bar{c}$ and to a lesser degree in $b\bar{b}$ spectroscopy are quite large such that higher order corrections [124] are important and must be included for quantitative conclusions.

A richer level structure than the one predicted by a simple $q\bar{q}$ model might of course exist. Quark pairs may bind to form quark molecules [127] $(c\bar{q})$ $(\bar{c}q)$ with a complex level scheme. The gluon field in a heavy $q\bar{q}$ system may have vibrational excitations [128] which may couple weakly to photons. So far there is no evidence for such states.

5.1. Hadrons with hidden charm

The parameters of $c\bar{c}$ vector states [129] observed in e^+e^- annihilation between 3.0 GeV and 4.5 GeV in c.m. are listed in table 5.1. The Novosibirsk group reports [129] a very precise value of the J/ψ and the ψ' mass, obtained by using a spin depolarizing resonance to calibrate the beam energy. The J/ψ, ψ' and ψ'' can be identified with the 1^3S_1, 2^3S_1 and 1^3D_1 levels, respectively, and there is general agreement on the resonance parameters of these states.

The situation above the ψ'', however, is still not settled. There are indications of a step in the cross section around 3.98 GeV, the DASP group observed two separate states at 4.04 GeV and 4.16 GeV, whereas there is general agreement on the existence of a state near 4.41 GeV. New data on the total cross section between 3.6 GeV and 4.5 GeV in c.m. by

Table 5.1
Resonance parameters of $c\bar{c}$ vector states

State	Mass (MeV)	Γ_{tot} (MeV)	Γ_{ee} (keV)
J/ψ	3096.93 ± 0.09	0.063 ± 0.009	4.8 ± 0.6
ψ'	3686 ± 0.15	0.215 ± 0.040	2.1 ± 0.3
ψ''	3768 ± 5	26 ± 5	0.27 ± 0.06
$\psi(4030)$	4030 ± 5	52 ± 10	0.75 ± 0.10
$\psi(4160)$	4159 ± 20	78 ± 20	0.78 ± 0.31
$\psi(4415)$	4415 ± 6	43 ± 20	0.43 ± 0.13

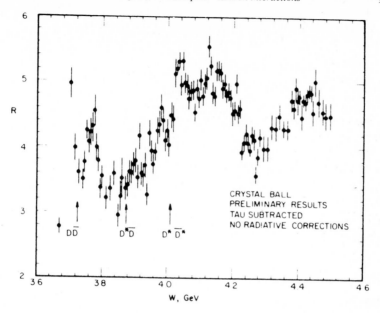

Fig. 5.2. The cross section for $e^+e^- \rightarrow$ hadrons as measured by the Crystal Ball collaboration, normalized to the point cross section.

the Crystal Ball collaboration [129] are shown in fig. 5.2. Plotted is the total annihilation cross section, normalized to the point cross section. The shown data are corrected for τ production but not for radiative effects. The radiative correction will enhance the peak structure. The data confirm the existence of states around 4.03 GeV and 4.16 GeV and indicate structure in the cross section below 4.0 GeV.

Several states with even charge conjugation have been observed in the decays J/ψ or ψ' into final states of γ + anything. The states [130] found at 3.41 GeV, 3.51 GeV and 3.55 GeV can be associated rather naturally with the $^3P_0$, $^3P_1$ and the $^3P_2$ levels, whereas the states reported at 2.82 GeV [131], 3.45 GeV [132] and 3.59 GeV [133] were not so easily fit into the $c\bar{c}$ scheme.

The Crystal Ball detector, designed to measure photons with good energy resolution over a large solid angle has produced a wealth of new data on the even charge conjugation states. Their main findings are [134]:

(i) They find no evidence for the states at 2.82 GeV and 3.45 GeV with a sensitivity well below that of the earlier experiments.

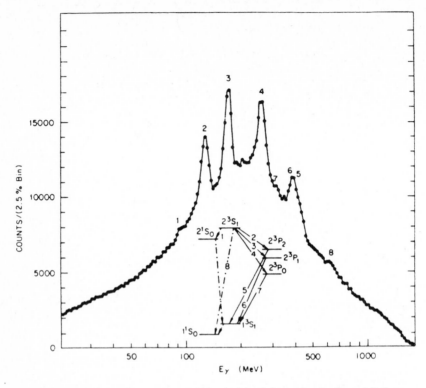

Fig. 5.3. The inclusive photon spectrum observed in ψ' decays by the Crystal Ball group.

(ii) They have observed the decays $\psi' \to \gamma^3 p \to \gamma\gamma J/\psi$ with much higher statistics than previous experiments. The inclusive photon spectrum from the ψ' after the final cuts is plotted in fig. 5.3.

The spectrum is very rich, and the connection between the observed structure and the level scheme is shown in the insert.

The parameters for the P states extracted from this spectrum and the cascade decay $\psi' \to \gamma\gamma J/\psi$ with $J/\psi \to e^+e^-$ $(\mu^+\mu^-)$ are listed in table 5.2. The values of $B(\psi' \to \gamma P)$ listed are from the Crystal Ball [134] and they are in agreement with earlier data [135]. The first two values for the cascade branching ratio are world averages [136], branching ratios for the cascade decay via the 0^{++} state are from the Crystal Ball [134–136]. The widths are in general agreement with the QCD expectation. Since a spin-1 particle cannot decay with the emission of two massless vector particles, one expects the 1^{++} state to be narrower than the 2^{++} and 0^{++}

Table 5.2
Parameters for the Charmonium P-states

State (J^{PC})	Mass (MeV)	Width (MeV)	$B(\psi' \to \gamma P)$ (%)	$B(\psi' \to \gamma P)(P \to \gamma J/\psi)$ (%)
2^{++}	$3553.9 \pm 0.5 \pm 4.0$	1.8 ± 0.6	$7.4 \pm 0.4 \pm 1.4$	1.16 ± 0.12
1^{++}	$3508.4 \pm 0.4 \pm 4.0$	< 1.5	$8.4 \pm 0.4 \pm 1.3$	2.34 ± 0.21
0^{++}	3413 ± 5.0	16.3 ± 3.6	$9.3 \pm 0.4 \pm 1.4$	$0.059 \pm 0.015 \pm 0.004$

states which can decay by two gluon emission. The values for $B(\psi' \to \gamma P)$ are below the theoretical predictions [137] by roughly a factor of two.

An interesting byproduct is the observation of the isospin-forbidden decay $\psi' \to \pi^0 J/\psi$. They find $B(\psi' \to \pi^0 J/\psi) = (0.09 \pm 0.02 \pm 0.01)\%$, consistent with the value $B(\psi' \to \pi^0 J/\psi) = (0.15 \pm 0.06)\%$ obtained [138] in an earlier measurement by the MARK II collaboration. The electromagnetic interaction predicts a rate which is more than one order of magnitude smaller than the observed rate such that this mechanism appears to be excluded. The decay might arise from an isospin breaking amplitude as in the decay $\eta \to 3\pi$.

(iii) The Crystal Ball collaboration has new data [134] on the resonance observed below the J/ψ mass in inclusive transitions [139] from the J/ψ and in exclusive [139, 140] final states both from the J/ψ and the ψ'. The state has now also been seen [134] in the inclusive photon spectrum obtained at the ψ'. It is natural to identify this state with the η_c, the lowest pseudoscalar $^1S_0(c\bar{c})$ state.

The photon spectra from these decays are plotted in fig. 5.4a, b, before and after background subtraction. A simultaneous fit to the most recent data gives the following results:

mass $= 2984 \pm 4$ MeV, width $= 12.4 \pm 4.1$ MeV,

$B(J/\psi \to \gamma\eta_c) = (1.13 \pm 0.33) \times 10^{-2}$,

$B(\psi' \to \gamma\eta_c) = (0.28 \pm 0.08) \times 10^{-2}$.

In the $c\bar{c}$ model the transition $B(\psi' \to \gamma\eta_c)$ is a forbidden spin flip transition and this may explain the low branching ratio. The width of the η_c is about a factor 200 larger than the width of the J/ψ, in qualitative agreement with the QCD prediction.

(iv) The Crystal Ball group has also observed [134] a new resonance in the inclusive photon spectrum of the ψ'. This state is just below the ψ' and is a candidate for η_c', the long sought 2^1S_0 state.

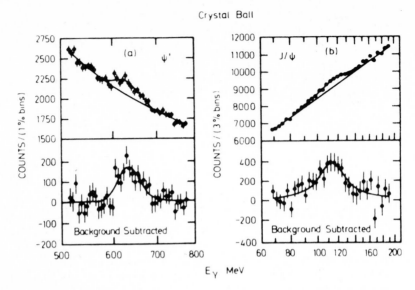

Fig. 5.4. Inclusive photon spectrum $\psi' \to \gamma X$ (a) and $J/\psi \to \gamma X$ (b) as measured by the Crystal Ball collaboration, shown both with and without a smooth background subtraction.

The photon spectrum from the ψ' for photon energies between 60 MeV and 110 MeV is plotted on a linear scale in fig. 5.5, with and without a polynomial background subtracted. The statistical significance of the peak is 4.4σ. Representing the resonance by a Gaussian gave a mass of $M = 3592 \pm 5$ MeV including statistical and systematic errors. The width is less than 8 MeV and the branching ratio is 0.2–1.3%. These are 95% confidence limits. The mass value agrees with an earlier measurement [133] by the DESY Heidelberg group, however, the branching ratio observed by Crystal Ball is substantially lower than that found by DESY–Heidelberg.

5.2. Hadrons with hidden beauty

The first evidence [141] for a new quark came from a lepton pair production experiment at FNAL. In this experiment they observed narrow states, the Υ states, in the e^+e^- mass spectrum produced in proton–nucleous collisions. The energy of DORIS was subsequently increased and the two lowest Υ states confirmed [142]. The measured

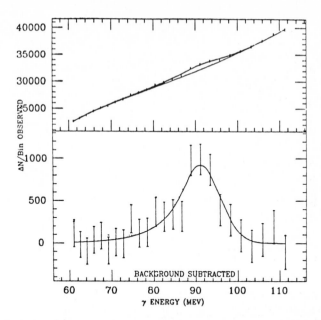

Fig. 5.5. Inclusive photon spectrum $\psi' \to \gamma X$ plotted on a linear scale for photon energies between 60 and 110 MeV. The spectrum is shown both with and without background subtraction.

[143, 144] Γ_{ee} widths showed [145] that these resonances are 1^{--} $b\bar{b}$ states, made of a quark b with charge $1/3e$.

The data have been extended to higher energies by the CUSB and CLEO collaborations at CESR. Some of the total cross section data [100, 146] obtained by these groups are plotted in figs. 5.6 and 5.7. Both experiments find clear evidence for two new 1^{--} states Υ'' and Υ''' which can be identified with the 3^3S_1 and the 4^3S_1 state. The $\Upsilon(3S)$ state, like the $\Upsilon(1S)$ and $\Upsilon(2S)$ states, has an observed width consistent with the energy spread of the beams, whereas the $\Upsilon(4S)$ state is considerably wider. It is therefore natural to assume that the threshold for $b\bar{b}$ production is located between the $\Upsilon(3S)$ and $\Upsilon(4S)$ state. The resonance parameters [146, 147] for the four observed $\Upsilon(1^{--})$ states are listed in table 5.3.

The groups at CESR have measured [100] the total cross section in fine steps for c.m. energies between the $\Upsilon(3S)$ and $\Upsilon(4S)$. The R_{visible} measured by CUSB is plotted in fig. 5.8 versus c.m. energy. No structure is

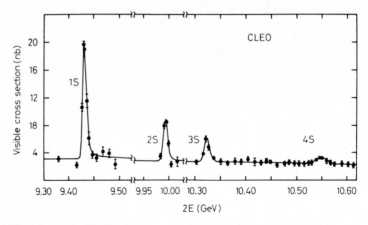

Fig. 5.6. The total hadronic cross section measured by the CLEO collaboration.

seen and the 90% upper confidence limit on Γ_{ee} for a narrow resonance extracted from these data are also shown. 1^{--} states resulting from the quantized excitations of the gluon string joining the quark and the antiquark had been predicted [128] to exist in this mass range, with a width $\Gamma_{ee} = 0.2 \pm 0.15$ keV. The sensitivity of the present data are close to the lower limit.

The cascade decays $\Upsilon(2S) \rightarrow \pi^+ \pi^- \Upsilon(1S)$ [100] and $\Upsilon(3S) \rightarrow \pi^+ \pi^- \Upsilon(1S)$ have been observed [100, 146, 147], and the data are listed in table 5.4.

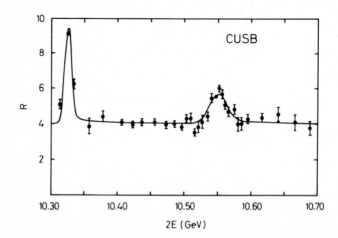

Fig. 5.7. The total hadronic cross section measured by the CUSB collaboration.

Table 5.3
Resonance parameters for the $\Upsilon(^3S_1)$ states

State	Mass (MeV)	Excitation energy (MeV)	Γ_{ee} (keV)	Group
$\Upsilon(1^3S_1)$	9462 ± 10	–	$1.29 \pm 0.09 \pm 0.13$	DESY
	9433 ± 28	–	1.19 ± 0.02	CLEO
		–	1.06 ± 0.04	CUSB
$\Upsilon(2^3S_1)$	–	553 ± 10	$0.58 \pm 0.08 \pm 0.26$	DESY
		560 ± 1	0.52 ± 0.02	CLEO
			0.51 ± 0.03	CUSB
$\Upsilon(3^3S_1)$	–	890 ± 1	0.39 ± 0.02	CLEO
			0.36 ± 0.03	CUSB
$\Upsilon(4^3S_1)$	–	1114 ± 2	0.31 ± 0.02	CLEO
			0.21 ± 0.02	CUSB

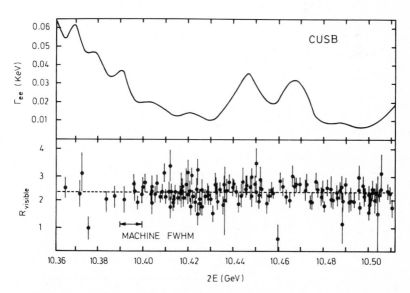

Fig. 5.8. R_{visible} measured by the CUSB collaboration in fine energy steps between the $\Upsilon(3S)$ and $\Upsilon(4S)$. The 90% upper confidence limit on the leptonic width Γ_{ee} for a narrow resonance is also shown.

Table 5.4
Data on the cascade decay

Group	$B(\Upsilon(2S) \to \pi^+\pi^- \Upsilon(1S))$ (%)	$B(\Upsilon(3S) \to \pi^+\pi^- \Upsilon(1S))$ (%)
LENA	27 ± 9	–
CLEO	19.1 ± 3.1	3.9 ± 1.9
CUSB	–	9.7 ± 4.3

CLEO observed the decay $\Upsilon(2S) \to \pi^+\pi^- \Upsilon(1S)$ by measuring the mass recoiling against all $\pi^+\pi^-$ pairs observed in the $\Upsilon(S)$ decays. The observed mass distribution, plotted in fig. 5.9, shows a clear peak at the mass of the $\Upsilon(1S)$.

The LENA group at DESY has searched [148] for the radiative decay $\Upsilon(2S) \to \gamma\gamma\Upsilon(1S)$ with $\Upsilon(1S) \to \mu^+\mu^-$ or e^+e^-. From the data they ex-

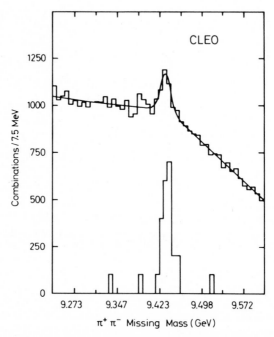

Fig. 5.9. The missing mass recoiling against all opposite charged pion pairs at the $\Upsilon(2S)$. The lower histogram is the $\pi^+\pi^-$ recoil mass for events of the type $\Upsilon(2S) \to \pi^+\pi^- e^+e^-$ ($\mu^+\mu^-$).

tracted an upper limit $B(\Upsilon(2S) \rightarrow \gamma\gamma\Upsilon(1S)) < 7\%$. The theoretical expectations range between 2% and 7%. In QCD (section 7) the $P_{2,0}$ states decay in leading order into two gluons. Since the photon in the cascade $\Upsilon \rightarrow \gamma P_{2,0}$ is soft, these gluons will result in two back to back jets of hadrons. This topology is different from the dominant decay $\Upsilon(^3S_1) \rightarrow 3$ gluons. The CUSB collaboration has observed [146] events with the two-jet topology at both the $\Upsilon(2S)$ and $\Upsilon(3S)$. They find from the observed thrust distribution:

$$B(\Upsilon(2S) \rightarrow 2 \text{ jets}) = 20 \pm 3\%,$$
$$B(\Upsilon(3S) \rightarrow 2 \text{ jets}) = 8 \pm 2\%.$$

Note that two-jet events may also result from $\Upsilon \rightarrow \gamma^* \rightarrow q\bar{q}$. However, these branching ratios would lead to less two-jet events than observed.

5.3. Hadrons with charm

5.3.1. D^0 and $D^\pm$ states

Possible transitions between the known D states and the corresponding Q values are summarized in fig. 5.10.

The ψ'' decays, to a good approximation, only into $D\bar{D}$ final states. The production cross sections for these mesons at the ψ'' resonance were determined [149] by the MARK II collaboration to:

$$\sigma(D^0) = 8.0 \pm 1.0 \pm 1.2 \text{ nb}, \qquad \sigma(D^+) = 6.0 \pm 0.7 \pm 0.1 \text{ nb}.$$

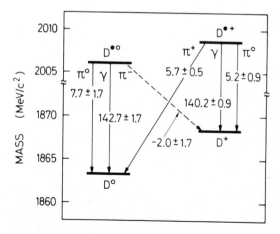

Fig. 5.10. Q-values for the $D^* \rightarrow D$ transition.

For comparison, the cross section predicted for $e^+e^- \to Z^0 \to c\bar{c}$ is on the order of 4 nb. An additional advantage is that the D's are produced almost at rest and the favourable kinematics allow a precise determination of the mass $M_0 = (E^2 - p^2)^{1/2}$ where E is the beam energy and p the D momentum. These nice features have been exploited by the groups working at SPEAR and have lead to a wealth of data [149] on the D states.

Mass spectra for $K^{\pm}\pi^{\mp}$, $K^0\pi^{\mp}\pi^{\pm}$ and $K^{\pm}\pi^{\mp}\pi^{\pm}$, measured at the ψ'' by the MARK II Collaboration [150] are shown in fig. 5.11. From such data, groups working at SPEAR found $m_{D^0} = 1864.3 \pm 0.9$ MeV and $m_{D^+} = 1868.4 \pm 0.9$ MeV.

A complete list of D branching ratios can be found in ref. [149]. Below we discuss a measurement of the lifetimes and a determination of the GIM mixing angles.

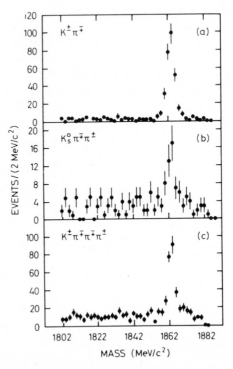

Fig. 5.11. Mass spectra for $\psi'' \to D^0\bar{D}^0$, candidate events with (a) $D \to K^{\pm}\pi^{\mp}$, (b) $D \to K^0_s\pi^{\mp}\pi^{\pm}$, (c) $D \to K^{\pm}\pi^{\mp}\pi^{\mp}\pi^{\pm}$. The data were obtained by the MARK II collaboration.

5.3.2. *Lifetime of charmed mesons*

Possible Cabibbo favoured decay modes of charmed mesons into light hadrons are shown in fig. 5.12. An assumption often made, was that the charmed quark would decay according to $c \to - d \sin \theta + s \cos \theta$ with the second quark merely acting as a spectator. This mechanism, shown in fig. 5.12a,c,d predicts that all charmed mesons should have the same lifetime. It has been pointed out by Pais and Treiman [151] that $\Gamma(D^+ \to \ell^+ \nu_\ell X) = \Gamma(D^0 \to \ell^+ \nu_\ell X)$ since $|\Delta I = 0|$ for the Cabibbo allowed decay $c \to \ell \nu_\ell s$. The semileptonic branching ratios can therefore be used to determine the ratios of the lifetimes:

$$\frac{B(D^+ \to \ell^+ \nu_\ell X)}{B(D^0 \to \ell^+ \nu_\ell X)} = \frac{\Gamma_{tot}(D^0 \to X)}{\Gamma_{tot}(D^+ \to X)} = \frac{\tau(D^+)}{\tau(D^0)}$$

Both the DELCO [152] and the MARK II [150] collaborations report results on this ratio. DELCO finds:

$$\tau(D^+)/\tau(D^0) = 4.3, \qquad B_e(D^+) = 22^{+4.4}_{-2.2}\%.$$

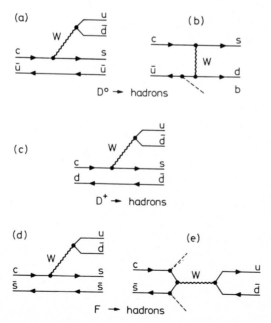

Fig. 5.12. Possible Cabibbo-favoured decay modes of charmed mesons into light hadrons.

Their results were extracted from a sample of $D\bar{D}$ events in which one or both of the charmed mesons decayed semileptonically.

The MARK II Collaboration determined [150] the branching ratios from an inclusive measurement of charmed meson production. The events were tagged by identifying either a charged or a neutral D meson and measuring the decay products of its partner. In this way the semileptonic branching ratios are determined directly and they find:

$$\tau(D^+)/\tau(D^0) = 3.1^{+4.6}_{-1.4}, \qquad B_e(D^+) = 18.6 \pm 6.4\%.$$

The lifetime of charmed particles has been determined [153] directly, using emulsions or high resolution bubble chambers. The results of the various experiments are summarized in table 5.5.

The lifetime of charged D mesons has also been measured [154] in a high energy photoproduction experiment at CERN. The experiment used a live target followed by a spectrometer. The target was made of 40 layers of 300 μm thick silicon disks spaced 100 μm apart. The energy loss in the target stack is proportional to the number of charged particles, i.e., it is possible to identify the production vertex and the secondary vertex from the decay of the charm particles, by the steps in the observed pulse height, and hence to measure the distance which the charmed particle travelled between production and decay. The measured time distribution of identified charmed particle decays is shown in fig. 5.13. Correcting for the D^0 contamination results in a lifetime $\tau_{D^+} = (9.5^{+3.1}_{-1.9}) \times 10^{-13}$ s.

The D^+ lifetime combined with the average branching ratio for $D^+ \rightarrow e^+ \nu_e X$ results in a semileptonic width

$$\Gamma(D^+ \rightarrow e^+ \nu_e X) \approx (1.8 \pm 1) \times 10^{11} \text{ s}^{-1}.$$

Table 5.5
Lifetimes of charged and neutral D mesons

Particle	Lifetime ($\times 10^{-13}$ s)	Events	Group
$D^\pm$	$8.0^{+4.9}_{-2.4}$	7	LEBC CERN
	$6.5^{+2.2}_{-1.0}$	4	SLAC hybrid
	$9.5^{+6.5}_{-3.3}$	6	FNAL emulsion
	$9.5^{+3.1}_{-1.9}$	(96)	NAI, CERN
D^0	$3.2^{+2.2}_{-1.0}$	6	FNAL
	$1.9^{+1.7}_{-0.6}$	4	SLAC hybrid
	$3.0^{+1.1}_{-0.7}$	17	FNAL emulsion
	$1.34^{+1.2}_{-0.9}$	5	WA 57 CERN
	$0.5^{+0.6}_{-0.3}$	3	WA 17 CERN

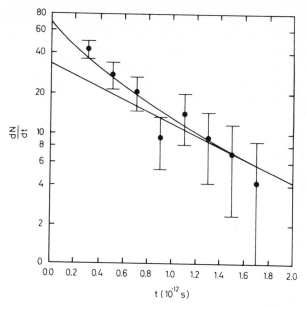

Fig. 5.13. Time distribution of identified charmed particle decays.

Cabibbo, Maiani and Corbo have evaluated [155] the semileptonic width in the spectator model and they find

$$\Gamma_{\mathrm{SL}} = \frac{G_{\mathrm{F}}^2 m_{\mathrm{c}}^5}{192\pi^3} g(\varepsilon)\left[1 - 2\alpha_{\mathrm{s}} f(\varepsilon)/B\pi\right],$$

with $\varepsilon = m_{\mathrm{s}}/m_{\mathrm{c}}$. Here $g(\varepsilon)$ is a phase space correction due to the finite mass of the s quark and the term in brackets is a QCD strong interaction correction. With $m_{\mathrm{c}} = 1.75$ GeV/c^2 they find $\Gamma_{\mathrm{SL}} = 1 \times 10^{11}$ s^{-1}, consistent with the experimental results. The width of the D$^+$ is therefore consistent with the prediction based on the spectator model. The experimental results on D$^+$ and D^0 lifetimes shows that D^0 must have additional decay modes.

Additional decay modes resulting from W$^+$ annihilation are indeed Cabibbo allowed both for D^0 and F$^+$ decays whereas they are Cabibbo forbidden for D$^+$ decays. These decay modes, shown in fig. 5.12b,e may account for a factor of five in the lifetime ratio $\tau(\mathrm{D}^0)/\tau(\mathrm{D}^+)$. To conclude, the lifetime of charmed hadrons cannot be reliably computed in the spectator model. Many mechanisms which lead to different lifetimes for different charmed particles have been proposed and the theoretical situation is reviewed in ref. [156].

5.3.3. The GIM mechanism

The GIM mechanism [79] predicts that a charmed quark decays predominantly into a strange quark, according to

$$c \rightarrow - d\sin\theta_B + s\cos\theta_B.$$

The mixing angle θ_B in the GIM model is identified with the familiar Cabibbo angle θ_C with $\sin\theta_C \cong 0.22$, determined [104, 105] from strange particle decays. Both angles θ_A and θ_B can be determined from a measurement of the two-body decay modes

$$D^0 \rightarrow K^-\pi^+, D^0 \rightarrow K^-K^+, D^0 \rightarrow \pi^-\pi^+$$

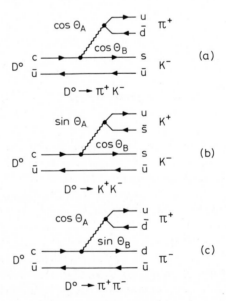

Fig. 5.14. Diagrams for two-body D^0 decays: (a) $D^0 \rightarrow \pi^+ K^-$, (b) $D^0 \rightarrow K^+ K^-$, (c) $D^0 \rightarrow \pi^+ \pi^-$.

as shown in fig. 5.14. These decays yield

$$\tan^2 \theta_A = 1.08 \frac{\Gamma(D^0 \to K^- K^+)}{\Gamma(D^0 \to K^- \pi^+)},$$

$$\tan^2 \theta_B = \frac{0.93\Gamma(D^0 \to \pi^+ \pi^-)}{\Gamma(D^0 \to K^- \pi^+)}.$$

Invoking SU(3) invariance and the GIM mechanism results in:

$$\tan^2 \theta_A = \tan^2 \theta_B = \tan^2 \theta_C = 0.05.$$

The MARK II collaboration has determined [157] these decay modes, making use of the fact that at the $\psi''D^0$'s are pair produced with a unique momentum of 288 MeV/c. The invariant mass spectra obtained for the twobody decay modes are shown in fig. 5.15. The peak occurs at the D^0 mass if the two particles are identified correctly, and shifted by about ± 120 MeV/c^2 if one of the particles is misidentified. Both a $\pi^+\pi^-$ and a K^-K^+ signal is observed yielding:

$$\frac{\Gamma(D^0 \to \pi^- \pi^+)}{\Gamma(D^0 \to K^- \pi^+)} = 0.033 \pm 0.015,$$

$$\frac{\Gamma(D^0 \to K^- K^+)}{\Gamma(D^0 \to K^- \pi^+)} = 0.113 \pm 0.030.$$

The data show that Cabibbo forbidden decays occur at roughly the level predicted. The interpretation, however, is complicated by strong interaction effects.

The $\bar{d}c$ coupling can also be determined, and perhaps more reliably, from a measurement [158] of $\nu + d \to \mu^- + c \to \mu^+ \mu^- X$. Using the Kobayashi–Maskawa formalism [33] (section 3) this process determines the value of $U_{c\bar{d}} \simeq \sin\theta_1 \cos\theta_2 \approx \sin\theta_C$. The reaction $\nu + s \to \mu^- + c$ will of course also yield opposite-sign dileptons. However, the two reactions will have different x dependence, since the d is a valence quark whereas the s is a sea quark. An analysis [159] of the observed dilepton rate of about 0.01 yields $0.19 < |U_{c\bar{d}}| < 0.34$, consistent with $\sin\theta_C \simeq 0.22$.

The $\bar{s}c$ coupling can be determined from the branching ratio of $D^+ \to e^+ \nu_e \bar{K}^0$ and the D^+ lifetime. An analysis [159] yields $|U_{sc}| = 0.66 \pm 0.33$.

The GIM mechanism is thus supported by the data.

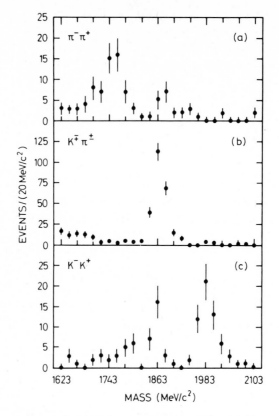

Fig. 5.15. Invariant mass distribution of candidate events for: (a) $D^0 \to \pi^- \pi^+$, (b) $D^0 \to K^{\mp} \pi^{\pm}$, (c) $D^0 \to K^- K^+$.

5.3.4. Evidence for the F meson

The DASP collaboration observed [160] a signal at 4.42 GeV in c.m. which they attribute to $F^+ \to \eta \pi^+$. A scatter plot of the $\eta \pi^+$ mass versus the fitted recoil mass, assuming $e^+ e^- \to F F^*$ is shown in fig. 5.16, for events at 4.42 GeV and events outside the region. At 4.42 GeV there is a cluster of six events, whereas at other energies the events have a smooth mass distribution. Of the six events observed at 4.42 GeV, less than 0.2 events can be ascribed to the background. This was estimated from the measured luminosity and the number of events observed outside of 4.42 GeV in the same mass region with the conservative assumption that all

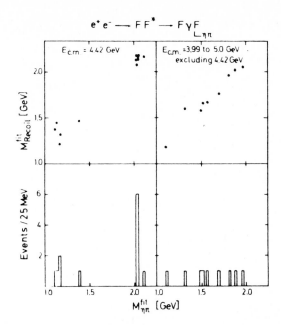

Fig. 5.16. Fitted mass versus fitted recoil mass, assuming $e^+e^- \to FF^*$, where $F^* \to \gamma F$ and $F \to \eta\pi$. (a) at 4.42 GeV, (c) all other energies excluding 4.42 GeV. Histograms (b) and (d) are the projections of (a) and (c) along the $M(\eta\pi)$ axis respectively. The data were obtained by the DASP collaboration.

these events are background events. The cluster at 4.42 GeV gives:

$$m_F = 2.03 \pm 0.06 \text{ GeV}/c^2, \qquad m_{F^*} = 2.14 \pm 0.06 \text{ GeV}/c^2,$$

including systematic uncertainties.

Supporting evidence for the F has come from a photoproduction experiment in the Ω spectrometer at the SPS and from emulsion exposures. The WA4 collaboration at the SPS observe [161] a signal in the decay modes $F \to \eta\pi$, $\eta 3\pi$, $\eta' 3\pi$ and $\phi\rho$. The results are shown in fig. 5.17 and summarized in table 5.6.

Three F candidate events are found in the emulsion data [153]. The observed decay modes and mass values in MeV/c^2 are:

$$\pi^-\pi^+\pi^-\pi^0 \ (2026 \pm 56),$$
$$K^+\pi^-\pi^+K^0 \ (2089 \pm 121),$$
$$\pi^+\pi^+\pi^-\pi^0 \ (2017 \pm 25).$$

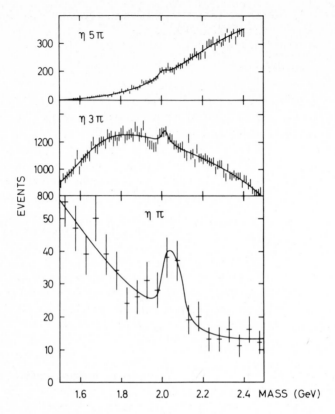

Fig. 5.17. Mass spectra of photoproduced $\eta + n\pi$ final states, measured by the WA4 collaboration at CERN.

Table 5.6
Evidence for F photoproduction

Decay mode	Mass (MeV)	$B\sigma$(nb)
$\eta\pi$	2047 ± 23	27 ± 7
$\eta 3\pi$	2021 ± 13	60 ± 15
$\eta' 3\pi$	2008 ± 20	20 ± 8
$\phi\rho'$	2049 ± 15	33 ± 10

The mass values observed in the new experiments are consistent with the values reported by the DASP collaboration.

5.4. Hadrons with beauty

In the standard model the b quark is the partner of a charge-$2/3$ quark in a left handed weak doublet. In the Kobayashi–Maskawa [33] scheme, the b quark decays weakly as indicated in fig. 5.18 via a flavour cascade b → c → s, whereas the direct decay b → u is strongly suppressed. The model [100] predicts that on the average 1.6 kaons, charged or neutral, are produced per b decay. The semileptonic branching ratio $B(b \to e\bar{\nu}_e$ hadrons$) = B(b \to \mu\bar{\nu}_\mu$ hadrons$)$ is predicted [162] to be in the range 11% to 13%.

The large width of the $\Upsilon(4S)$ compared to the width of the adjacent $\Upsilon(3S)$ state shows that the lightest B meson must have a mass between 5.18 GeV/c^2 and 5.28 GeV/c^2 using the DORIS energy scale. The CUSB collaboration [100, 146] has made an unsuccessful search for narrow photon lines in the debris of the $\Upsilon(4S)$. Such lines would have been a signature of B* → γB and the negative result yields the limit:

$$\frac{\Upsilon(4S) \to B^*B}{\Upsilon(4S) \to all} \leqslant 0.20.$$

With the assumption that $\Upsilon(4S)$ is below B*B threshold and with the observed $\Upsilon(4S)$ width as an input, a theoretical estimate [146] gives $m_B = 5.26 \pm 0.01$ GeV/c^2 as the mass of the lightest B meson.

The CUSB and CLEO collaborations [100] have measured the yield of mixed electron–hadron events in the vicinity of the $\Upsilon(4S)$. Both experiments only accept electrons with momenta above 1 GeV/c. This cut strongly reduces the number of events resulting from charm production,

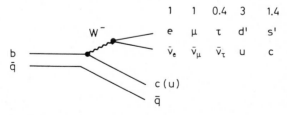

Fig. 5.18. b decays in the spectator model. The relative strength of the various decay modes is indicated.

whereas 2/3 of the B decays survive. Both experiments show a strong increase in the inclusive electron yield at the $\Upsilon(4S)$ resonance demonstrating that the $\Upsilon(4S)$ is indeed disintegrating into weakly decaying particles. The normalized cross sections for inclusive electron–hadron events and for hadron production measured by the CUSB collaboration are plotted in fig. 5.19. From these data they find $B(\text{B} \rightarrow e\bar{\nu}_e X) = 13.1 \pm 2.5 \pm 3.0\%$, in agreement with the value of $B(\text{B} \rightarrow e\bar{\nu}_e X) = 13.6 \pm 2.1 \pm 1.7\%$ measured by the CLEO group. The first error is the statistical uncertainty, the second the systematic. The CLEO collaboration has also measured [100] the branching ratio for $\text{B} \rightarrow \mu\bar{\nu}X$ and they find $B(\text{B} \rightarrow \mu\bar{\nu}_\mu) = 10.0 \pm 1.3 \pm 2.1\%$.

The electron momentum spectrum observed in mixed electron–hadron events at the $\Upsilon(4S)$, and corrected for the continuum contribution is plotted in fig. 5.20 for the CUSB (a) and the CLEO (b) data. The momentum spectra from both experiments favours a hadronic recoil mass of the order of 2 GeV/c^2. A hadronic recoil mass around 1 GeV/c^2 would result in a lepton spectrum much harder than the spectrum observed.

From the observed multiplicity distributions in $\text{B}\bar{\text{B}}$ decays the CLEO collaboration finds an average charged multiplicity of $\langle N_{\text{ch}} \rangle = 3.50 \pm 0.35$

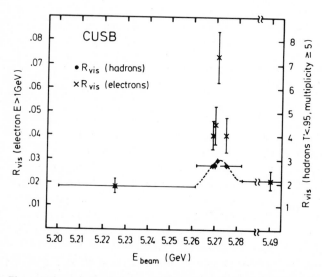

Fig. 5.19. The visible normalized cross section for inclusive electron ($p_e > 1.0$ GeV/c) production and the visible normalized total hadron cross section. The data are from CUSB.

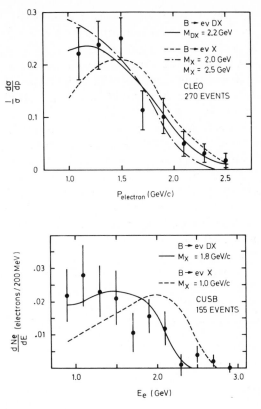

Fig. 5.20. The electron momentum spectrum from $B \to e\bar{\nu}X$. Data obtained by CLEO (upper figure), and by CUSB (lower figure).

in semileptonic B decays and $\langle N_{\text{ch}} \rangle = 6.31 \pm 0.35$ in nonleptonic B decays. Subtracting the electron they find that on the average 2.5 ± 0.3 hadrons are produced per semileptonic B decay. Groups at SPEAR find [163] that an average number of 2.5 ± 0.1 charged hadrons are produced in the decays of an equal mixture of D and D* mesons.

The semileptonic branching ratios, the lepton momentum spectrum, and the average charged multiplicities observed in semileptonic B decays are therefore in agreement with predictions based on the standard model.

Since c quarks predominantly decay into s quarks, the observed kaon yield can be used to determine the relative strengths of $b \to W\bar{c}$ and $b \to W\bar{u}$. Neutral kaons are identified in CLEO for momenta above 0.3

GeV/c by demanding a secondary vertex at least 7 mm from the beam line. The mass is reconstructed assuming the particles to be pions. CUSB identifies neutral kaons using topology only. CLEO identifies charged kaons for momenta between 0.5 GeV/c and 1.0 GeV/c by time of flight. To be less sensitive to normalization uncertainties the groups compare the ratios of kaons per event at the $\Upsilon(4S)$ to the number of kaons per event in the continuum.

The yield of neutral kaons observed by the CUSB collaboration normalized to the hadronic cross section is plotted in fig. 5.21 versus c.m. energy. It is clear that relatively more K^0 events are produced in $\Upsilon(4S)$ decays than in the continuum. The same behaviour is also observed for charged kaons and the results are summarized in table 5.7 (from reference [100]).

Monte Carlo calculations predict $R = 1.8$ for b → W$\bar{c}$, in agreement with the observed ratio $R \approx 1.9$. The data therefore are consistent with this decay mode only. To set a limit on the fraction $(1 - f)$ of the decay b → W$\bar{u}$ they evaluate

$$R_{measured} = fR(b → W\bar{c}) + (1 - f)R(b → W\bar{u}).$$

This expression gives $f = 1.12 \pm 0.25$ and a 90% upper confidence limit of

$$|b → W\bar{u}|^2/|b → W\bar{c}|^2 < 0.4.$$

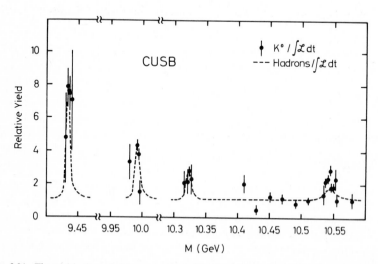

Fig. 5.21. The yield of neutral kaons measured by CUSB, as a function of center of mass energy.

Table 5.7
Comparison of the *observed* kaon yield with Monte Carlo predictions (taken from ref. [100])

Data	Number of kaons observed per event		
	K^0 CLEO	K^0 CUSB	K^+ CLEO
Continuum	0.73 ± 0.05	0.82 ± 0.08	1.12 ± 0.16
B$\bar{\text{B}}$ decay	1.43 ± 0.25	1.52 ± 0.20	2.02 ± 0.25
$R = \dfrac{\text{B}\bar{\text{B}} \text{ decays}}{\text{continuum}}$	1.96 ± 0.34	1.85 ± 0.30	1.80 ± 0.32
	Monte Carlo predictions for $\frac{1}{2}(K^\pm + K^0)$		
Continuum	0.90		
B$\bar{\text{B}}$: b $\rightarrow$ W$\bar{\text{c}}$	1.60, $R = 1.78$		
B$\bar{\text{B}}$: b $\rightarrow$ W$\bar{\text{u}}$	0.90, $R = 1.00$		

The CLEO collaboration has measured the cross section for $e^+e^- \rightarrow e(\mu)K X$ and the data are plotted in fig. 5.22. A strong peak is seen at $\Upsilon(4S)$. From these data and the average number of kaons per B decay it is possible to extract the average number of kaons per B decay for both semileptonic and nonleptonic decays and 0.8 ± 0.7 kaon per semileptonic B decay.

About 1.0 kaons are expected per semileptonic decay for b $\rightarrow$ W$\bar{\text{c}}$ and approximately no kaons for b $\rightarrow$ W$\bar{\text{u}}$. The number of kaons in semi-

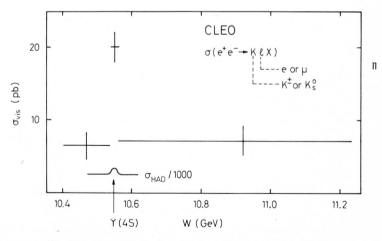

Fig. 5.22. The yield of $e^+e^- \rightarrow e\,(\mu) + K\,X$, measured by CLEO.

leptonic decays is therefore a sensitive measure of the relative strength of these transitions, however, more data are clearly needed.

The data on B decays are in agreement with the standard model and rule out many of the non standard models. It is indeed very likely that the t quark exists.

6. Hadron production in e^+e^- annihilation

At the parton level, e^+e^- annihilation proceeds [7–10] to lowest order by the first graph shown in fig. 6.1. The neutral timelike current, electromagnetic or weak, couples directly to a pair of quarks. At small distances the quarks behave as if they were free, resulting in a total cross section which is proportional to the muon cross section. Neglecting the weak contribution:

$$R = \sigma(e^+e^- \to q\bar{q} \to \text{hadrons})/\sigma(e^+e^- \to \mu^+\mu^-) = 3\sum_i e_i^2.$$

The factor of three accounts for the number of colours.

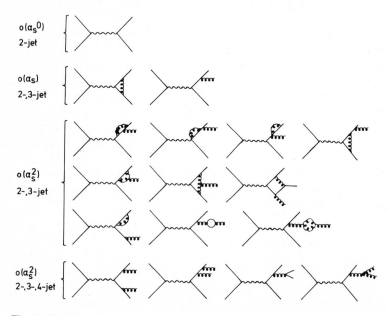

Fig. 6.1. Leading and next to leading order QCD diagrams for hadron production.

The cross section for pointlike scalar quarks would be a factor of four smaller. The spin-1/2 nature of the quarks is also reflected in the angular distribution of $(1 + \cos^2 \theta)$ with respect to the beam axis, compared to a $\sin^2 \theta$ distribution for scalar quarks.

However, quarks are confined, and as they move apart new quark pairs are created. At distances greater than 10^{-13} cm the quarks have condensed into hadrons, such that quark pairproduction materializes as two back to back hadron jets travelling along the direction of the primary quarks. The hadrons in the jet have typical transverse momenta of 300 MeV with respect to the jet axis, independent of energy, and large and increasing momenta along the jet axis.

To first order this picture will be modified in any field theory [12] of strong interactions by the graphs shown in fig. 6.1. The produced quarks radiate [11] a field quantum, which materializes as a hadron jet in the final state, leading to planar three-jet events. With increasing energy, multiple gluon emission with four or more partons in the final state will become visible. This final state is in general not planar. Higher order gluon emission is particularly interesting since these processes can be used to identify the nature of the theory. For example, the last Feynman graph in fig. 6.1 depicts the gluon self coupling. This coupling is an essential property of Quantum ChromoDynamics [14] (QCD), the leading candidate for a strong interaction theory, and is absent in Abelian field theories like QED.

We might arbitrarily divide the physics which can be extracted from a study of hadron production in e^+e^- annihilation into two classes. First one might try to identify the process on the parton level from the observed final state hadrons. Secondly one can study the mechanism by which a struck quark or a radiated gluon converts into a jet of hadrons. This is presumably a non-perturbative process and so far there is little theoretical guidance.

In this section we first discuss some data which directly support the simple picture outlined above and then the more general properties of the hadronic final state.

The evidence for gluons and their properties as determined in e^+e^- annihilation will be discussed in section 7.

6.1. *The total cross section*

The normalized (electron–positron) annihilation cross section R is a well defined quantity in QCD and can be evaluated [164] to all orders in the coupling constant α_s/π defined in eq. 5.2.

$$R = 3\sum_i e_i^2 1 + \alpha_s/\pi + C_2(\alpha_s/\pi)^2 + \cdots. \tag{6.1}$$

R is the sum of the squares of the quark charges, as in the naive quark model, with a small correction due to emission and absorption of gluons. The first term (α_s/π) contributes on the order of 5%. The size of the second order term depends on the renormalization scheme, but its value is always less than the first order term. For example, in the $\overline{\text{MS}}$ scheme $C_2 = 1.99 - 0.12 N_f$, where N_f is the number of quark flavours, i.e., $C_2(\alpha_s/\pi) = 0.02$.

The R values determined [165] by the PETRA groups are plotted in fig. 6.2, together with some lower energy data [166, 167]. Only statistical errors are included and the systematic uncertainties vary between 5% and 10% for the different experiments. The small angle luminosity monitors, the radiative corrections and the acceptance corrections make the largest contribution to the systematic error.

The general features of the observed cross section are in striking agreement with the quark model predictions, both the narrow vector states and the steps in R associated with the liberation of new quark flavours are clearly seen. R does not change between thresholds and its value is roughly given by eq. 6.1.

The MARK I Collaboration has measured [167] R for c.m. energies between 5.5 GeV and 7.5 GeV with good statistics and an estimated systematic uncertainty of 10%. A careful comparison [168] of these data with QCD shows that the experimental values are on the average 16% above the theoretical prediction.

A precise measurement of R is in many ways easier to do at high than at low energies. As shown below, the events are jetlike, with a high multiplicity and an angular distribution of the jet axis with respect to the beam axis of $1 + \cos^2\theta$, resulting in a large, well defined acceptance on the order of 70–80%. The contamination from tau pair production is reduced by a cut on multiplicity and the two-photon contributions by a cut on visible energy to a very low level.

The PETRA groups find R to be constant within errors above $b\bar{b}$ threshold and the data exclude the production of a new charge-2/3 flavour in this energy region. Above $b\bar{b}$ threshold the quark–parton model predicts $R = 11/3$ for u, d, s, c and b quarks and this prediction is shown as the solid line in fig. 6.2. The data seem to be slightly above this value.

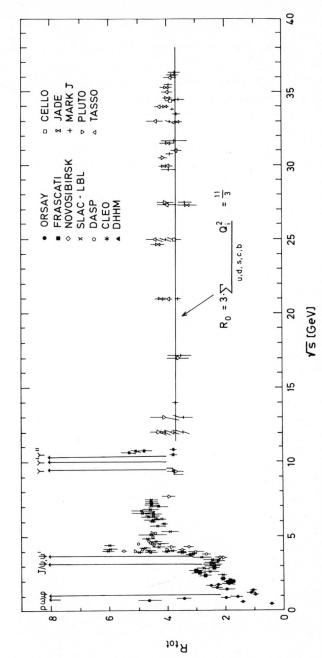

Fig. 6.2. R_{tot} the cross section for $e^+e^- \rightarrow$ hadrons, normalized to the point cross section $e^+e^- \rightarrow \mu^+\mu^-$, plotted versus c.m. energy. Only statistical errors are shown. The solid line indicates the quark model prediction for u, d, s, c and b quarks.

Since the data from the various groups agree within the statistical errors they can be averaged, and the resulting R value is plotted [165] on an expanded scale in fig. 6.3. The naive quark–parton model is shown as the dotted line. The QCD predictions [169], with and without weak effects (section 3) included, are also plotted. Weak effects are negligible at the present PETRA energy for $\sin^2 \theta_W = 0.23$. However, note that the total cross section data do constrain the value of $\sin^2 \theta_W$ as discussed in section 3.3.

The data clearly favour the presence of a QCD correction term, but only on the one standard deviation level. To turn a measurement of the total cross section into a determination of α_s requires that the combined statistical and systematic errors are kept on the 2% level.

The proliferation of leptons and quarks has led to suggestions [170] that these particles are not elementary, but composite, made of new entities. In this case the quark will have a size and this will modify the value of R.

Two different fits have been made to limit the quark size from present data.

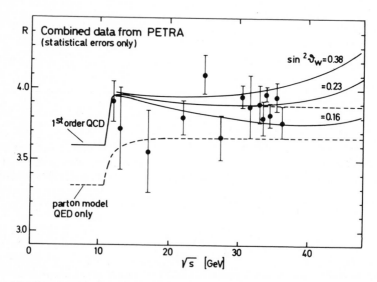

Fig. 6.3. The R values obtained by combining the data from the various PETRA groups are plotted versus $\sqrt{s}$. The lower dashed line is the quark model QED prediction. The upper, solid lines show the first order QCD prediction with $\Lambda = 300$ MeV for various values of $\sin^2 \theta_W$.

Söding and Wolf [171] introduce an electric $G_E(s)$ and a magnetic $G_M(s)$ form factor of the quarks and this yields:

$$R = 3\sum_i e_i^2 \left[\frac{2m_q^2}{s}|G_E(s)|^2 + |G_M(s)|^2 \right]. \tag{6.2}$$

However, since $m_q^2 \ll s$, the data can be expressed in terms of $G_M(s)$ only. Assuming $G_M = (1 - s/M^2)^{-1}$ leads to a limit $M > 124$ GeV, whereas a dipole form factor $G_M(s) = (1 - s/M_D^2)^{-2}$ leads to $M_D > 176$ GeV. Both limits are 90% confidence lower limits.

The MARK J group fit their data [36] to the form

$$R = R_0 \left[1 \mp s/(s - \Lambda^2) \right]. \tag{6.3}$$

The value for R_0 includes both the gluon correction and effects due to the neutral weak current. They find $\Lambda_+ > 180$ GeV and $\Lambda_- > 285$ GeV. We may thus conclude that quarks are pointlike down to distances of 10^{-16} cm.

6.2. Hadron jets

The jet structure is generally analyzed in terms of sphericity [172] or thrust [173].

The sphericity S is defined as:

$$S = \tfrac{3}{2} \min\left[\sum_i (p_T^i)^2 / \sum_i (p^i)^2 \right]. \tag{6.4}$$

Here p^i is the momentum and p_T^i the transverse momentum of a track with respect to a given axis. The jet axis is defined as the axis which minimizes transverse momentum squared. Sphericity measures the square of δ, the jet cone opening angle: $S = \tfrac{3}{2} \langle \delta^2 \rangle$, is 0 for a perfect jet and 1 for a spherical event.

Thrust T is defined as:

$$T = \max\left(\sum_i |p_\parallel^i| / \sum_i |p^i| \right). \tag{6.5}$$

Here p^i is the momentum of a track and $p_\parallel^i$ its projection along a given axis. The jet axis is defined as the axis which maximizes the directed momentum. Expressed in terms of δ, $T \approx (1 - \langle \delta^2 \rangle)^{1/2}$ and it approaches 1 for a perfect jet event and 1/2 for an isotropic event.

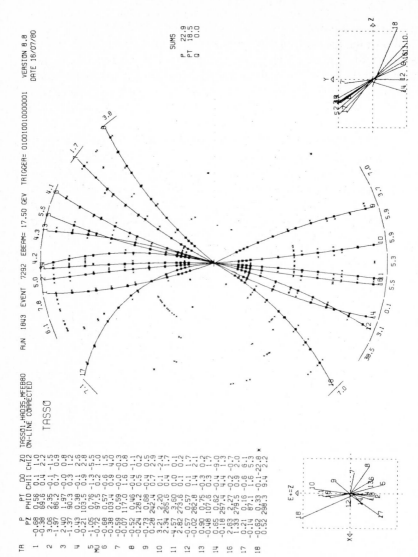

Fig. 6.4. Two-jet event observed by the TASSO group. The event is viewed along the beam direction.

The two-jet structure of hadronic events was first seen [174] by the MARK I group at SPEAR in a sphericity analysis of their data. At PETRA/PEP energies the jet structure is already visible in the raw data. Fig. 6.4 shows a typical two-jet event observed by the TASSO detector.

The average sphericity [175] measured at DORIS and at PETRA is plotted versus the c.m. energy W in fig. 6.5. The sphericity decreases proportionally to $W^{-1/2}$, demonstrating that the jets indeed become more collimated at high energies. The jet cone half opening angle shrinks from 31° at 4 GeV to 17° at 36 GeV, i.e., the hadrons in an 18 GeV jet occupy only 2% of 4π.

The spin of the quark is reflected in the angular distribution of the jet axis with respect to the beam axis. A spin-1/2 quark has $d\sigma/d\Omega \sim 1 + \cos^2\theta$. The angular distribution was first measured [174] by the MARK I collaboration at SPEAR, and they confirmed that the quarks are fermions. The angular distribution of two jet events measured at high energies by the TASSO groups is plotted in fig. 6.6. The data are in good agreement with the predicted $1 + \cos^2\theta$ distribution shown as the solid line.

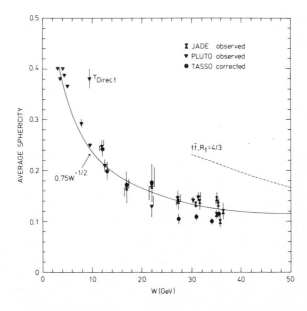

Fig. 6.5. The average sphericity plotted versus the c.m. energy.

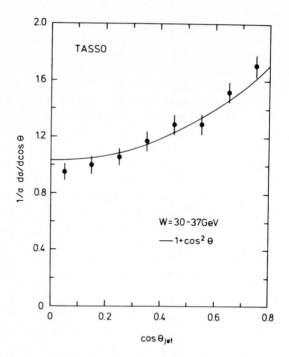

Fig. 6.6. The angular distribution of the jet axis in multihadron events with respect to the beam axis. The jet axis was determined by the TASSO group using the sphericity method. The solid line shows the $1 + \cos^2 \theta$ distribution expected from pair production of fermions.

6.3. Charge correlation

The back to back produced quarks have opposite charge. According to the standard picture they will fragment into hadrons by a neutral quark–gluon cascade conserving the initial charge. Therefore, apart from fluctuation, the resulting jet should remember the charge of the primary quark, so that the charge found in one jet should be correlated [176] with the charge of the other jet. Furthermore one expects this long range correlation to be found among the fast particles, and that the slow particles should exhibit short range correlations only. The TASSO group has found evidence [177] for these correlations using two different methods.

The first method is based on a suggestion [178] by Field and Feynman to use the momentum-weighted particle charge as a definition of the jet charge:

$$q_{\text{jet}}(\gamma) = \sum_{i=1}^{n} e_i x_i^{\gamma}. \tag{6.6}$$

Here e_i is the particle charge and $x_i = p_i / p_{\text{beam}}$ its relative momentum. The sum is over all particles in the jet. γ must be positive since the primary quark is expected to be a constituent of a particle with large x_i. A slightly modified definition of the charge is used in order to cancel the strong γ dependence:

$$q'_{\text{jet}}(\gamma) = q_{\text{jet}}(\gamma) \sum_i x_i^{\gamma} \Big/ \sum_i x_i^{2\gamma}. \tag{6.7}$$

With this definition of a jet charge they define a correlation:

$$P'(\gamma) = -\langle q'_1(\gamma) q'_2(\gamma) \rangle. \tag{6.8}$$

$P'(\gamma)$ is evaluated for all events which have at least one particle with $x_i \geqslant 0.35$ in each jet. The result, plotted in fig. 6.7 versus γ, shows a strong positive correlation as expected if long range charge correlations are present.

A long range charge correlation is also trivially imposed by the fact that the event as a whole must be neutral, i.e., if one jet is negatively charged, then the other must be positively charged by charge conservation. This effect has been evaluated by randomly redistributing all the charges in the jet and then evaluating $P'(\gamma)$. The result is shown as the

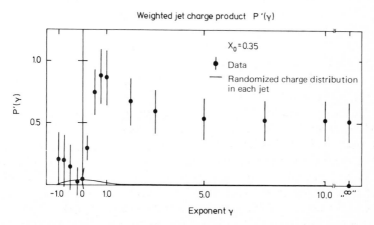

Fig. 6.7. The value of the momentum-weighted charge correlation product $P'(\gamma)$ as a function of γ, for events with at least one particle with $x_1 > 0.35$ in each jet. The solid line shows the charge correlation for a randomized charge distribution. The data were obtained by TASSO.

solid line. The observed charge correlation cannot be explained by the charge conservation mentioned above.

The charge correlation has been further investigated by evaluating the function:

$$\bar{\phi}(y, y') = -\frac{1}{\Delta y \Delta y'}\left\langle 1/n \sum_{k=i}^{n}\sum_{i \neq k} e_i(y)e_k(y')\right\rangle. \qquad (6.9)$$

In this expression, $e_i(y)$ is the charge of a particle i at rapidity y in the interval Δy, and $e_k(y')$ is the charge of a particle k at a rapidity y' in the interval $\Delta y'$. The rapidity is defined as

$$y = \tfrac{1}{2}\ln\left[(E + p_{\|})/(E - p_{\|})\right], \qquad (6.10)$$

where $p_{\|}$ is the particle momentum along the jet axis. The function $\bar{\phi}(y, y')$ is related to the probability that the particles i and k have opposite-sign charges minus the probability that the charges have the same sign. Since the event as a whole is neutral, the function $\bar{\phi}(y, y')$ simply shows how the charge of particle i at rapidity y is being compensated. The normalization is chosen such that $\int \bar{\phi}(y, y')\mathrm{d}y' = 1$. In fig. 6.8 the ratio $\tilde{\phi}(y, y') = \bar{\phi}(y, y')/\int \bar{\phi}(y, y')\mathrm{d}y$ is plotted versus y with the test particle in various rapidity intervals y'.

The distribution for $-0.75 \leqslant y' \leqslant 0$, i.e., a slow particle, is shown in fig. 6.8a. This distribution peaks at small negative values of y and shows that the charge of a slow particle is indeed compensated locally, as expected if only short range correlations are present. The data are in good agreement with the standard Monte Carlo calculations (see below) based on $q\bar{q}$ and $q\bar{q}g$ production shown as the solid line.

As the test particle becomes faster the distribution becomes increasingly skewed with a tail extending to positive y values. In fig. 6.8d the correlation function is plotted for $-5.5 \leqslant y' \leqslant -2.5$. Although the bulk of the charge is still compensated locally, there is now a significant signal at the opposite end of the rapidity plot. The probability that the charge of a particle with $y' < -2.5$ is compensated by a particle in the opposite jet with $y > 1$ is $(15.4 \pm 2.6)\%$.

This long range correlation is also reproduced by the standard Monte Carlo program (see section 7). However, this correlation is not present if the initial partons are neutral. This was demonstrated by using the same M.C. program, but assuming the initial quark to be neutral. The results, shown as the dashed line in fig. 6.8d, fail to reproduce the long range correlation observed.

The charges were also distributed at random among the particles in an event and the resulting correlation function evaluated. The ensuing

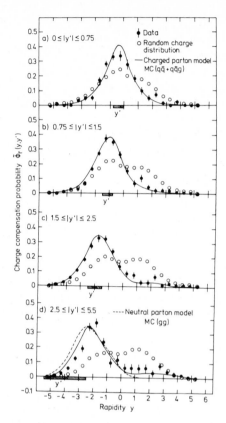

Fig. 6.8. The charge compensation probability plotted versus rapidity y for various rapidity intervals $\Delta y'$ of the test particle. The prediction for charged and neutral parton models is shown by the solid and the dashed lines, respectively. The charge compensation probability for a randomized charge distribution is shown by the open circles. The data were obtained by TASSO.

correlation functions, shown as the open circles in fig. 6.8 are much wider than the data. Thus the TASSO group has demonstrated that a long range charge correlation between particles in opposite jets exists and that the initial partons are charged.

6.4. The gross properties of the final state

The data above confirm the basic mechanism $e^+e^- \rightarrow q\bar{q}(g) \rightarrow$ hadrons, i.e., the hadrons result from the fragmentation of quarks and gluons. We next discuss some of the gross properties of the final state.

6.4.1. The neutral energy fraction

The JADE and the CELLO collaborations have determined ρ_γ, the fraction of total energy converted into photons, by a direct measurement of the photon energy deposited in lead glass counters surrounding the detector. JADE has also determined the total neutral energy fraction ρ_N by a measurement of the energy carried away by charged particles and subtracting this from the known c.m. energy. The results [165] are plotted versus energy in fig. 6.9 and show that both the neutral energy fraction and the gamma energy fraction are constant within errors, for energies between 12 GeV and 35 GeV. The data are also in agreement with a measurement [179] of the neutral energy fraction by the Crystal Ball collaboration.

The JADE group has also evaluated the neutrino energy fraction ρ_ν

$$\rho_\nu = \rho_N - \rho_\gamma - \rho_{K^0_L,n}.$$

The energy fraction $\rho_{K^0_L,n}$, carried off by K^0_L and neutrons, was estimated. They find $\rho_\nu < 10\%$ with 95% C.L. The Pati–Salam model [180] with integer-charge quarks predicts that between 18% and 28% of c.m. energy is carried off by neutrinos, in disagreement with the experimental results.

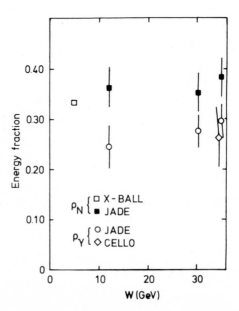

Fig. 6.9. The fraction of total energy into photons, ρ_γ, and neutral energy, ρ_N, plotted versus c.m. energy.

6.4.2. Charged multiplicity

The averaged charged multiplicity observed at high energies [165, 181] is plotted together with lower energy data [182] in fig. 6.10. The high energy data points from various groups are in reasonable agreement and well above the multiplicity predicted by extrapolating lower energy data according to the $(a + b)\ln s$ dependence predicted in the naive quark–parton model. In the leading log QCD, the averaged charged multiplicity n_{ch} is expected [183] to increase as

$$\langle n_{ch} \rangle = n_\rho + a \exp\{b[\ln(s/\Lambda^2)]^{1/2}\}, \quad b = 1.77. \tag{6.11}$$

Indeed the data are well represented [165] in the entire region by this form with $n_\rho = 2.0 \pm 0.2$, $a = 0.027 \pm 0.01$ and $b = 1.9 \pm 0.2$. The fit is shown as the dashed curve marked A in fig. 6.10. However, most of the sharp increase in multiplicity seems to be unrelated to multiple gluon emission. This is demonstrated by a computation of the charged multiplicity expected for quark pairproduction with subsequent fragmentation. The result of this computation which does not include gluon emission is shown as the solid line marked q$\bar{q}$ in fig. 6.10. A similar computation

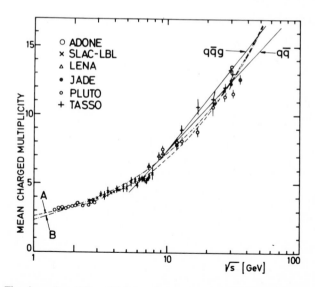

Fig. 6.10. The charged particle multiplicity plotted versus c.m. energy. The dashed line (A) indicates a fit using the QCD form $\langle n_{ch} \rangle = n_0 + a \exp b[\ln(s/\Lambda^2)]^{1/2}$. The dashed–dotted line (B) indicates a fit of the form $\langle n_{ch} \rangle \sim s^{1/4}$. The solid lines marked q$\bar{q}$g and q$\bar{q}$ are Monte Carlo results from a standard fragmentation model with and without gluons.

including gluon emission predicts a charged multiplicity higher by only one unit of charge at the highest PETRA energy. The bulk of the strong increase in multiplicity is therefore apparently a phase space effect; the full cascade process can only develop at c.m. energies above 10 GeV.

A simple statistical model [184] predicts the multiplicity to rise proportional to $s^{1/4}$. This dependence, shown as the curve marked (B) in fig. 6.10, represents the data rather well. The model, however, fails to fit [165] the charged particle dispersion:

$$D_{ch} = \left(\langle n_{ch}^2 \rangle - \langle n_{ch} \rangle^2 \right)^{1/2}.$$

6.5. Particle ratios and momentum spectra

Hadrons have now been identified over a large range in momentum and the available data with references are summarized in table 6.1.

Table 6.1
Data on identified hadrons at high energies

Particles	Group	Technique	Momentum range (GeV/c)	Ref.
π^0	TASSO	Liquid argon	0.5–4.0	[185]
$\pi^\pm$	JADE	dE/dx	< 0.7, 2–7	[165][a]
	TASSO	TOF Čerenkov	0.4–10.0	[186]
$K^\pm$	JADE	dE/dx	< 0.7	[165][a]
	TASSO	TOF	< 1.1	[186]
$K^0, \overline{K}^0$	JADE	$K_s^0 \to \pi^+\pi^-$	all momenta	[165][a]
	PLUTO	$K_s^0 \to \pi^+\pi^-$	all momenta	[187]
	TASSO	$K_s^0 \to \pi^+\pi^-$	all momenta	[188, 189]
	MARK II	$K_s^0 \to \pi^+\pi^-$	all momenta	[37][a]
$\Lambda, \overline{\Lambda}$	JADE	$\overline{\Lambda} \to \overline{p}\pi^+$ ($\overline{p}$ and π^+ by dE/dx)	0.4–1.1	[190]
	TASSO	vertex fits to pairs with opposite charges	1–10	[189]
	MARK II	vertex fits to pairs with opposite charges	1–10	[37][a]
$p\overline{p}$	JADE	dE/dx	0.3–0.9	[165][a]
	TASSO	TOF	0.5–2.2	[186]
	MARK II	TOF	0.5–2.0	[37][a]

[a] Preliminary data.

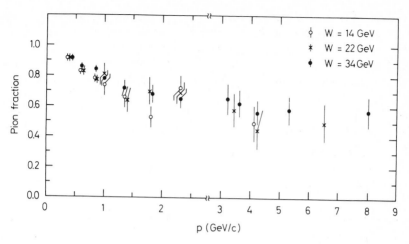

Fig. 6.11. The pion fraction in $e^+e^- \to$ hadrons, as a function of momentum for various c.m. energies. The data are from the TASSO collaboration.

The pion fraction, measured by the TASSO collaboration, is plotted versus momentum for various c.m. energies in fig. 6.11. The pion fraction is nearly independent of c.m. energy, and decreases slowly with increasing pion momentum, from nearly 100% at low momenta to around 50–60% at the highest momenta measured.

The cross section $d\sigma/dp$ to produce $\pi^\pm$, π^0, $K^\pm$, $K^0\overline{K}^0$, p$\overline{\text{p}}$ and $\Lambda\overline{\Lambda}$ at a c.m. energy of 30 GeV is plotted in fig. 6.12 versus momentum. The data show that the cross section for $\pi^+ + \pi^-$ production is equal to twice the cross section for π^0 production. Also, the cross sections for charged and neutral kaon production are in rough agreement in the momentum region where they overlap. Although pions dominate at low energies, other particle species become increasingly important with momentum. For example, at 10 GeV a ratio of $\pi^\pm : K^0\overline{K}^0 : \Lambda\overline{\Lambda}$ of 90:50:20 is observed. Thus a substantial number of heavy particles is produced, and a typical event at 30 GeV in c.m. contains 11 $\pi^\pm$, 5.5 π^0, 1.4 $K^0\overline{K}^0$ (1.4 K^+K^-) $\geqslant 0.4$ p$\overline{\text{p}}$ and 0.3 $\Lambda\overline{\Lambda}$ in the final state. The number of neutral kaons is about a factor of three larger than the yield observed in pp interactions.

The invariant cross section for neutral kaon production is plotted [191] in fig. 6.13 for data obtained at different c.m. energies. The constituent strange quarks in the K^0 result from primary s$\overline{\text{s}}$ production, weak decays of c$\overline{\text{c}}$ and b$\overline{\text{b}}$ quarks, plus the contribution from s$\overline{\text{s}}$ pairs created from the

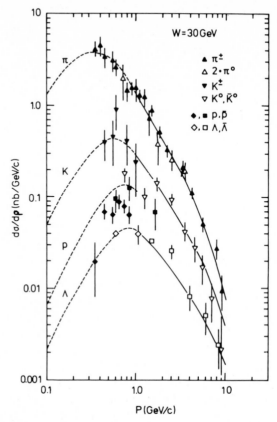

Fig. 6.12. Differential cross sections $d\sigma/dp$ for inclusive particle production, $e^+e^- \to hX$, for $\pi^+ + \pi^-$ (TASSO, preliminary), $2\pi^0$ (TASSO), $K^+ + K^-$ (TASSO), $K^0 + \overline{K}^0$ (TASSO), $p + \overline{p}$ (JADE, TASSO), $\Lambda + \overline{\Lambda}$ (JADE, TASSO). The dashed curves are of the form $(E^2/p^2)d\sigma/dp \sim \exp(-bE)$ with $b = 3.6$ GeV. The solid curves are drawn just to guide the eye.

vacuum. The contribution from these sources and their sum have been evaluated using the standard Monte Carlo model and are plotted in fig. 6.14. The Monte Carlo computations are in fair agreement with the data and indicate that the sea contribution is the dominant contribution even up to large values of x.

To form a baryon requires three quarks in a colour singlet. One might therefore naively expect baryon production to be strongly suppressed relative to meson production. Maybe the most surprising result obtained

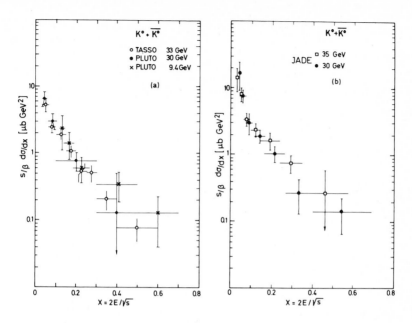

Fig. 6.13. The invariant cross section $(s/\beta)\,d\sigma/dx$ for inclusive $K^0 + \overline{K}^0$ production versus x for various c.m. energies, obtained by TASSO and PLUTO (a) and JADE (b).

in the inclusive measurements has been the large cross section for baryon production. This may reflect the fact that the quarks have baryon charge 1/3. The measured cross sections $d\sigma/dp$ for inclusive p, $\bar{p}$ production around 30 GeV in c.m. are plotted [37] in fig. 6.15. The total cross section for $p\bar{p}$ production is plotted in fig. 6.16 versus energy, together with data at lower energies. The error in the proton data is quite large since the cross sections have only been measured over a limited momentum range. The data indicate that about 0.7 nucleon pairs are produced per event.

Inclusive $\Lambda\overline{\Lambda}$ production however, can be measured over a wide momentum range since the Λ's can be identified via the decay mode $\Lambda \to p\pi$. The invariant cross section for $\Lambda\overline{\Lambda}$ production measured by the TASSO collaboration is plotted in fig. 6.17, and compared to the invariant cross sections for inclusive charged pion and neutral kaon production measured by the same group. The invariant cross section for $\Lambda\overline{\Lambda}$ production normalized to the invariant cross sections for kaon or pion production is apparently independent of momentum, i.e., the relative probability of creating a colourless diquark system and a colourless three quark system is independent of momentum.

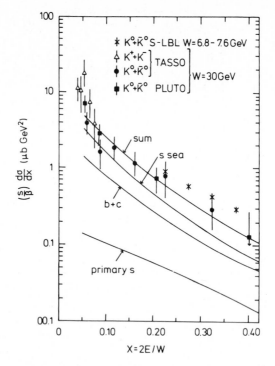

Fig. 6.14. The invariant cross section $(s/\beta)d\sigma/dx$ for inclusive $K^0 + \overline{K}^0$ production, plotted versus x. The estimated contributions from primary $s\bar{s}$, $c\bar{c}$ and $b\bar{b}$ production and from $s\bar{s}$ pair creation from the vacuum, together with their sum, are also shown.

This has led many authors [192, 193] to consider diquark production, i.e., not only $(q\bar{q})$ quark pairs but also $(qq \cdot \bar{q}\bar{q})$ states are produced in the quark cascade.

In fact, the available data can be well fit assuming the relative probability for creating a diquark is 0.075%. The prediction [193] of such a simple model is shown in fig. 6.18 (solid and dashed lines). More detailed models of baryon production based on the string model and incorporating diquarks have also been given. The dashed–dotted curves in fig. 6.18 show the predictions [192] by the Lund group based on such a model. The model fits the proton data at low momenta but decreases faster with momentum than the observed cross section. The model fails to predict the Λ data by a factor of 2 to 3.

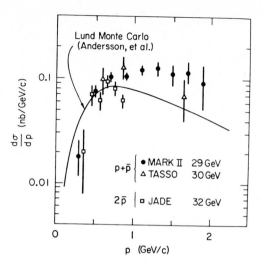

Fig. 6.15. The cross section $d\sigma/dp$ for inclusive $p+\bar{p}$ production versus momentum.

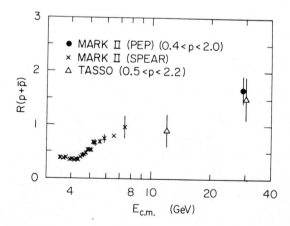

Fig. 6.16. The total cross section for inclusive $p+\bar{p}$ production, normalized to the muon pair production cross section, versus c.m. energy.

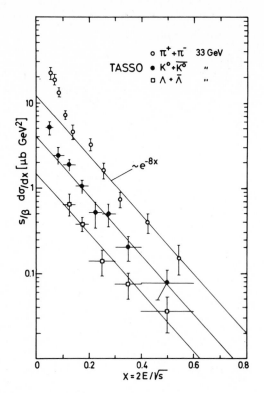

Fig. 6.17. The invariant cross section $(s/\beta)\,\mathrm{d}\sigma/\mathrm{d}x$ for inclusive $\pi^+ + \pi^-$, $K^0 + \overline{K}^0$ and $\Lambda + \overline{\Lambda}$ production versus x.

6.6 Scaling violations

The differential cross section for producing a hadron with energy E, momentum p and mass m at an angle θ with respect to the beam axis, can be written [195] in terms of two structure functions W_1 and W_2 as:

$$\frac{\mathrm{d}\sigma}{\mathrm{d}x\,\mathrm{d}\Omega} = \frac{\alpha^2\beta x}{s}\left(mW_1 + \tfrac{1}{4}\beta^2 x\nu W_2 \sin^2\theta\right), \tag{6.12}$$

with $\beta = p_h/E_h$, $x = 2E_h/\sqrt{s}$ and $\nu = (E_h/m\sqrt{s})$. ν is the energy of the virtual photon viewed in the rest system of h. Integrating this expression

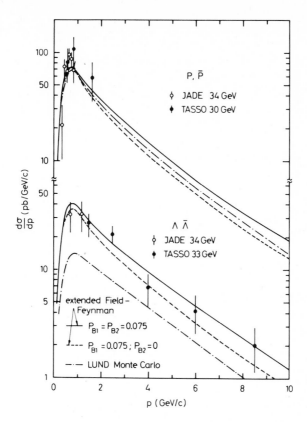

Fig. 6.18. The cross section $d\sigma/dp$ for inclusive production of $p+\bar{p}$ and $\Lambda + \bar{\Lambda}$ at c.m. energies between 30 and 34 GeV. The solid and dashed curves show the predictions based on the extended Field–Feynman model of Meyer. The dashed-dotted curves show the prediction of the Lund model.

over angles, and retaining only the leading terms yields:

$$\frac{d\sigma}{dx} \simeq \frac{4\pi\alpha^2}{s} \beta x m W_1.$$ (6.13)

The quark models predict W_1 to scale, i.e. to be a function of x only. This leads to:

$$\frac{d\sigma}{dx}(e^+e^- \to hX) = \sum_q \sigma(q\bar{q}) \left[D_q^h(x) + D_{\bar{q}}^h(x) \right]$$ (6.14)

$D_q^h(x)$, the fragmentation function gives the probability that a primary quark q yield a hadron h with relative momentum x.

In QCD $s/\beta \, d\sigma/dx$ no longer scales; soft collinear gluon emission (fig. 6.1) will deplete the particle yield at large values of x and enhance the yield at small values of x, since the energy is now shared between the quark and the gluon. In general—since q^2 is very large compared to Λ^2—the effects are rather small: of the order of 10–20% in the PETRA/PEP energy range. To see a clear effect, data in the SPEAR and DORIS energy range must be compared with data obtained at higher energies at PETRA and PEP.

The scaled cross sections $s \, d\sigma/dx$ for inclusive charged particle production as measured by DASP at DORIS, by MARK I at SPEAR and by TASSO, JADE and CELLO at PETRA are plotted [165] in fig. 6.19 versus y. The solid line ($s \, d\sigma/dx = 23 \exp(-8x_p) \, \mu b \; GeV^2$) is drawn to guide the eye. The cross section at $x < 0.2$ increases dramatically with energy and shows that the observed increase in the multiplicity is due to slow particles. The low energy cross section is above the high energy cross section at larger values of x, as expected in QCD. In order to reduce the

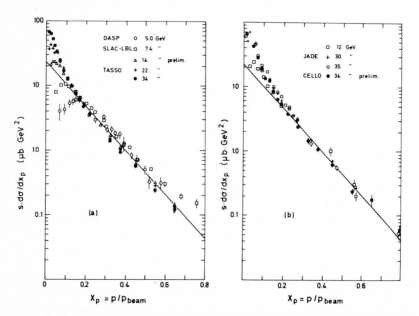

Fig. 6.19. The invariant cross section $s \, d\sigma/dx_0$ plotted versus x, for c.m. energies between 5.0 and 34 GeV. The solid line, $s \, d\sigma/dx = 23 \exp(-8x_p) \; GeV^2 \, \mu b$, is drawn to guide the eye.

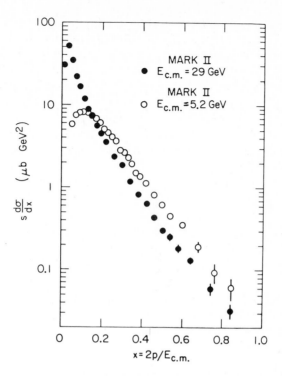

Fig. 6.20. Comparison of the inclusive cross sections $s\,\mathrm{d}\sigma/\mathrm{d}x$ for $e^+e^- \to h^{\pm} X$ measured by the MARK II collaboration at 5.2 and 29 GeV.

systematic uncertainties it is advantagous to compare data collected with the same detector over a large range in energies. In fig. 6.20, the inclusive cross section $s\,\mathrm{d}\sigma/\mathrm{d}x$ measured at 5.2 GeV and at 29 GeV using the MARK II detector is shown. The low energy data are clearly above the high energy data at $x \geqslant 0.3$.

In fig. 6.21, the invariant cross section is plotted versus the c.m. energy for fixed values of x. The energy dependence is consistent with that expected in QCD. A direct comparison with QCD might be premature since effects caused by crossing the $c\bar{c}$ and the $b\bar{b}$ threshold and the fact that the hadrons are not identified still remains to be studied.

The invariant cross section $(s/\beta)\,\mathrm{d}\sigma/\mathrm{d}x$ for identified charged pions is plotted in fig. 6.22 versus x for c.m. energies between 4.5 GeV and 34 GeV. Also, these data, which came from the DASP and the TASSO collaborations, indicated a sizable scaling violation.

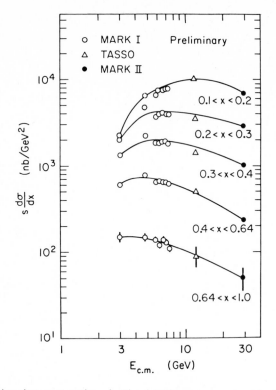

Fig. 6.21. The invariant cross section $s\,d\sigma/dx$ plotted versus c.m. energy for fixed values of $x = 2P/\sqrt{s}$. The solid lines are drawn just to guide the eye.

7. The gluon

The leading candidate for a field theory of strong interactions is, at present Quantum ChromoDynamics (QCD) [14]. In QCD, the strong force is mediated by eight massless coloured vector particles, the gluons, which couple directly to colour. The strong charge, or coupling strength α_s (eq. 5.2) is independent of quark flavour, but depends on the momentum transfer Q in the process, resulting in a coupling which is strong at small values of Q^2 (large separations) and weak at large values of Q^2 (small separations). The Q^2 dependence of the coupling constant reflects the fact that gluons are coloured objects with self interactions (fig. 6.1).

To demonstrate that QCD is indeed the correct theory of strong interactions, one must show that gluons with the properties listed above

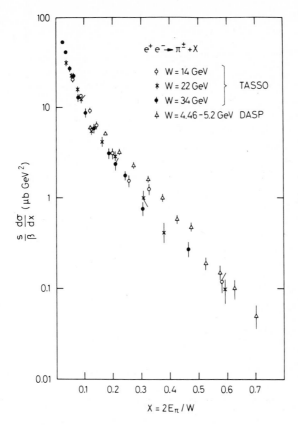

Fig. 6.22. The invariant cross section $(s/\beta)\,d\sigma/dx$ for inclusive $\pi^+ + \pi^-$ production plotted versus $x = 2E_\pi/\sqrt{s}$.

exist. The first evidence for gluons came from deep inelastic lepton–nucleon experiments and will be discussed in section 9. Electron–positron interaction has given direct evidence for gluons from a study of heavy quarkonium states and from quark–gluon bremsstrahlung.

7.1. Quarkonium

7.1.1. The three-gluon decay of $\Upsilon(1S)$

In QCD, a 1^{--} state is expected to decay predominantly via a three-gluon intermediate state.

The energy distribution of the gluons in $1^{--} \to ggg$ can be written [196] as:

$$\frac{1}{\Gamma} \frac{d\Gamma}{dx_1 dx_2} = \frac{6}{\pi^2 - 9} \frac{x_1^2(1-x_1)^2 + x_2^2(1-x_2)^2 + x_3^2(1-x_3)^2}{x_1^2 x_2^2 x_3^2},$$

(7.1)

where $x_i = E_i/E_b$ is the scaled energy.

This should lead to planar events, defined by the three gluons fragmenting into three jets of hadrons. However, Υ decays yield, in general, two energetic gluon jets plus one low energy gluon jet. The symmetric case $x_1 = x_2 = x_3 = 2/3$ is rather unlikely, and even in this case each jet will, on the average, have an energy of only 3 GeV. The data on $e^+e^- \to$ hadrons do not show a clear jet structure at 6 GeV in c.m.

PLUTO has analyzed the hadron data [197] from Υ decays using triplicity [198], a generalization of thrust to three axes. In this method the final state hadrons with momenta $p, p_2 \ldots p_N$ are grouped into three classes C_1, C_2 and C_3 with momenta $p(C_N) = \Sigma|p_i|$ where the sum is over all particles assigned to class C_N. Triplicity T_3, is then defined as:

$$T_3 = \left(\sum_i p_i\right)^{-1} \max[|p(C_1)| + |p(C_2)| + |p(C_3)|].$$

(7.2)

T_3 is 1 for a perfect three-jet event and $3(\sqrt{3}/8)$ for a spherical event.

The momenta of the three jets are given by $p_1 = p(C_1), p_2 = p(C_2)$ and $p_3 = p(C_3)$, and the angles between these vectors θ_1, θ_2, and θ_3 are the angles between the three jets. The triplicity and the thrust distribution for events on the Υ are plotted in fig. 7.1. The data include both neutral and charged tracks and they have been corrected for the continuum contribution $e^+e^- \to q\bar{q}$ and for the vacuum polarisation. The triplicity and thrust distributions for the following decay modes have been evaluated:

(i) $e^+e^- \to \Upsilon \to$ hadrons, where the hadrons are distributed according to phase space.

(ii) $e^+e^- \to \Upsilon \to q\bar{q}$. Such a distribution might be expected if the Υ decays via a one-gluon intermediate state. Also $e^+e^- \to \Upsilon \to gg$ would lead to a similar topology.

(iii) $e^+e^- \to \Upsilon \to ggg$. This is the lowest order diagram in QCD.

The results are also plotted in fig. 7.1 and it is clear that the three-gluon distribution agrees well with the data, whereas neither phase space nor two-body $q\bar{q}(gg)$ decays fit the observed distributions. The thrust distribution for events collected in the continuum adjacent to the Υ is plotted

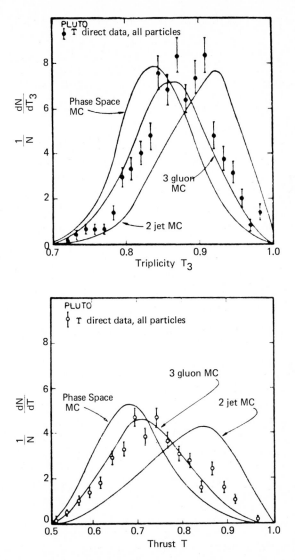

Fig. 7.1. Triplicity and thrust distributions observed by the PLUTO group for events at the $\Upsilon(1S)$ resonance. Monte Carlo predictions based on a phase space model, two-body $q\bar{q}$ and three-body ggg final states are also shown.

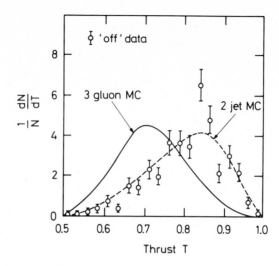

Fig. 7.2. The thrust distribution in the continuum, observed by the PLUTO group at energies adjacent to the $\Upsilon(1S)$ resonance.

in fig. 7.2. This distribution is peaked at large values of thrust and is well fit by $e^+e^- \to q\bar{q}$ but not by $e^+e^- \to ggg$.

With a colourless gluon, Υ would decay [199] to lowest order via a one-gluon intermediate state, producing a final state with two collinear hadron jets resulting in similar thrust distributions on and off resonance. The marked difference in event topology observed on and off the resonance is strong indirect evidence that the gluon *has* colour.

The angular distribution of the thrust axis observed in the continuum [148, 200] and at the $\Upsilon(1S)$ is plotted in fig. 7.3. The continuum events clearly favour the $1 + \cos^2\theta$ distribution predicted for $e^+e^- \to q\bar{q}$, whereas the data at the $\Upsilon(1S)$ favour a somewhat flatter distribution.

A fit to the data of the form $1 + \alpha_T \cos^2\theta$ gives [148] $\alpha_T = 0.33 \pm 0.16$, consistent with $\alpha_T = 0.39$ predicted [196, 201] for the decay $\Upsilon(1S) \to ggg$ with vector gluons, and disagrees strongly with the value $\alpha_T = -1.0$ predicted for scalar gluons. The CLEO collaboration has carried out similar measurements at CESR and finds [100] $\alpha_T = 0.35 \pm 0.11$, consistent with the PLUTO value. QCD also predicts that the normal to the plane defined by the three gluons should be distributed as $1 + \alpha_N \cos^2\phi$ with $\alpha_N = -0.33$ for vector gluons, and $\alpha_N = \pm 1$ for $0^\pm$ gluons. From a fit to its data CLEO obtains [100] $\alpha_N = -0.26 \pm 0.03$, in good agreement with the vector hypothesis.

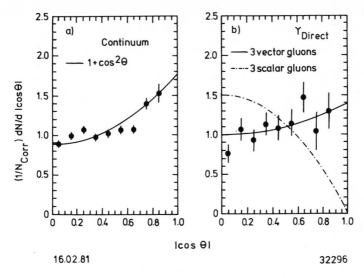

16.02.81   32296

Fig. 7.3. Data on the distribution of the thrust axis with respect to the beam axis, observed for continuum and $\Upsilon(1S)$ events by the LENA group. The data are compared to theoretical predictions.

Independent information on the gluon spin comes from a measurement [100, 200] of the ratio

$$A = \frac{\Gamma(\Upsilon' \to \pi^+ \pi^- \Upsilon)}{\Gamma(\psi' \to \pi^+ \pi^- J/\psi)}.$$

For vector gluons, this ratio depends on the size of the system, whereas for scalar gluons the value is independent of size. The value of A has been evaluated by Gottfried [202] and he finds $A = 0.1$ or 1 for vector and scalar gluons, respectively. Experimentally the early LENA result [200] of $A = 0.09 \pm 0.05$ has been confirmed by the CLEO group [100], which finds $A = 0.085 \pm 0.058$.

7.1.2. $Q\bar{Q}$ spectra and decay widths

The observed [146, 203] $c\bar{c}$ and $b\bar{b}$ states identified by the spectroscopic notation $n^{2S+1}L_J$ are plotted in fig. 7.4 as a function of excitation energy. The level ordering is in agreement with the spectrum shown in Fig. 5.1 for two non-relativistic fermions bound in a steeply rising potential. The excitation energy E^* measured with respect to the 1^3S_1 state, and the ratio $R_{ee} = \Gamma_{ee}(n^3S_1)/\Gamma_{ee}(1^3S_1)$ of the leptonic widths are listed in table

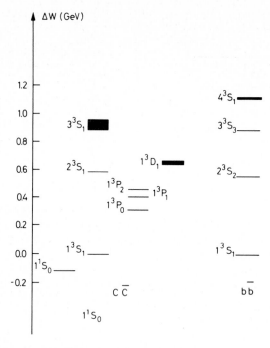

Fig. 7.4. Observed $c\bar{c}$ and $b\bar{b}$ states plotted versus excitation energy. The states are identified using the spectroscopic notation $n^{2S+1}L_J$.

7.1. It is hoped that some of the uncertainties [204] caused by higher order QCD radiative effects will cancel and that R_{ee} can be compared directly to first order calculations.

Several attempts [205–209] have been made to extract information on the strong force from the data on the $c\bar{c}$ and the $b\bar{b}$ system.

One approach is to derive [205] the potential directly from the observed levels, using the inverse scattering formalism. It has also been shown [208] that a rather simple potential of the form $V(r) = A + Br^\nu$, with the parameters determined by the lowest levels, can fit both the $c\bar{c}$ and $b\bar{b}$ states using identical parameter values.

In a complementary approach, one tries to derive a potential from first principles. Richardson [207] has obtained a QCD potential of the form

$$V(Q^2) = -\frac{4}{3} \frac{12\pi}{33-2N_f} \cdot \frac{1}{Q^2} \cdot \frac{1}{\ln(1+Q^2/\Lambda^2)}, \qquad (7.3)$$

Table 7.1

Properties of the $c\bar{c}$ and $b\bar{b}$ system

State	Measurement		Emp. potential [208]		QCD potential [209]	
	E* (MeV)	R_{ee}	E* (MeV)	R_{ee}	E* (MeV)	R_{ee}
ψ'	589.1 ± 0.1	0.44 ± 0.06	589	0.35	589	0.45
ψ''	671 ± 2		705		715	
$^3P_{2,1,0}$	426 ± 4		425		425	
$\Upsilon(2S)$	560.0 ± 0.3	0.46 ± 0.03	560	0.43	550	0.45
$\Upsilon(3S)$	890.3 ± 0.4	0.33 ± 0.03	890	0.28	890	0.32
$\Upsilon(4S)$	1113.0 ± 1.0	0.23 ± 0.02	1120	0.20	1160	0.26

written in momentum space. The potential has only one parameter Λ which may be identified with the characteristic strong interaction mass. The Fourier transform results in a potential which is proportional to $1/r$ at small distances, as expected for the exchange of a massless vector particle, and proportional to r at large distances, as expected for the confining term. The potential, with second order QCD corrections [209] included, is plotted in fig. 7.5 versus r.

The average r values for the J/ψ and the Υ are indicated. It is clear that both the $c\bar{c}$ and the $b\bar{b}$ family are mainly sensitive to the intermediate part of the potential between the Coulombic and the linear part. A fit to the observed spectrum gives $\Lambda_{\overline{MS}} = 0.508$ GeV which is consistent but

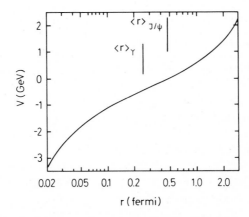

Fig. 7.5. The modified Richardson potential versus r.

larger than the value of α_s determined in other processes. The results are also listed in table 7.1.

Both QCD and simple potential models reproduce the available data on $c\bar{c}$ and $b\bar{b}$ rather well using the same values of the parameters for both flavours. This shows that the strong force does not depend on the quark flavour.

7.1.3. Determination of the coupling constant α_s

The data on the Υ states are in good agreement with predictions based on QCD and can therefore be used to determine α_s. The ratio of the gluonic and the leptonic widths of the $\Upsilon(1S)$ is given in leading order by eqs. (5.1) and (5.3):

$$\frac{\Gamma_{3g}}{\Gamma_{ee}} = \frac{10(\pi^2 - 9)}{81\pi e_b^2} \frac{\alpha_s^3(M)}{\alpha^2}. \tag{7.4}$$

Lepage and Mackenzie have calculated [210] the first order QCD correction to this ratio. They find, using the $\overline{MS}$ scheme:

$$\frac{\Gamma_{3g}}{\Gamma_{ee}} = \frac{10(\pi^2 - 9)}{81\pi e_b^2} \frac{\alpha_s^3(M)}{\alpha^2} \left[1 + (9.1 \pm 0.5) \frac{\alpha_s(M_\Upsilon)}{\pi} + \cdots \right] \tag{7.5}$$

The coefficient in front of $\alpha_s(M_\Upsilon)/\pi$ depends on renormalization scheme and scale.

The renormalization scale can be chosen such that eq. (7.4) is correct, and the authors find this to be the case for $M = 0.48 M_\Upsilon$.

Experimentally the three gluon width is related to the total width by:

$$\Gamma_{3g} = \Gamma_{tot} - \Gamma_{ee} - \Gamma_{\mu\mu} - \Gamma_{\tau\tau} - R\Gamma_{\mu\mu} - \Gamma_{\gamma 2g}$$

$$= \Gamma_{tot} - (3 + R + \Gamma_{\gamma 2g}/\Gamma_{ee})\Gamma_{ee}. \tag{7.6}$$

The term $R\Gamma_{ee}$ is the contribution from the vacuum polarization and $\Gamma_{\gamma 2g}/\Gamma_{ee}$ is a small correction for the decay $\Upsilon \to \gamma gg$.

$$\Gamma_{3g}/\Gamma_{ee} = \frac{1 - (3 + R + \Gamma_{\gamma 2g}/\Gamma_{ee})B_{ee}}{B_{ee}} \tag{7.7}$$

The values for Γ_{ee} are listed in table 5.3.

New data on the leptonic branching ratio of the $\Upsilon(1S)$ resonance have been reported [100, 146] by the CLEO and CUSB collaborations from a measurement of the cascade decay:

$$\Upsilon(2S) \to \pi^+\pi^- \Upsilon(1S) \to \pi^+\pi^- e^+ e^- (\mu^+\mu^-).$$

The CLEO and the CUSB collaborations find $B_{ee} = 3.6 \pm 0.4\%$, and $B_{ee} = 3.2 \pm 0.8\%$, respectively, using the known [100, 148] branching ratio $B(\Upsilon(2S) \to \pi^+\pi^- \Upsilon(1S)) = 19.1 \pm 3.1\%$. Combining results with earlier value of $B_{ee} = 3.0 \pm 0.8\%$ measured by the DORIS groups give a world average of $B_{ee} = 3.3 \pm 0.5\%$.

With these values Lepage and McKenzie obtain:

$$\alpha_s(0.48M_p) = 0.152^{+0.012}_{-0.010}, \qquad \Lambda_{\overline{MS}} = 100^{+34}_{-25} \text{ MeV}$$

The value for α_s is in good agreement with the value determined from quark-gluon bremsstrahlung.

7.2. Search for gluonium states in J/ψ radiative decays

Two or more gluons are expected [211] to form gluonium states, bound colourless particles with a mass spectrum starting perhaps around 1–1.5 GeV. Replacing one of the gluons in the decay $J/\psi \to ggg$ by a photon, results in a final state which consists of one photon and two gluons. This decay mode, observed [212] by the MARK II collaboration with a branching ratio consistent with the QCD prediction, seems to be well suited to search for gluonium, since the two gluons recoiling against the colourless photon must be in a colour singlet state. At present, two possible candidates for gluonium, $\iota(1440)$ and $\theta(1640)$, have been observed in radiative J/ψ decays.

7.2.1. $J/\psi \to \gamma\iota(1440)$

The decay $J/\psi \to \gamma\iota(1440)$ was first observed [212] by the MARK II collaboration in the channel $K_s K^\pm \pi^\mp$. The $K_s K^\pm \pi^\mp$ mass distribution for candidate events, plotted in fig. 7.6, shows a clear peak centered at 1440 MeV. Selecting events with $m_{K\overline{K}} < 1.05$ GeV enhances the signal as shown by the shaded distribution. The resonance parameters determined from the 5C mass distribution are listed in table 7.2.

The $\iota(1440)$ has also been observed [134] by the Crystal Ball collaboration in $J/\psi \to \gamma K^+ K^- \pi^0$, and the $K^+ K^- \pi^0$ mass distribution for candidate events is plotted in fig. 7.7. The shaded distribution corresponds to events with $M_{K\overline{K}} < 1125$ MeV. The mass, width and cascade branching ratio determined from these data are also listed in table 7.2. The two data sets are in excellent agreement. In particular, both groups find a large cascade branching ratio and this makes it tempting [213] to associate the $\iota(1440)$ with a gluonium state.

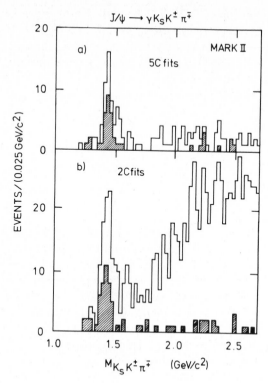

Fig. 7.6. The $K_s K^{\pm} \pi^{\mp}$ mass spectrum for $J/\psi \to \gamma K_s K^{\pm} \pi^{\mp}$ candidate events which satisfy (a) a 5C fit to this hypothesis, or (b) a 2C fit (observation of the photon not required). Events in the shaded region have $m_{K\bar{K}} < 1.05$ GeV. The data were obtained by the MARK II collaboration.

Table 7.2
The $\iota(1440)$

Quantity	MARK II	Crystal Ball	E meson
Mass (MeV)	1440^{+10}_{-15}	1440^{+20}_{-15}	1418 ± 10
Width (MeV)	50^{+30}_{-20}	60^{+20}_{-30}	50 ± 10
$B(J/\psi \to \gamma\eta) \times B(\eta \to K\bar{K}\pi)$	$(4.3 \pm 1.7) \times 10^{-3}$	$(4.0 \pm 1.2) \times 10^{-3}$	

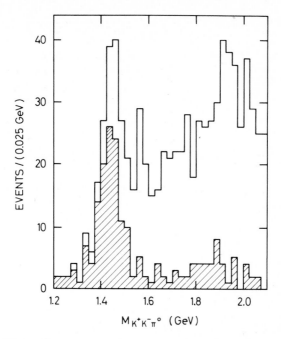

Fig. 7.7. The $K^+ K^- K^0$ mass spectrum for $J/\psi \to \gamma K^+ K^- \pi^0$ candidate events. Events in the shaded region have $m_{K\overline{K}} < 1125$ MeV. The data were obtained by the Crystal Ball collaboration.

The values for the mass and the width of the $\iota(1440)$ are consistent with the values reported [214] for the well established E meson. However, the spin parity of the E meson has been found to be 1^- and this makes it a natural candidate for the axial nonet together with the D, A, and Q_A mesons. The question whether ι is the E particle or not has now been decided by a measurement of the quantum numbers of the ι.

The Crystal Ball collaboration has determined [134] the quantum numbers of the $\iota(1440)$ to be 0^- from a spin parity analysis. The analysis was carried out in 100 MeV steps for an invariant $K\overline{K}\pi$ mass between 1300 and 1800 MeV. The analysis included $K\overline{K}\pi$ (phase space), $\delta^0\pi^0(0^-)$, $\delta^0\pi^0(1^+)$, $K^*\overline{K}+C.C.(0^-)$ and $K^*\overline{K}+C.C.(1^+)$.

The $K\overline{K}\pi$ phase space contribution was assumed to be incoherent, whereas the other partial waves were allowed to interfere with arbitrary phase. The ι and K^* helicities were allowed to vary. Significant contributions to the $K\overline{K}\pi$ yield came only from $KK\pi$ (phase space), $\delta^0\pi^0(0^-)$

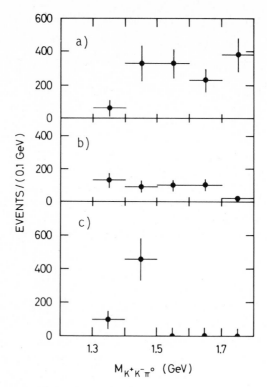

Fig. 7.8. The contributions for various partial waves as a function of $K\overline{K}$ mass: (a) phase space ($K\overline{K}\pi$ flat), (b) $K^*K^- + C.C.$ in $J^P = 1^+$, (c) $\delta\pi$ in $J^P = 0^-$. The analysis was made by the Crystal Ball collaboration.

and $K^*\overline{K} + C.C.(1^+)$. These contributions, corrected for the detection efficiency, are plotted versus $K^+K^-\pi^0$ mass in fig. 7.8. The $K^*\overline{K} + C.C.(1^+)$ partial wave contribution is small and independent of mass. The $\delta\pi(0^-)$ contribution, however, shows a clear resonance structure around the ι mass. The spin parity of the ι is therefore 0^-. Many cross checks have been made [134] and they all support the results of the partial wave analysis.

The analysis also finds:

$$\frac{B(\iota \to K^*\overline{K} + C.C.)}{B(\iota \to K^*\overline{K} + C.C. + \delta\pi)} < 0.25 \quad (90\% \text{ C.L.}).$$

Note that Dionisi et al. [215] determine

$$\frac{B(E \to K^*\overline{K} + C.C.)}{B(E \to K^*\overline{K} + C.C. + \delta\pi)} = 0.86 \pm 0.12.$$

The ι spin can be determined from the angular distributions with the assumption that the ι decays predominantly into $\delta\pi$. A fit to the full three-dimensional decay distribution strongly favours spin 0. If the relative probability for spin 0 is taken to be 1 the spin 1 and spin 2 hypotheses have relative probabilities of 10^{-4} and 0.008 respectively.

The experimentalists therefore conclude that the $\iota(1440)$ is a new pseudoscalar meson which cannot be assigned to the ground state nonet. It has been suggested that the $\iota(1440)$ is a radially excited $q\bar{q}$ state and such states have been predicted [216] to exist in this mass range.

It is of course still tempting to identify the $\iota(1440)$ with a gluonium resonance.

7.2.2. $J/\psi \to \gamma\theta(1640)$

The Crystal Ball collaboration has observed [134] a new resonance $\theta(1640)$ in the decay $J/\psi \to \gamma\eta\eta$. The $\eta\eta$ invariant mass distribution for events which satisfy the 5C fit to this decay mode is plotted in fig. 7.9. A

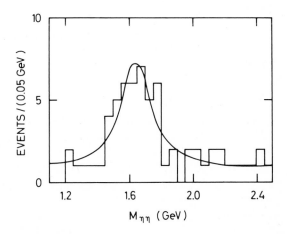

Fig. 7.9. The $\eta\eta$ mass distribution for $J/\psi \to \gamma\eta\eta$ candidate events. The data are from the Crystal Ball collaboration, and the solid line is the result of a Breit–Wigner plus a flat background fit.

clear resonance signal is seen in the data and a fit to a Breit–Wigner resonance plus a flat background results in a mass $M = 1640 \pm 50$ MeV and a width $\Gamma = 220^{+100}_{-70}$ MeV. The cascade branching ratio $B(\mathrm{J}/\psi \to \gamma\theta)$ $\times B(\theta \to \eta\eta) = (4.9 \pm 1.4 \pm 1.0) \times 10^{-4}$ is nearly an order of magnitude smaller than the corresponding cascade branching ratio for the $\iota(1440)$. A search for the decay mode $\theta \to \pi^0\pi^0$ was negative and the resulting 90% upper C.L. is:

$$B(\mathrm{J}/\psi \to \gamma\theta) \times B(\theta \to \pi^0\pi^0) < 6 \times 10^{-4}.$$

The $\theta(1640)$ must have $J^{PC} = 0^{++}, 2^{++}$, since it is produced in J/ψ radiative decays and decays into $\eta\eta$. A spin-parity analysis favours 2^{++}.

The $\theta(1640)$ does not fit into the ground state 2^{++} nonet and the mass is probably too low for the $\theta(1640)$ to be a radial excitation. Another possibility is that the θ is a $(q\bar{q}q\bar{q})$ four quark state. However, such a state is expected to have a rather large width.

The $\theta(1640)$ has the quantum number of the bound two-gluon ground state. Although its mass is somewhat larger than expected [217] for the gluon ground state, this assignment is still a possibility.

7.3. Quark–gluon bremsstrahlung

A quark which is being accelerated will radiate gluons analogous to normal bremsstrahlung. Quark–gluon bremsstrahlung [12, 218], shown to first-order in the coupling constant in fig. 7.10a, is expected to occur in any field theory of strong interaction. The first order cross section in QCD is given by:

$$\frac{1}{\sigma_0} \frac{\mathrm{d}\sigma(q\bar{q}g)}{\mathrm{d}x_1 \mathrm{d}x_2} = \frac{2\alpha_s}{3\pi} \frac{x_1^2 + x_2^2}{(1-x_1)(1-x_2)}. \tag{7.8}$$

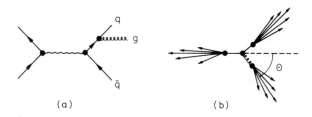

(a)   (b)

Fig. 7.10. (a) Quark–gluon bremsstrahlung to first order in α_s. (b) Final state in quark–gluon bremsstrahlung.

x_1 and x_2 are the fractional quark energies, $x_i = 2E_i/\sqrt{s}$, and σ_0 is the cross section. The energy and the angular distribution of the radiated gluon is equal to that of bremsstrahlung photons:

$$\frac{d\sigma}{dk\,d\Omega} \sim \frac{\alpha_s}{k\sin\theta}, \tag{7.9}$$

where k is the energy of the gluon and θ its angle with respect to the jet axis as defined in fig. 7.10b. The gluons are coloured and will materialize as a jet of hadrons with a mean p_T on the order of 300 MeV/c, similar to the value observed in quark fragmentation, and a normal multiplicity distribution.

Quark–gluon bremsstrahlung has well defined experimental signatures. The transverse momentum with respect to the jet axis will grow with energy:

$$\langle p_T^2 \rangle \sim Q^2/\ln(Q^2/\Lambda^2). \tag{7.10}$$

In some rare cases the gluon will be emitted with a transverse momentum which is large compared to 300 MeV/c. Then the event will consist of three well defined jets defining a plane.

Events with these properties were found [219] almost immediately at the turn on of PETRA, and at the Batavia Conference all groups [220–223] working at PETRA reported data.

Examples of three-jet events are shown in fig. 7.11. Before reviewing the evidence for gluon bremsstrahlung, I will briefly describe the Monte Carlo models used [224] to compare theory and data.

7.3.1. Monte Carlo simulation

All groups have made extensive Monte Carlo computations to confront the various production mechanisms with the data. The inputs to these calculations are summarized below:

(a) Quark pairs are pair-produced with a cross section proportional to e_i^2.

(b) The basic gluon bremsstrahlung process (fig. 6.1) is treated to first order in the strong coupling constant α_s by Hoyer et al. [225], whereas the computation by Ali et al. [227] includes all second order diagrams except those with internal gluon lines.

(c) The formalism of Field and Feynman [178] or the one set up by the Lund group [228] is then used to compute the fragmentation of the constituents.

In the Field–Feynman model, quark fragmentation is a cascade process where light quark pairs $u\bar{u}$, $d\bar{d}$ and $s\bar{s}$ are created by the colour field

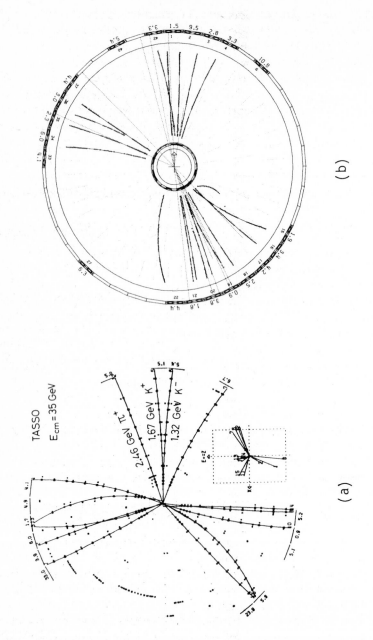

Fig. 7.11. Examples of three-jet events observed by (a) TASSO and (b) JADE.

from the vacuum in the ratio $2:2:1$. The $q\bar{q}$ pairs recombined with their neighbours to form mesons: i.e. $q \rightarrow q + q_1\bar{q}_1 \rightarrow (q\bar{q}_1) + q_1 \rightarrow M_1 + q_1 \rightarrow M_1 + q_1 + \bar{q}_2 q_2 \rightarrow M_1 + (q_1\bar{q}_2) + q_2 \rightarrow M_1 + M_2 + q_2 \ldots$. The process stops when the "free" quark reaches a threshold energy E_0; the quark then combines with the left-over quark from the other jet to form a wee hadron.

The cascade process is described by three parameters: a_F, σ_q and $P/(P+V)$.

(i) $a_F : f_q^h(z)$ is the probability that a quark q with energy E_q fragments into a hadron h with relative energy $z = (E+p)_h/(E+p)_q$. This probability function is given by: $f_q^h(z) = 1 - a_F + 3a_F(1-z)^2$. a_F is assumed to be the same for u, d and s quarks and its value is determined experimentally. For heavy quarks $f_q^h(z)$ is assumed to be constant.

(ii) σ_q: The quarks in the $q\bar{q}$ pair are created with equal and opposite transverse momenta according to a Gaussian with a rms width σ_q. The quarks from two different $q\bar{q}$ pairs are combined, yielding a meson with a net transverse momentum with respect to the jet axis.

(iii) $(P/P+V)$: Only pseudoscalar P $(\pi, K\ldots)$ and vector mesons $V(\rho, K\ldots)$ are produced in the primordial chain. The program has lately been extended to include [193] baryon production. This is done by assigning a small probability to the production of diquarks $(q_1\bar{q}_1 q_1\bar{q}_1)$ in the fragmentation chain.

The fragmentation of gluons is treated as a two step process, in which the gluon first fragments into a $q\bar{q}$ pair and then quark subsequently fragments into hadrons as outlined above. In the Hoyer et al. program, the gluon impacts all its energy to one of the quarks, i.e., in this model quark and gluon fragmentation are identical. Ali et al. describe $g \rightarrow q\bar{q}$ by the splitting function [229] $f(z) = z^2 + (1-z)^2$ with $z = E_g/E_q$.

Field and Feynman found [178] that deep inelastic lepton–hadron interactions and also hadron–hadron interactions are simultaneously described by the following values of the parameters

$$a_F = 0.77, \qquad \sigma_q = 0.30 \text{ GeV}/c, \qquad P/(P+V) = 0.5.$$

The TASSO group has made a fit to their data in a region dominated by two jet events. Simultaneous fits, varying a_F, σ_q and $P/(P+V)$ were made to the x distribution $(x = p_h/E_b)$, the mean $\langle p_T^2 \rangle_{out}$ distribution normal to the event plane (see below), and the charged multiplicity distribution. The best fits were obtained [230] for $a_F = 0.57 \pm 0.20$, $\sigma_q = 0.32 \pm 0.04$ GeV/c and $P/(P+V) = 0.56 \pm 0.15$, in agreement with the values found in lepton–hadron and hadron–hadron interactions. The quality of the fits is shown in fig. 7.12.

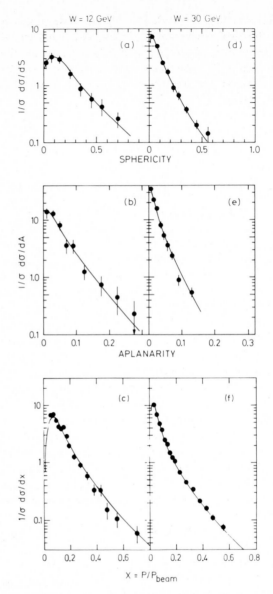

Fig. 7.12. Comparison of the data with the QCD model (solid lines) at 12 GeV and 30 GeV in c.m.

The Lund model [227] is based on the massless string model [228]. In this model the two quarks are no longer independent objects, and the string fragments as a whole. The Lund model predicts that quarks and gluons fragment differently and that the jets do not longer coincide directly with the parton momenta. The model has been used extensively by the JADE group and agrees with the data.

7.3.2. *Event topology*

The production mechanism can be delineated from the event shape. For example, light $q\bar{q}$ production results in two well defined jets along the same axis with small momenta transverse to this axis, $q\bar{q}g$ results in planar events, four parton events like $q\bar{q}gg$ define two planes, heavy $q\bar{q}$ production gives spherical events. By now, several methods are used to determine the shape and the topology of an event. Some of these methods are briefly discussed below.

The *shape* of an event is conveniently evaluated by constructing the second rank tensor [172, 231]

$$\mathbf{M}_{\alpha\beta} = \sum_{j=1} p_{j\alpha} \cdot p_{j\beta} \quad (\alpha, \beta = x, y, z) \tag{7.11}$$

where $p_{j\alpha}$ and $p_{j\beta}$ are momentum components along the α and the β axes for the jth particle in the event. The sum is over all charged particles in the event. Let $\mathbf{n}_1$, $\mathbf{n}_2$ and $\mathbf{n}_3$ be the unit eigenvectors of the tensor, associated with the normalized eigenvalues Q_i, where $Q_i = \Sigma(\mathbf{p}_j \cdot \mathbf{n}_j)^2 / \Sigma \mathbf{p}_j^2$.

These eigenvalues are ordered such that $Q_1 \leqslant Q_2 \leqslant Q_3$ and are normalized, with $Q_1 + Q_2 + Q_3 = 1$. The principal jet axis is then the $\mathbf{n}_3$ direction. The event plane is spanned by $\mathbf{n}_2$ and $\mathbf{n}_3$, and $\mathbf{n}_1$ defines the direction in which the sum of the squares of the momentum components is minimized. Every event can be represented in a two-dimensional plot of aplanarity $A = \frac{3}{2}Q_1$ (i.e. normalized momentum squared out of the event plane), versus sphericity $S = \frac{3}{2}(Q_1 + Q_2)$. In such a plot, two-jet events will cluster at small values of A and S, planar events have small values of A, whereas for spherical events both A and S will be large. This method has been used by TASSO [220] and JADE [223].

MARK J [221, 232] uses a linear method based on energy flow, where the coordinate system is defined as follows: the $\mathbf{e}_1$ axis coincides with the thrust axis, which is defined as the direction of maximum energy flow. They next investigate the energy flow in a plane perpendicular to the thrust axis. The direction of maximum energy flow in that plane defines a direction e_2 with a normalized energy flow

$$major = \sum_i |\boldsymbol{p}_i \cdot \boldsymbol{e}_2|/E_{vis}, \tag{7.12}$$

where $E_{vis} = \sum |\boldsymbol{p}_i|$. The third direction, e_3, is orthogonal to both the thrust and the *major* axis, and it is very close to the minimum of the momentum projection along any axis, i.e.,

$$minor = \sum_i |\boldsymbol{p}_i \cdot \boldsymbol{e}_3|/E_{vis}. \tag{7.13}$$

The PLUTO group [233] has developed a two step cluster method to determine the event *topology*. The first stage associates all particles into preclusters, irrespective of their momenta. Particles belong to the same precluster if the angles between any two tracks are less than a limiting angle α. The momentum of a precluster is the sum of the momenta of all the particles assigned to that precluster. The preclusters are then combined to clusters if the angle between the momentum vectors is less than a given value β. The number of clusters n is defined as the minimum number of clusters which fulfill the inequalities:

$$\sum_{i=1}^{n} E_{ci} > E_{vis}(1-\epsilon), \tag{7.14}$$

where E_{ci} is the cluster energy and ϵ a small number. If the energy of a cluster, defined as the sum of the energies of all particles assigned to the cluster, exceeds a threshold energy E_{th} then the cluster is called a jet. Typical values for the various parameters are $\alpha = 30°$, $\beta = 45°$, $\epsilon = 0.1$ and $E_{th} = 2.0$ GeV. A similar cluster method [234] has also been used by MARK II.

7.3.3. The evidence for gluons
In this part we review the evidence [37, 171, 235, 236] for the process $e^+e^- \rightarrow q\bar{q}g$ (fig. 6.1).

(A) The average transverse momentum of the hadrons with respect to the jet axis will grow with energy. Normalized transverse momentum distributions, measured by TASSO and evaluated with respect to the sphericity axis, are plotted in fig. 7.13 versus p_T^2 for different c.m. energies. The observed p_T^2 distribution clearly broadens with energy. In QCD the growth is explained as hard non-collinear gluon emission. Fits based on this mechanism are shown in fig. 7.13. However, it is also possible to fit the data up to moderate values of p_T^2 by increasing σ_q as a function of c.m. energy.

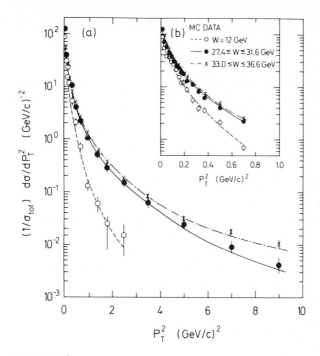

Fig. 7.13. $(1/\sigma)d\sigma/dp_T^2$ at 12 GeV, 27.4–31.6 GeV as a function of p_T^2. The curves are QCD fits to the data with $\sigma_q = 320$ MeV/c.

(B) Planarity: regardless of the value of σ_q (or the mean p_T), hadrons resulting from the fragmentation of a quark must, on the average, be uniformly distributed in azimuthal angle around the quark axis. Therefore, apart from statistical fluctuations, the two-jet process $e^+e^- \to q\bar{q}$ will not lead to planar events, whereas the radiation of a hard gluon, $e^+e^- \to q\bar{q}g$ will result in an approximately planar configuration of hadrons with large transverse momentum *in* the plane and small transverse momentum with respect to the plane. Thus the observation of such planar events, at a rate significantly above the rate expected from statistical fluctuations of the $q\bar{q}$ jets, shows in a model-independent way that there must be a third confined particle in the final state. The third particle is not a quark since it has baryon number zero and cannot have 1/2 integer spin.

We first compare the distribution of $\langle p_T^2 \rangle_{\text{out}}$, the momentum component normal to the event plane squared, with that of $\langle p_T^2 \rangle_{\text{in}}$, the momentum component in the event plane perpendicular to the jet axis.

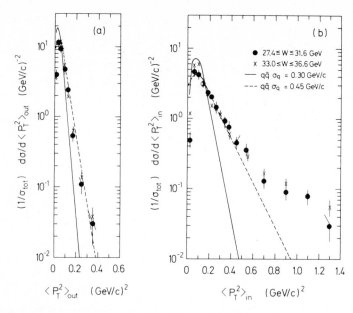

Fig. 7.14. Distributions of mean transverse momentum squared per event for charged particles, (a) normal to and (b) in the event plane, measured by the TASSO collaboration at low and high energies. The curves are the predictions for a $q\bar{q}$ final state with $\sigma_q = 300$ MeV/c (solid lines) and $\sigma_q = 450$ MeV/c (dashed lines).

The data obtained by the TASSO group are plotted in fig. 7.14 and fig. 7.15 for c.m. energies between 12 GeV and 36.6 GeV. The distribution of $\langle p_T^2 \rangle_{\text{out}}$ changes little with energy in contrast to the distribution of $\langle p_T^2 \rangle_{\text{in}}$, which grows rapidly with energy, in particular there is a long tail of events not observed at lower energies. Fits to the data, assuming $e^+e^- \to q\bar{q}$ and $\sigma_q = 300$ MeV/c (solid curves) or $\sigma_q = 450$ MeV/c (dashed curves), are shown in fig. 7.14. The $\langle p_T^2 \rangle_{\text{out}}$ distribution at high energies is not fit by $\sigma_q = 300$ MeV/c, however a good fit can be obtained by increasing σ_q to 450 MeV/c. The $q\bar{q}$ model however, completely fails to reproduce the long tail observed in $\langle p_T^2 \rangle_{\text{in}}$ at high energies. This discrepancy cannot be removed by increasing the mean transverse momentum of the jet. Fig. 7.14b shows a fit assuming $\sigma_q = 450$ MeV/c (which gives a good fit to $(1/\sigma)d\sigma/dp_T^2$ and to $\langle p_T^2 \rangle_{\text{out}}$). The agreement is poor. We therefore conclude that the data include a number of planar events not reproduced by the $q\bar{q}$ model, independent of the average p_T assumed.

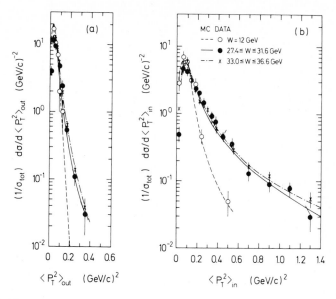

Fig. 7.15. The same plot as in fig. 7.14. The curves are the second order QCD predictions with $\alpha_s = 0.17$ and $\sigma_q = 320$ MeV/c.

Gluon bremsstrahlung offers a natural mechanism to explain the observed planarity of the events. Fig. 7.15 shows a second order QCD fit to the data using the Monte Carlo method outlined above. The fit assumed a constant value of $\sigma_q = 320$ MeV and $\alpha_s = 0.17$ (see below). The long tail in $\langle p_T^2 \rangle_{in}$ is reproduced in this model. Note, that the growth in $\langle p_T^2 \rangle_{out}$ is explained by the occurrence of a small fraction of four (or more) jet events which, in general, are not planar.

The data from PLUTO [222] and JADE [223], analyzed in a similar manner, are in full agreement with the findings of the TASSO group.

The planarity of the events is also observed [226, 232] by the MARK J group, using a different technique. They divided each event into two hemispheres using the plane defined by the *major* and the *minor* axis (see above) and analyzed the energy distribution in each hemisphere as if it resulted from a single jet. The jet with the smallest transverse momentum with respect to the thrust axis is defined as the narrow jet. The other as the broad jet. The oblateness, defined as $O = major - minor$ is a measure of the planarity of the event and is zero for phase space and two-jet events and finite for three-jet final states. The normalized event distribu-

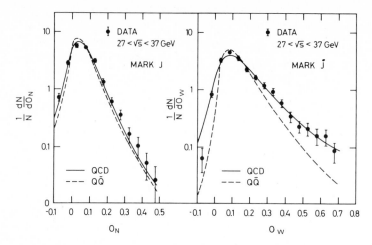

Fig. 7.16. The distribution $(1/N)dN/dO$ determined by the MARK J collaboration as a function of oblateness O_N for the narrow and O_W for the wide jet, separately. The solid curves are predictions based on $e^+e^- \to q\bar{q}g$, the dashed curves show the prediction for $e^+e^- \to q\bar{q}$.

tion measured for c.m. energies between 27 and 37 GeV is plotted versus oblateness in fig. 7.16 for the narrow and the wide jet separately and compared to the predictions for $e^+e^- \to q\bar{q}$ (dashed line) and $e^+e^- \to q\bar{q}g$ (solid line). A good fit is obtained with the $q\bar{q}g$ final state whereas the $q\bar{q}$ final state does not fit the oblateness distribution for the broad jet.

The data discussed above demonstrate that the observed planar events cannot result from fluctuations in the quark pair production with a Gaussian distribution in transverse momentum around the jet axis defined by the hadrons. Each PETRA group has now observed more than 1000 planar events with an estimated background from fluctuations of two-jet events of about 20%. Wide-angle gluon bremsstrahlung $e^+e^- \to q\bar{q}g$ naturally results in planar events. The observed rate for such events is consistent with the QCD predictions. Besides this source there are two ad hoc possibilities: a flat phase space of unknown origin, or the transverse momentum distribution of the quark fragmentation having a long non-Gaussian tail. The first possibility can be excluded by observing events with three axes, the second by excluding the possibility that the three axes are defined by two multiparticle jets and a single high momentum particle at a large angle with respect to the jet axes.

(C) *Properties of planar events.* The TASSO collaboration uses a generalization [237] of sphericity to define three-jet events. In this method

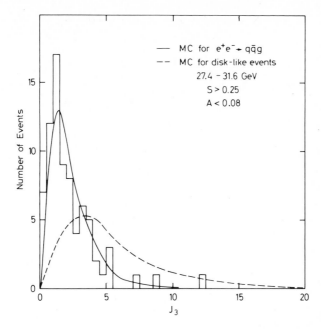

Fig. 7.17. Planar events ($S > 0.25$, $A < 0.08$) measured by the TASSO collaboration and plotted versus the tri-jettiness J_3. The Monte Carlo predictions for $e^+e^- \to q\bar{q}g$ (solid line) and for $e^+e^- \to$ hadrons (phase space, dashed line) are also shown.

the tracks are projected on to the event plane defined by n_2 and n_3 (see above). The projections are divided into three groups, and the sphericity for each group, S_1, S_2 and S_3, is determined. The three axes and the particle assignment to the three groups are defined by minimizing the sum of S_1, S_2 and S_3. This defines the direction of the three jets and assigns the particles to these jet directions.

In fig. 7.17 the TASSO events are plotted versus tri-jettiness J_3, defined as

$$J_3 = \langle p_T^2 \rangle_{\text{in}} / \tfrac{1}{2} \left(300 \text{ MeV}/c^2\right)^2.$$

Here $\langle p_T^2 \rangle_{\text{in}}$ is evaluated for all charged tracks in an event, with respect to their assigned axes. Thus for three-jet events with a mean transverse momentum of 300 MeV with respect to the jet axis we expect to find the events clustered around $J_3 = 1$, compared with a wide distribution in J_3 in case of a flat phase space distribution. The data agree with the expecta-

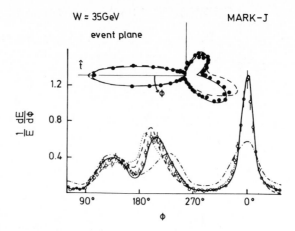

Fig. 7.18. A plot of the energy distribution in the plane defined by the thrust and the *major* axis, for events with oblateness greater than 0.3. The curve shows fits according to: (i) a QCD calculation with $\alpha_s = 0.18$ (solid line), (ii) a two jet $q\bar{q}$ model with a Gaussian (dashed line) or exponential (dotted line) p_T distribution, (iii) a pure phase space distribution (dashed–dotted line). The results were obtained by MARK J collaboration.

tions for $e^+e^- \rightarrow q\bar{q}g$, shown as the solid line and disagree with the phase space distribution plotted as the dashed line.

The MARK J group observes [232, 238] a three jet structure in the energy flow analysis discussed above. To enhance effects resulting from gluon emission they select planar events with oblateness $O > 0.3$. The result of superimposing events with the trust axis pointing to the left and with the event plane in the plane of the paper is shown in fig. 7.18. An antenna pattern is clearly visible. The energy flow is projected on to the event plane and its polar angle distribution is compared with various models. Due to the event selection criteria both phase space models and $q\bar{q}$ models show three lobes. These models however, do not fit the observed distribution, independent whether the fragmentation is assumed to have a Gaussian or an exponential p_T distribution. The data are well fit by the QCD model shown as the solid line.

The three jet structure is also seen [223, 239] in the JADE data. They selected three-jet events by demanding $Q_2 - Q_1 > 0.07$. The event plane is defined by the two least energetic jets and the 0° direction by the most energetic jet. The tracks are then projected on to the event plane and the resulting momentum flow, $(1/\Sigma_i p_i)\mathrm{d}p/\mathrm{d}\theta$ plotted. Superimposing the

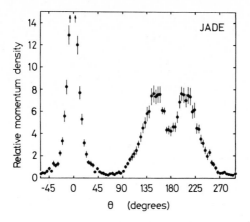

Fig. 7.19. Relative momentum flow in the event plane. The data are from the JADE group.

events leads to the momentum distribution shown in fig. 7.19. There is a clear dip at 180°, opposite the fast jet, and two adjacent peaks demonstrating the three lobe structure. A good fit can be obtained using QCD, whereas $q\bar{q}$ production fails to reproduce the data.

The same conclusion is also reached using the cluster analysis with no assumption on the number of jets. This analysis has been done [233] by the PLUTO group using the cluster method described above. The distribution of the observed number of jets per event are listed in table 7.3 and compared to the predictions based on $q\bar{q}, q\bar{q} + q\bar{q}g$ ($\alpha_s = 0.15$) and phase space. The models are all normalized to the number of observed events. The data clearly favour a clustering of the particles around three axes.

Table 7.3
Number of clusters

n_j	Data	phase space	$q\bar{q}$	$q\bar{q} + q\bar{q}g$
1	2	1	3	3
2	551	30	680	567
3	249	154	152	247
4	53	306	23	46
5	3	268	1	2
6	1	86	–	–
7	–	14	–	–

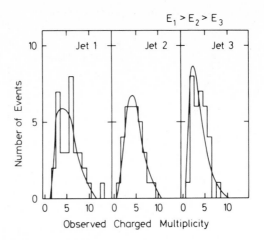

Fig. 7.20. The charged-particle multiplicity distribution for each of the three jets in a planar event, with the jets ordered according to energy. The QCD prediction is shown as the solid line. The data are from the TASSO collaboration.

The remaining question is then to decide whether the third jet is defined by a single particle or by a group of particles. This can be done by examining the events. Fig. 7.11 shows typical candidates for three jet events as observed by JADE and TASSO. Note, that several tracks cluster around each axis.

The multiplicity distributions as measured by the TASSO Collaboration for each of the three jets are plotted in fig. 7.20. The jets are ordered according to energy, $E_1 > E_2 > E_3$. The energies of the jets were computed from the observed opening angles between the jets, neglecting parton masses.

It is obvious that, in general, each jet contains several charged particles, and that the observed multiplicity distribution is reproduced by the QCD calculation shown as the solid line in fig. 7.20.

The TASSO group has also evaluated the transverse momentum of charged particles in planar events with respect to the jet axis to which they were assigned. The jet direction and the assignment of particles to one of the three jets were determined using the generalized sphericity method. The distribution $(1/N)\,\mathrm{d}N/\mathrm{d}p_T^2$ is plotted versus p_T^2 as the solid circles in fig. 7.21. The distribution is in agreement with the QCD prediction shown as the solid line. Also plotted (open circles) in fig. 7.21 is the p_T^2 distribution measured with respect to the jet axis in two-jet

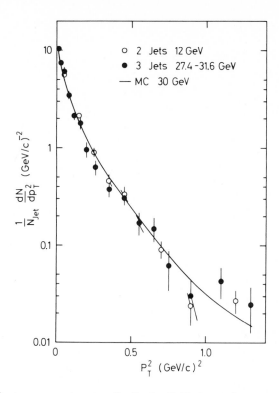

Fig. 7.21. The transverse momentum distribution $(1/N)\,dN/dp_T^2$ of the hadrons in planar events with respect to the three-jet axis is shown as the solid circles. The open circles represent the transverse momentum distribution with respect to the jet axis in two-jet events at lower energies. The solid line represents the Monte Carlo QCD prediction. The data were obtained by the TASSO group.

events at 12 GeV. The distributions are in excellent agreement and the data support the conjecture that the mean p_T occurring in the fragmentation of a parton is independent of energy.

The JADE group [240] uses an independent method suggested by Ellis and Karliner [241] to demonstrate the existence of three-jet events. From the data taken at c.m. energies around 30 GeV they select planar events which satisfy the condition $Q_2 - Q_1 > 0.1$ and determine the thrust axis. The event is then divided into two jets by a plane normal to the thrust axis, and p_T computed separately for each jet; the jet with the smallest p_T is called the slim jet, the other the broad jet. The broad jet is then transformed into its own rest system. If the broad jet consists of two jets

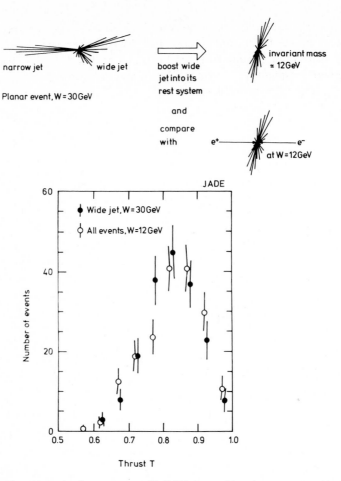

Fig. 7.22. The wide jet in planar events at 30 GeV is boosted into its own rest system. The solid circles show the thrust distribution of the wide jet in this system. The open circles show the thrust distribution of two-jet events at 12 GeV. The data are from the JADE collaboration.

they will now appear as two back to back jets along the new thrust axis T^*. The distribution of T^* in this system is plotted in fig. 7.22 together with the thrust distribution of two-jet events measured at 12 GeV. The two distributions are in excellent agreement. Also other quantities like the invariant mass, mean p_T and charge multiplicity evaluated for the broad jet in its own rest system are in agreement with the same quantities evaluated for a two jet event at 12 GeV.

In conclusion: The data on $e^+e^- \to$ hadrons at high energies show clear evidence for a three jet structure resulting from the fragmentation of a new light parton with baryon number zero. This observation is naturally explained in any field theory of strong interactions as quark–gluon bremsstrahlung. Actual fits based on QCD are in excellent agreement with the data.

7.3.4. The spin of the gluon

The spin of the gluon can be determined from the angular correlation observed between the three partons in $e^+e^- \to q\bar{q}g$ events. It is convenient to describe this process using the variables $x_i = E_i / E_b$, where the energy carried off by the quark or the gluon E_i is measured in units of the beam energy E_b. The three jet event is defined in fig. 7.23a with the variables ordered such that $x_1 > x_2 > x_3$. The thrust of the $q\bar{q}g$ event is given by x_1 with $x_1 + x_2 + x_3 = 2$. The variable x_i is related to the angles between the partons, θ_i (fig. 7.23), as

$$x_i = \frac{2\sin\theta_i}{\sin\theta_1 + \sin\theta_2 + \sin\theta_3}. \tag{7.15}$$

The distribution of the events as a function of x_i, averaged over production angles relative to the incident e^+e^- directions, can be written [241]

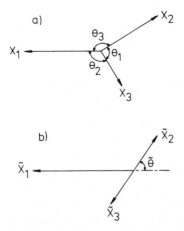

Fig. 7.23. (a) Kinematic variables to describe $e^+e^- \to q\bar{q}g$. (b) Definition of the angle $\tilde{\theta}$.

as:

$$\frac{1}{\sigma_0}\left(\frac{d\sigma}{dx_1 dx_2}\right)_V = \frac{2\alpha_s}{3\pi}\left(\frac{x_1^2 + x_2^2}{(1-x_1)(1-x_2)}\right.$$

$$\left. + \text{cyclic permutations of } 1,2,3\right) \qquad (7.16)$$

for the vector case, and

$$\frac{1}{\sigma_0}\left(\frac{d\sigma}{dx_1 dx_2}\right)_S = \frac{\tilde{\alpha}_s}{3\pi}\left(\frac{x_3^2}{(1-x_1)(1-x_2)}\right.$$

$$\left. + \text{cyclic permutations of } 1,2,3\right) \qquad (7.17)$$

for the scalar case.

TASSO multihadron events collected at c.m. energies between 25 GeV and 36.6 GeV have been analyzed [235, 242] using these variables, with the angles θ_i determined from the jet directions defined by charged tracks. Monte Carlo computations show that the x_i values determined in this manner agree well with those of the parent partons. The resolution in x_1 was found to be on the order of 6% with systematic uncertainties of the order of 1%. The observed distribution is shown in fig. 7.24. Collinear two jet events cluster along the base line with $x_1 = 1$ and dominate the event sample. Well defined three jet events occur for $x_1 < 0.9$ with symmetrie three star events at the top of the diagram.

The TASSO group has determined the spin using the variables $\cos\tilde{\theta} = (x_2 - x_3)/x_1$ suggested by Ellis and Karliner [241]. $\tilde{\theta}$ is the angle between parton 1 and the axis of the parton 2 and 3 system boosted to its own rest frame, as defined in fig. 7.23b.

To ensure that the spin analysis is not affected by higher order terms one should avoid x_1 close to 1. Furthermore, for x_1 close to 1 the distributions vary rapidly so that smearing effects caused by the hadronization of gluons and quarks are important. For these reasons only events with $x_1 < 0.9$ are used in the analysis. In this case the lowest-energy jet has a mean energy around 6 GeV and the smallest opening angle between any two partons is 70°.

The distribution [236, 242] of the events as a function of $\cos\tilde{\theta}$ is plotted in fig. 7.25 and compared with the distributions predicted for vector

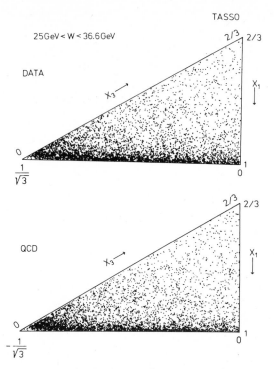

Fig. 7.24. Dalitz plot of $e^+e^- \rightarrow$ hadron events at 33 GeV using the normalized variables $x_i = E_i/E_b$ where E_i is the jet- and E_b the beam energy. For comparison the QCD prediction with $\alpha_s = 0.17$ is also shown. Both distributions contain the same number of events.

(solid) and scalar (dashed–dotted) gluons. The prediction was made using the model of Hoyer et al. [225] and including hadronization according to Field–Feynman [178]. The distribution predicted for scalar gluons on the parton level without hadronization is also plotted (dashed curve). A comparison shows that the distribution is rather insensitive to hadronization effects. Note, that the distributions are normalized to the number of events in the plot, i.e., the scalar and vector cases are discriminated using the shape only. The data clearly favour the vector case. To avoid binning effects the mean value of $\cos \tilde{\theta}$ was evaluated and compared to the theoretical prediction. The experimental value of $\langle \cos \tilde{\theta} \rangle = 0.3391 \pm 0.0079$ can be compared to the values $\langle \cos \tilde{\theta} \rangle_V = 0.341 \pm 0.003$ for vector gluons and $\langle \cos \tilde{\theta} \rangle_S = 0.298 \pm 0.003$ for scalar gluons. The

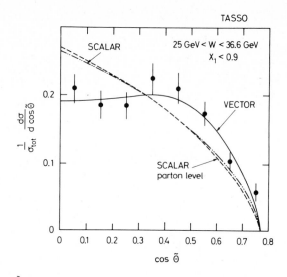

Fig. 7.25. The $\cos\tilde{\theta}$ distribution of events with $x_1 < 0.9$. The solid line and the dashed-dotted line show the distributions predicted for vector gluons and scalar gluons respectively. The predictions include hadronization. For comparison the prediction for a scalar gluon on the parton level is shown as the dashed line. The distributions are normalized to the number of observed events.

comparison yields:

$$\langle\cos\tilde{\theta}\rangle_{exp} - \langle\cos\tilde{\theta}\rangle_{QCD} = (2\pm8)\times10^{-3},$$

$$\langle\cos\tilde{\theta}\rangle_{exp} - \langle\cos\tilde{\theta}\rangle_{S} = (41\pm8)\times10^{-3}.$$

The data are thus in excellent agreement with the QCD prediction but differ from the scalar prediction by roughly five standard deviations, corresponding to a confidence level of 10^{-6}. The result is remarkably insensitive both to the exact value of α_s and the details of the fragmentation. Varying the value of α_s by $\pm20\%$ changes the computed value of $\langle\cos\tilde{\theta}\rangle$ by about 1%. Evaluating $\langle\cos\tilde{\theta}\rangle$ in the parton model without fragmentation leaves the scalar prediction unchanged and increases the predicted value for a vector gluon by about 2%.

The gluon spin can also be determined from the x_1 distribution, i.e. the distribution of the most energetic jet in three-jet events. The results obtained [243] by the CELLO group are shown in fig. 7.26. The data clearly favour vector gluons ($\chi^2 = 5.1$) over scalar gluons ($\chi^2 = 30.8$).

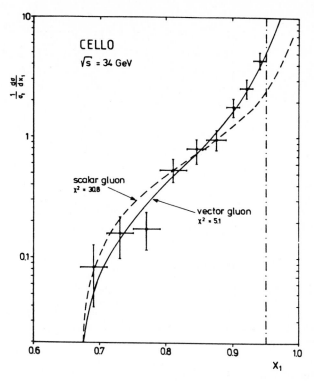

Fig. 7.26. The thrust distribution observed by the CELLO collaboration in three-jet events. The predictions for vector and scalar gluons are shown by the solid and the dashed line respectively.

The data obtained by PLUTO and by MARK II also support a vector gluon.

Independent information on the gluon spin can be obtained from the angular distribution of the thrust axis with respect to the beam axis. According to QCD the angular distribution is given by $1 + a(T)\cos^2\theta_T$, where the coefficient $a(T)$ decreases with the thrust. The TASSO group finds [236] $a(T) = 1.00 \pm 0.11$ for events with thrust between 0.9 and 1.0 and $a(T) = 0.75 \pm 0.18$ for events with thrust below 0.9. The QCD predictions of $a(T) = 0.97$ respectively 0.75 are in good agreement with the data but better statistics is clearly needed.

7.3.5. Alternative sources of three-jet events

Several alternative mechanisms have been proposed to explain the occurence of three-jet events. These models have been examined in detail by Söding [235], using the TASSO data. He finds that although all models can explain some of the observed features none of the models can explain all. In particular, the models proposed do not reproduce the fractional energy distributions, i.e. the angular correlations between jets.

The constituent-interchange model [244] explains three-jet events as a $e^+e^- \rightarrow q\bar{q}M$ process, where the gluon is replaced by a meson M emitted at large angles with respect to the quark direction. The observed multiplicity distribution shows that single pion production must be strongly suppressed and M must stand for a series of high mass mesons. This higher twist contribution results in a constant value of $\langle p_T^2 \rangle$ independent of c.m. energy W, whereas both the data and QCD show that $\langle p_T^2 \rangle$ increases as $W^2/\ln(W^2/\Lambda^2)$. To account for this observation, the effective quark coupling constant must increase by a factor of four between 12 GeV and 33 GeV. However, the model, even with the coupling constant adjusted, fails to fit the observed x_1 distribution, as shown in fig. 7.27. This has also been noted in an earlier analysis [245] by the PLUTO group. Söding estimates that not more than 5% of the cross section in the three jet region can be attributed to higher twists.

It has been suggested that three-jet events may be described as a $e^+e^- \rightarrow q\bar{q}$ process in which the hadronization is given a long exponential tail. Such models can be made to fit the p_T^2 distribution and the $\langle p_T^2 \rangle$ distribution in and normal to the event plane. However, they completely fail to reproduce the x_1 distribution.

It has also been suggested [246] that the long p_T^2 tail observed at high energies is due to the fragmentation of heavy quarks which may have a much larger intrinsic value of σ_q. Assuming a Gaussian distribution in p_T^2 with $\sigma_q = 310$ MeV$/c$ for u, d and s quarks and $\sigma_q = 800$ MeV$/c$ for the c and the b quark leads to distributions in p_T^2, $\langle p_T^2 \rangle_{in}$ and $\langle p_T^2 \rangle_{out}$ in accordance with the data. However, this model again fails to reproduce the x_1 distribution, and the predicted transverse distribution of momentum with respect to the three axes is to broad.

7.3.6. Determination of the quark–gluon coupling constant α_s

Attempts to extract the quark–gluon coupling constant α_s from $e^+e^- \rightarrow$ hadrons meet with problems of both experimental and theoretical nature:

(i) The omission of neutrals in some experiments. It has been shown that neutral and charged particles behave similarly. Using only charged

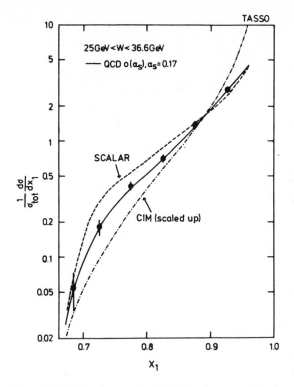

Fig. 7.27. The thrust distribution observed by TASSO and compared to the prediction of scalar gluons (dashed line), vector gluons (solid line) and the constituent interchange model (CIM) scaled up to agree with the p_T^2 distribution (dashed–dotted line).

particles will therefore not change the mean value of a quantity like thrust but will increase its error.

(ii) Apparent multi-jet contributions from b decays.

(iii) QED corrections [247], in particular hard photon emission in the initial state.

(iv) Fluctuations in the hadronization process, which may cause events to be improperly classified, i.e., an event with two or four primary partons might be classified as a three-jet event, due to fluctuations or to overlap between the final state jets. These effects may not be crucial as long as the minimum angle between any two partons is large compared to the opening angle of the jet and a jet is defined using Sterman–Weinberg criteria [248].

(v) Contributions from higher order QCD diagrams (fig. 6.1). These corrections have been evaluated by various groups with apparently contradictory results. Fabricius et al. [249] computed the thrust distribution to order α_s^2 and found this correction to be small and in general negative. Ellis et al. [250] and independently Vermaseren et al. [251] found a positive α_s^2 correction of the order of 30–40%. It was later realized [218,252] that both results may be correct, since the groups compute different quantities. Consider the diagram for two gluon emission depicted in fig. 7.28. The effective mass squared of the two partons i, j is given by $s_{ij} = (p_i + p_j)^2 = y_c s$, where p_i, p_j are the four-momenta of the partons, y_c is a numerical parameter and s is the total c.m. energy squared. Ellis et al. and Vermaseren et al. calculate the three- and four-jet cross sections of the parton level, i.e., a four parton state remains a four parton state down to extremely small values of $y_c \approx 10^{-5}$, i.e., for a typical PETRA value of $s = 1000$ GeV^2, $s_{ij} \approx 10^{-5}s \approx 10^{-2}$ GeV^2. The result of such a computation is very sensitive [253] to the cut off mass. However, since a typical jet has a mass around 6 GeV the two partons would appear experimentally as one jet, i.e. as one primary parton, due to the non-perturbative hadronization process.

This problem is avoided in the computation by Fabricius et al. They start with the three- and four-parton bare cross sections, but they use a Sterman–Weinberg definition of an observable jet. Two partons cannot be separated if the opening between the partons is less than δ or if either parton carries less than εE_b of the energy. Fabricius et al. define an event as a three jet event if all but a fraction $\varepsilon/2$ of its total energy is contained within three separate cones of full opening angle δ. For the calculation they use $\varepsilon = 0.2$ and $\delta = 45°$. Although this calculation is better suited to compare theory and experiment there may still be problems due to ambiguities arising from emission of soft final state partons.

The value of α_s has been determined by several groups, using a variety of methods, and the results are listed in table 7.4. Some of the results

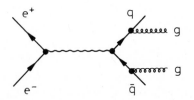

Fig. 7.28. Double gluon bremsstrahlung leading to four-jet events.

Table 7.4
Determination of α_s

Group	Ref.	$\alpha_s{}^a$	Method
CELLO	[254]	$0.16 \pm 0.02 \pm 0.03$	Three-jet events, momentum tensor (α_s)
JADE	[239]	$0.18 \pm 0.03 \pm 0.03$	Three-jet events, momentum tensor (α_s)
MARK J	[238]	0.19 ± 0.02	Three-jet events, energy flow (α_s^2)
MARK II	[37]	$0.18 \pm 0.015 \pm 0.03$	Energy–energy correlation (α_s)
PLUTO	[233]	$0.15 \pm 0.03 \pm 0.02$	Three-jet events, cluster analysis (α_s)
PLUTO	[255]	0.18 ± 0.02	Event shapes (α_s)
TASSO	[230]	$0.17 \pm 0.02 \pm 0.03$	Three-jet events, momentum tensor (α_s)^2

^{a} Value $\pm$ statistical error $\pm$ systematic uncertainty.

were obtained in leading order (α_s), others were evaluated using the Ali et al. program [227] which includes some of the next to leading order diagrams (α_s^2). The methods used by various groups to extract a value of α_s are discussed in some detail below.

The strong coupling constant α_s is directly related to the number of three-jet events. After choosing a minimum opening angle between any pairs of partons (q, $\bar{\text{q}}$ or g), the QCD cross section can be integrated and normalized to the total e^+e^- annihilation cross section. This ratio depends only on α_s and can be compared directly to the experimental ratio of three-jet events to the total number of hadronic events. In practice, corrections must be made for the effects discussed above. An example of the quality of such fits is shown in fig. 7.12. The data are from TASSO [230]. Note that the fragmentation parameters were determined at low energies. Similar fits have been made by JADE [233, 239] and CELLO [254]. PLUTO determined [233] the number of three-jet events from a cluster analysis.

MARK J has determined [238] α_s from the number of three-jet events defined by $O_b > 0.3$ or $O_b \geqslant O_n \geqslant 0.3$, with $O_n > 0$. O_b and O_n are the oblateness (fig. 7.16) for the broad and the narrow jet respectively. The group has also determined α_s from the average oblateness $\langle O_b \rangle$.

The value of α_s has also been determined from a measurement of the energy–energy correlation [256, 257]. This correlation is defined as

$$\frac{1}{\sigma} \frac{d\Sigma_E}{d\theta} = \frac{2}{\sigma} \sum_{j,k} \int \frac{d^3\sigma}{dz_j \, dz_k \, d\theta} z_j z_k \, dz_j \, dz_k, \qquad (7.18)$$

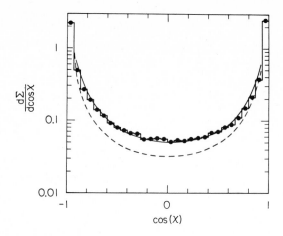

Fig. 7.29. The angular dependence of the energy–energy correlation function. The dashed line shows the parton level QCD prediction, the solid line includes hadronization effects. The data are from the MARK II collaboration.

where $z_j = E_j/E_{beam}$ is the fractional energy of particle j, and θ is the opening angle between the momentum vectors of particles j and k. The sum is over all pairs j and k which satisfy this condition, and the result is averaged over all events. This expression has been calculated to first order in QCD, neglecting hadronization.

The energy–energy correlation has been evaluated by various groups. The data obtained by the MARK II collaboration [37] at PEP are plotted in fig. 7.29 ($\chi \equiv \theta$). The dashed curve shows the perturbative QCD prediction, the solid line gives the prediction including hadronization. It is obvious that non-perturbative effects are still very important at 30 GeV. To extract a value of α_s which is less dependent on nonperturbative effects the groups consider the difference

$$A(\theta) = \frac{1}{\sigma} \left(\frac{d\Sigma_E}{d\theta}(\pi - \theta) - \frac{d\Sigma_E}{d\theta}(\theta) \right). \qquad (7.19)$$

To first order, non-perturbative effects cancel and the observed effect should mainly result from gluon bremsstrahlung. Data from PLUTO [258], MARK II [37] and CELLO [224] are plotted in fig. 7.30, and compared with the QCD prediction for $\alpha_s = 0.18$. The data are in agreement with the predictions.

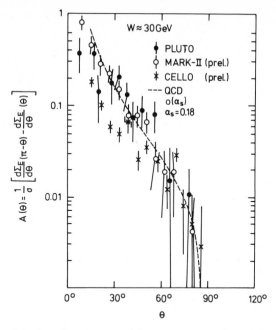

Fig. 7.30. Forward–backward asymmetry of the energy–energy correlation function. The QCD prediction for $\alpha_s \sim 0.18$ is shown as the dashed line. The data were obtained by PLUTO, CELLO and MARK II.

The PLUTO group has investigated the uncertainties resulting from detector imperfection, radiative corrections and hadronization effects. They find that these effects introduce uncertainties on the order of 10–20% for θ between 45° and 90°. The corrections are larger at smaller angles.

PLUTO has also determined [255] the value of α_s using the following shape dependent observables:

– $\langle 1 - T \rangle$, where T is the thrust.

– The energy-weighted jet broadness, $\langle \sin^2 \eta \rangle = \langle (E_i / W) \sin^2 \delta_i \rangle$, where E_i and δ_i are the energy and the angle of a track with respect to the jet axis, i.e., η is an energy-weighted jet opening angle.

– The squared invariant mass [259] of the wider of the two jets normalized to s, the total energy squared.

– The integral over the energy–energy correlation function in the large angle region $60° < \theta < 120°$.

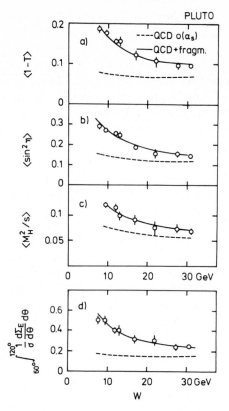

Fig. 7.31. The energy dependence of the average thrust, energy weighted jet opening angle, invariant mass of the broad jet and the integral over the energy–energy correlations for θ between 60° and 120°. The dashed line shows the perturbative QCD contribution $a/\ln(W^2/\Lambda^2)$, the solid line includes hadronization effects (b/W). The data are from the PLUTO collaboration.

Effects due to perturbative QCD have an energy dependence proportional to $1/\ln(W^2/\Lambda^2)$, whereas non-perturbative effects will decrease with energy as $1/W$. The PLUTO data span a large energy range and can be used to separate perturbative and nonperturbative effects by the different dependence on energy. The data and the results of the fits are plotted in fig. 7.31. The perturbative contribution is shown by the dashed line, the solid curve includes both effects. The fits are rather good and all consistent with the value $\alpha_s = 0.18 \pm 0.02$.

The value of α_s has thus been extracted by several methods with different sensitivities to theoretical problems like higher order QCD corrections and to experimental effects like b decays. It is quite remarkable that all determinations listed in table 7.4 are consistent with $\alpha_s = 0.17$. The same value $\alpha_s = 0.17$ is also found by Fabricius et al. using the full second order calculation. With the second order formalism developed by Ellis et al., Ali [260] finds $\alpha_s = 0.12 - 0.13$.

One of the crucial predictions of QCD is that the strength of the coupling α_s decreases logarithmically with increasing value of the momentum transfer squared [eq. (5.2)]. The MARK II collaboration at PEP has done an interesting attempt to determine this Q^2 dependence from an analysis of the $p_\perp$ distribution within a jet. The method is based on work by Konishi et al. [261]. They compute the energy-weighted cross section for particles within a jet using a quark–gluon cascade model. The relevant momentum transfer entering the formula for α_s depends on the jet cone opening angle 2δ as $Q^2\delta^2$, i.e., the energy-weighted cross section for single-jet production can be used to determine α_s at various values of the momentum transfer. The method is discussed in detail by Hollebeek [37]. The results shown in fig. 7.32 are quite intriguing but more work, both experimentally and theoretically, is needed to ensure the α_s is indeed running.

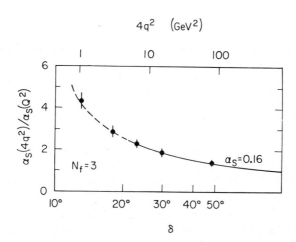

Fig. 7.32. The dependence of α_s as a function of the jet opening angle δ or $4q^2$. The data are from the MARK II collaboration.

7.3.7. Do quarks and gluons fragment differently?

One might expect on general grounds that gluons and quarks hadronize into different final states. A gluon may fluctuate into pairs of quarks and gluons. Furthermore the ggg coupling is 9/4 times stronger than the $q\bar{q}g$ coupling, so that gluon emission will be more frequent for gluons than for quarks. This leads us to expect that a gluon and a quark will fragment into hadrons differently—the hadron spectrum from a gluon fragmentation will be softer, with a correspondingly higher multiplicity. Computations [261, 262] show that at asymptotic energies, a gluon will fragment into a jet of hadrons with multiplicity and opening angle which are larger by a factor 9/4 than the corresponding quantities for a quark jet of the same energy.

Using the string model Anderson, Gustafson and co-workers have predicted [227] that the yield of low-energy particles emitted at large angles with respect to the jet axis depends on the jet being a result of the fragmentation of a quark or a gluon. The JADE group [239] has carried out this analysis using charged and neutral particles. Planar events with $Q_2 - Q_1 > 0.10$ were divided into a slim jet and a broad jet by the plane normal to the thrust axis. The broad jet is then boosted into its own rest system, and the particles assigned to the two subjets. The softest jet is called the gluon jet. Monte Carlo calculations show that with the cuts used, this is true more than 50% of the time and it simply reflects the softness of a bremsstrahlung spectrum. All the particles are projected on to the plane defined by T, the thrust axis of the event and T^* the thrust axis of the boosted two jet system. They then plot the particle densities between the gluon jet and the slim jet and between the quark jet and the slim jet in terms of normalized angles θ_i/θ_{max}, where θ_{max} is the opening angle between the gluon jet and the slim jet or the quark jet and the slim jet, respectively. The data plotted in fig. 7.33 show that the density of tracks is larger by a factor of two between the slim jet and the quark jet. The result of a Monte Carlo computation [225] based on similar fragmentation functions for quarks and gluons fails to reproduce the dip observed in the particle density distribution between the quark and the slim jet, as shown by the dotted histogram in fig. 7.33. The data, however, are reproduced in the Lund Monte Carlo program [227], where the quark has a harder fragmentation function than the gluon jet. The fit using the Lund Monte Carlo is shown by the solid histogram. These findings have been confirmed by a recent analysis [224].

The JADE group has determined [236] the $p_\perp$ growth displayed by the individual jets in planar three-jet events. In this analysis, planar events

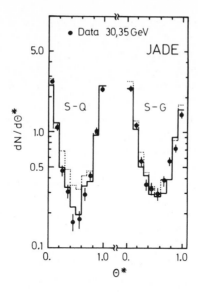

Fig. 7.33. Angular distribution of charged particles between the slim jet and the gluon jet (S–G), and between the slim jet and the quark jet (S–Q) as a function of the normalized angles θ/θ_{max}.

were selected by demanding $Q_2 - Q_1 > 0.07$ and $Q_1 < 0.06$ where Q_j are the normalized eigenvectors of the sphericity tensor, with $Q_3 > Q_2 > Q_1$. The three jet directions and the particle assignments to the jets were determined by maximizing triplicity, i.e. the directed momentum along three axes. The jet energies E_j are computed from the direction of the jets, ordered such that $E_1 > E_2 > E_3$. Monte Carlo studies show that in this case the gluon jet can be assigned to jet 1, 2, or 3 with the probability 12%, 22% or 51%. In 16% of the cases no gluon was emitted, when a two jet event was mislabelled as a three jet event due to fluctuations in the hadronization process. The mean transverse momentum of the three jets is plotted in fig. 7.34 as a function of jet energy. The data show that the softest jet has the largest $p_\perp$ with respect to its jet axis for a given energy, whereas there is little difference between the two quark jets 1 and 2.

The JADE group also finds that the jet broadening is more important to the event plane than normal to the plane. Both findings are reproduced in the LUND model but not by the Hoyer et al. simulation, in which quarks and gluons have identical hadronization.

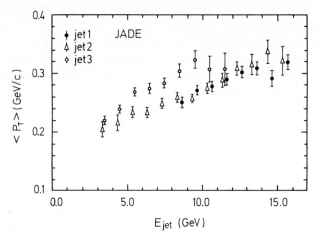

Fig. 7.34. The mean transverse momentum of particles with respect to their assigned jet axis in planar three jet events as a function of jet energy. The jets are ordered according to energy as $E_1 > E_2 > E_3$.

The results are very interesting, however, the findings of the JADE group are not confirmed by similar analyses carried out by the MARK J [236] and the MARK II Collaboration [37].

7.4. Summary

To demonstrate that QCD is the correct theory of strong interaction one has to show that gluons, i.e. massless, coloured vector particles with flavour-neutral couplings exist. Furthermore the gluons should have self coupling leading to a running coupling constant.

(i) Gluons exist. They offer the only consistent explanation of all the features observed in e^+e^- annihilation into hadrons at high energies, in particular the occurrence of three-jet events. There is also no alternative explanation of the properties of the final state hadrons in the decay $\Upsilon \to$ hadrons.

(ii) Gluons have spin one. Maybe the cleanest observation is from the angular correlation observed between the jets in three-jet events. Both $\Upsilon \to$ hadrons and $\Upsilon' \to \pi^+\pi^- \Upsilon$ are in agreement with vector spin predictions and disagree with predictions based on scalar gluons.

(iii) The gluon is flavour neutral. The $c\bar{c}$ and $b\bar{b}$ mass spectrum and the leptonic widths can be explained using the same potential for both c and b quarks.

(iv) Gluons may be coloured. This would explain why $\Upsilon \to g \to$ hadrons is forbidden.

(v) Gluon self interaction, a crucial feature of QCD, has not yet been observed convincingly. Some early evidence of an analysis of the angular width of gluon jets may indicate that α_s is running. Furthermore there are candidates for gluonium.

8. Two-photon interactions

Electron–positron collisions are a prolific source of photon–photon collisions [263, 264] as shown in fig. 8.1, where Q^2 and ν of the spacelike photon are determined from a measurement of energy and angle of the scattered lepton. These processes offer a unique opportunity to vary the mass of the target and the projectile over a wide range, from collisions of two nearly real photons, via deep inelastic electron scattering on a photon target, to collisions of two heavy photons. The interest in these studies is enhanced by the dual rôle of the photon as a hadron and as a pointlike particle which can initiate hard scattering processes even on its mass shell.

Experimentally, two-photon events are separated from annihilation events by a cut on the observed energy. The c.m. energy of the $\gamma\gamma$ system is, in general, much lower than the available energy reflecting the product of two bremsstrahlung spectra. In some cases one or (rarely) both of the scattered electrons are detected in shower counters mounted at small angles with respect to the beam axis. The background from beam-gas events is low and can be measured from the number of events which satisfy the selection criteria but originate outside the interaction volume.

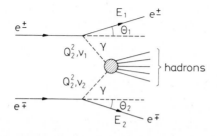

Fig. 8.1. Hadron production in $e^+e^- \to e^+e^- X$.

Another potential source of background results from inelastic Compton scattering. However, estimates of this process find it to be negligible.

The mass of the produced hadron system can be computed from the energy and angles of the scattered electrons. However, tagging both electrons leads to a large reduction in rate, so that in practice only one or none of the electrons is detected. This has no drawbacks in the case of real photons and a simple final state, like $e^+e^- \to \gamma\gamma e^+e^- \to \pi^+\pi^- e^+e^-$, where the energy can be determined from the final state pions. In general, however, there are both charged and neutral final state particles and they are travelling at small angles with respect to the beam direction. It is therefore difficult to detect and measure the momenta of all particles in the final state. Thus the measured visible mass of the hadron system W_{vis} can only be related to the true mass W of the produced hadron system by a Monte Carlo calculation. Furthermore, tagging is required in order to study the Q^2 dependence of a process.

During the past few years several groups have reported experimental results on photon–photon interactions. An up to date discussion of the data and the theoretical aspects of $\gamma\gamma$ collisions including a complete set of references can be found in the talks given by Wedemeyer [265] and Bardeen [266] at the Bonn meeting.

8.1. Resonance production

All hadrons with even charge conjugation and spin different from one can be produced [267] in $\gamma\gamma$ collisions. The corresponding cross section can be written as

$$\sigma(e^+e^- \to e^+e^- X) = \left(2\alpha \ln s/m_e^2\right)^2 f(x)(2J+1)\Gamma(X \to 2\gamma)/M_X^3,$$

$$(8.1)$$

with $x = m_X^2/s$ and $f(x) = \frac{1}{2}(2+x)^2\ln(1/x)-(1-x)(3+x)$. Thus such a measurement determines the partial width $\Gamma(X \to \gamma\gamma)$ and if the branching ratio $B(X \to 2\gamma)$ is known, Γ_{tot}, the total width of the resonance.

The data available on $\eta'(958)$, $f^0(1270)$ and $A_2(1310)$ are listed in Table 8.1. These data were all collected in the no tag mode. Beam-gas background, annihilation events and cosmic ray events were removed by kinematical cuts. The results obtained by the various groups are in agreement within the quoted errors.

Table 8.1
Data on $\Gamma_{\gamma\gamma}(x)$

Particles (J^{PC})	Group	Decay mode	$\Gamma_{\gamma\gamma}$ (keV)[a]	Ref.
$\eta'(958)$	MARK II	$\rho^0\gamma$	$5.9 \pm 1.6 \pm 1.2$	[268]
(0^{-+})	JADE	$\rho^0\gamma$	7.5 ± 0.7	[265]
$f^0(1270)$	PLUTO	$\pi^+\pi^-$	$2.3 \pm 0.5 \pm 0.35$	[269]
(2^{++})	TASSO	$\pi^+\pi^-$	$3.2 \pm 0.2 \pm 0.6$	[270]
	MARK II	$\pi^+\pi^-$	$3.6 \pm 0.3 \pm 0.5$	[271]
	CELLO	$\pi^+\pi^-$	$3.6 \pm 0.2 \pm 0.7$	[272]
	Crystal Ball	$\pi^0\pi^0$	$2.9 ^{+0.55}_{-0.39} \pm 0.6$	[273]
A_2	JADE	$\rho^\pm\pi^\mp$	1.2 ± 0.4	[265]
(2^{++})	Crystal Ball	$\eta\pi^0$	$0.77 \pm 0.18 \pm 0.27$	[273]

[a] width $\pm$ systematic uncertainty $\pm$ statistical error.

As an example of such an analysis let us consider the reaction $e^+e^- \to f^0 e^+e^- \to \pi^+\pi^- e^+e^-$ as measured [270] by the TASSO collaboration. They selected two positively charged tracks originating from the interaction region. To reduce the background from beam-gas events the tracks should be coplanar to within $10°$ with respect to the beam axis and each track should have a transverse momentum with respect to the beam axis of 0.3 GeV/c or more. Two-prong cosmic rays and annihilation events are rejected by requiring the two tracks to be non-collinear by more than $7.5°$ and the sum of the magnitude of the momenta to be less than 20% of the beam energy. The transverse momentum imbalance of the event with respect to the beam axis should be less than 0.3 GeV/c.

The mass distribution of the remaining events, assigning pion masses to the particles, is shown in fig. 8.2. The data represent a total integrated luminosity of 9240 nb^{-1} at c.m. energies between 22.4 GeV and 36.7 GeV. The contribution [274] from $e^+e^- \to e^+e^- e^+e^-$ and $e^+e^- \to e^+e^+\mu^+\mu^-$ is shown as the solid line. The data are in good agreement with the prediction except for an excess of events near 1.3 GeV.

The invariant mass distribution after subtracting the QED contribution is shown in fig. 8.3. It is natural to identify the observed mass peak near 1.27 GeV with the $f^0(1270)$. A spin-two resonance like the f^0 can be produced with helicity amplitudes 0, 1 or 2. To determine the dominant

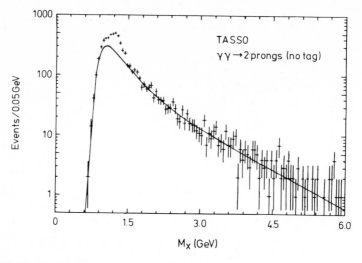

Fig. 8.2. Untagged two prong events from TASSO plotted versus the pair mass. The QED contribution is represented by the solid line.

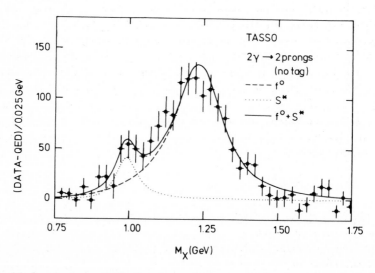

Fig. 8.3. Invariant mass distribution of $\gamma\gamma \rightarrow 2$ prongs, assuming the tracks to be pions.

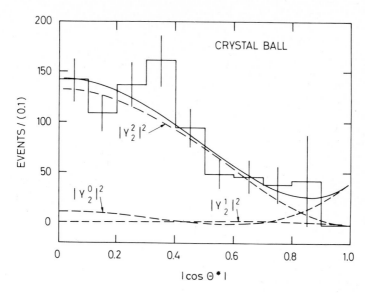

Fig. 8.4. The decay angular distribution of $f^0 \to \pi^0\pi^0$. The data—corrected for the acceptance—are from the Crystal Ball collaboration.

amplitude the data are plotted as a function of $\cos\theta^*$ where θ^* is the angle between the beam axis and one of the charged particles in the events. Fig. 8.4 shows data [273] from Crystal Ball which have the largest acceptance in $\cos\theta^*$. The angular distributions expected for the three helicity amplitudes are shown. The data, in agreement with theoretical expectations and the findings of other groups, clearly favour the helicity two amplitude. In fact, all the groups use only this amplitude to determine $\Gamma_{\gamma\gamma}$. The observed mass distribution in fig. 8.3 can be fit to the f^0 using the standard values of f^0 mass and the width plus an additional term resulting either from $\pi^+\pi^-$ continuum production or from a lower mass resonance.

8.2. Exclusive channels

Exclusive particle production in $\gamma\gamma$ collisions [264, 275] has recently received some attention.

At the Wisconsin Conference the TASSO collaboration reported [276] the first data on $e^+e^- \to \pi^+\pi^-\pi^+\pi^-e^+e^-$. They found the four pion channel to be dominated by $\rho^0\rho^0$ production, with a cross section which

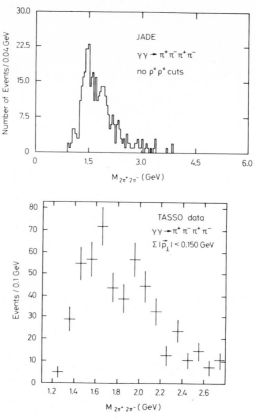

Fig. 8.5. Invariant four pion mass spectra observed by TASSO and by JADE.

rises sharply at threshold and reaches a level well above the cross section estimated from the Vector Dominance Model (VDM).

New data are reported by several groups. The four-pion invariant mass spectra observed by the TASSO group [265] and by the JADE group [265] are shown in fig. 8.5a, b. These data may contain some structure and a more refined analysis including $\rho^0\rho^0$, $\rho^0\pi^+\pi^-$ and $\pi^+\pi^-\pi^+\pi^-$ phase space production are now underway. MARK II [277] and CELLO [272] have determined the $\rho^0\rho^0$ cross section using a less complex analysis and the data are shown in fig. 8.6. The data agree both in shape and magnitude with the old TASSO data.

Recently the TASSO collaboration reported [278] data on $e^+e^- \rightarrow p\bar{p}e^+e^-$ at a c.m. energy of 2.0–2.6 GeV. They find a cross section of

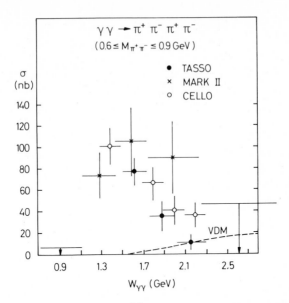

Fig. 8.6. The cross section for $e^+e^- \to \rho^0\rho^0$ as measured by TASSO, MARK II and CELLO, plotted versus the c.m. energy. The vector dominance prediction is shown by the dashed line.

$4.5 \pm 1.6 \pm 0.8$ nb, where the first error is statistical and the second systematic. An estimate of the cross section, using the reaction $p\bar{p} \to \rho\rho$, $p\bar{p} \to \rho\omega$ and the vector dominance model, gives values between 0.2 nb and 1.0 nb.

8.3. The total cross section for $\gamma\gamma \to$ hadrons

The amplitude for $\gamma\gamma \to$ hadrons will presumably contain both the hadronic piece [279] and the pointlike piece [280] shown in fig. 8.7. In the

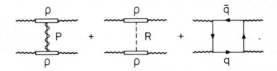

Fig. 8.7. The hadronic and the pointlike contribution to the total cross section for $\gamma\gamma \to$ hadrons.

hadronic piece the photons convert into vector mesons which subsequently interact, producing a final state similar to that observed in
hadron–hadron collisions, where the secondary hadrons tend to be
produced with low transverse momenta with respect to the beam axis. In
addition, however, the photon has a pointlike piece where the photon
couples directly to a quark pair initiating a hard scattering process. In
this case the secondary hadrons will appear as two jets of hadrons
distributed roughly as $1/p_T^4$ with respect to the beam axis. Although the
hadronic piece will dominate at small p_T, the pointlike contribution with
its slower p_T dependence will be dominant at large values of p_T.

The total cross section for $\gamma\gamma \rightarrow$ hadrons can be estimated from the
imaginary part of the elastic scattering amplitude:

$$\sigma(\gamma\gamma \rightarrow \text{hadrons}) = 240 \text{ nb} + \frac{270 \text{ nb GeV}}{W} + \sigma_{\gamma\gamma}^P. \tag{8.2}$$

The first term results from Pomeron exchange and is estimated using
the factorization relation $\sigma_{\gamma\gamma} \cdot \sigma_{pp} = (\sigma_{\gamma p})^2$. The second term involves both
f and A_2 exchange, and leads to a cross section which decreases as $1/W$
where W is the $\gamma\gamma$ c.m. energy. The pointlike contribution can be crudely
estimated using the box diagram in fig. 8.7. This process is analogous to
the QED process $e^+e^- \rightarrow \mu^+\mu^-e^+e^-$ yields

$$\sigma_{\gamma\gamma}^P = \frac{4\pi\alpha^2}{W^2} 3\sum_i e_i^4 \ln(W^2/m_i^2), \tag{8.3}$$

where e_i denotes the charge and m_i the mass of the i-th quark. This cross
section decreases roughly as $1/W^2$. Including only u, d and s quarks
with a mass of 100 MeV results in a cross section value of 650 nb at a $\gamma\gamma$
c.m. energy of 1 GeV.

The PLUTO group has measured [281] the cross section for $e^+e^- \rightarrow$
$\gamma^*\gamma^*e^+e^- \rightarrow$ hadrons $+ e^+e^-$. The data were obtained by requiring that
one of the electrons is scattered between 23 mrad and 55 mrad, and
deposits at least 4 GeV in the shower counter. The second electron is not
detected, so that its photon is nearly real. The process can be considered
as inelastic electron scattering on a real transversely polarized photon,
i.e., $e\gamma \rightarrow e'X$. The kinematics is defined above in fig. 8.1 and the cross
section $d\sigma(e\gamma \rightarrow eX)$ can be written in terms of the transverse and
longitudinal cross section σ_t and σ_ℓ as:

$$d\sigma(e\gamma \rightarrow eX) = \Gamma_t \left[\sigma_t(Q^2, W) + \varepsilon\sigma_\ell(Q^2, W) \right] d\Omega \, dE_1'. \tag{8.4}$$

In this expression, Γ_t is the flux and ε the polarization of the virtual

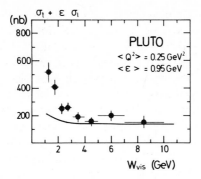

Fig. 8.8. The cross section $\gamma^*e \to e'X$ plotted versus the visible energy W_{vis}. The VDM prediction is shown as the solid line. The data were obtained by the PLUTO group.

photon. These quantities are given by

$$\Gamma_t = \frac{\alpha E'\left[1+(1-y)^2\right]}{2\pi^2 Q^2 y}, \qquad \varepsilon = 2(1-y)/\left[1+(1-y)^2\right], \qquad (8.5)$$

where $y = 1 - (E_1'/E)\cos^2\theta/2$ is the relative energy of the virtual photon.

The resulting cross section is plotted in fig. 8.8 as a function of W_{vis} at an average $\langle Q^2 \rangle$ of 0.25 GeV/c^2. The VDM prediction is shown as the solid line.

The observed energy W_{vis} was converted into the c.m. energy W using a multi-pion phase space model with limited transverse momentum. The corrections are on the order of 15–20%. The longitudinal cross section was neglected in the rest of the analysis in agreement with the data on electroproduction on a hadron target. The Q^2 dependence of the transverse cross section was taken from the vector dominance model using the ρ pole only:

$$\sigma_t(Q^2, W) = \sigma_{\gamma\gamma} \cdot \left(\frac{m_\rho^2}{Q^2 + m_\rho^2}\right)^2 \qquad (8.6)$$

The data at small Q^2 are indeed consistent with this model as shown in fig. 8.8.

The total cross section $\sigma_{\gamma\gamma}$ for two real photons was obtained by extrapolating to $Q^2 = 0$ and correcting for the difference between W and

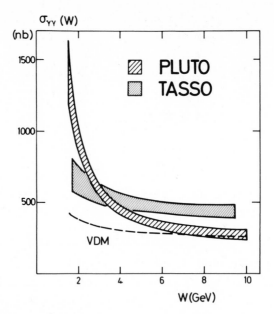

Fig. 8.9. The total cross section for $\gamma\gamma \rightarrow$ hadrons as measured by PLUTO, plotted versus the $\gamma\gamma$ c.m. energy. Preliminary TASSO data are also shown.

W_{vis}. The cross sections are well represented by the expression

$$\sigma_{\gamma\gamma} = A\left(240 \text{ nb} + \frac{270 \text{ nb GeV}}{W}\right) + \frac{B}{W^2}\left(\frac{m_\rho^2}{m_\rho^2 + Q^2}\right), \qquad (8.7)$$

with $A = 0.97 \pm 0.16$ and $B = 2250 \pm 500$ nb GeV^2.

This fit, with one standard deviation error bars, is plotted in fig. 8.9. The cross section agrees with the VDM prediction for $W \geq 6$ GeV. The rapid rise at small W might be indicative of a pointlike contribution.

Preliminary data [282] from the TASSO group are also shown. Note, that these data have a systematic error of 25% in addition to the statistical error shown.

8.4. Evidence for a pointlike coupling in $\gamma\gamma \rightarrow$ hadrons

The pointlike contribution (fig. 8.7) might show up more clearly in the transverse momentum distribution of the hadrons at large values of p_T where the hadronic contribution is very small.

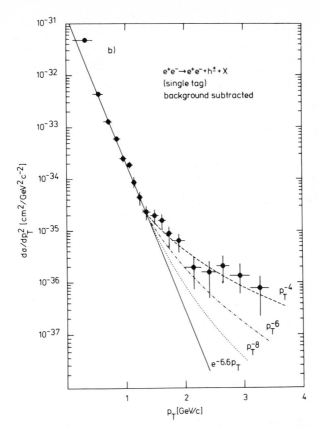

Fig. 8.10. The differential cross section $d\sigma/dp_T^2$ plotted versus p_T. The data are from TASSO.

The TASSO group [283] selected events with a tagged electron between 24 and 60 mrad depositing at least 4 GeV in the shower counter. At least three charged particles should be observed in the inner detector, one with $p_T > 0.3$ GeV/c, two and more with $p_T > 0.2$ GeV/c. A total of 1125 events with an average c.m. energy of 6.1 GeV satisfied the criteria.

The cross section $d\sigma/dp_T^2$ is plotted in fig. 8.10. The data show a steep exponential drop at small p_T which flattens out for $p_T \gtrsim 1.5$ GeV/c.

A fit to the data of the form $C_1\exp(ap_T)+C_2(p_T^b)$ gave $a = -7.4\pm0.3$ GeV and $b = -3.87\pm0.6$ GeV. This should be compared to pp interactions, where the produced pion spectrum falls exponentially as $\exp(-ap_T)$ with 6 GeV^{-1} $\leqslant a \leqslant 7$ GeV^{-1}. This cross section always falls faster than p_T^{-8} and approaches this value only at large c.m. energies.

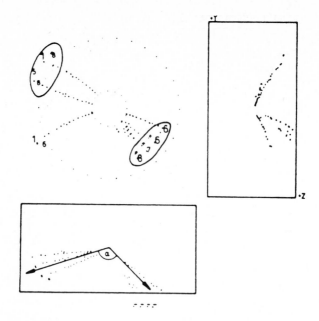

Fig. 8.11. A candidate event for $\gamma\gamma \to q\bar{q} \to$ hadrons, observed by PLUTO.

Secondary hadrons from $\gamma\gamma$ collisions and pp interactions have a similar steep exponential p_T dependence at low values of p_T. However, the two distributions are markedly different at large values of p_T, the p_T distributions of hadrons from $\gamma\gamma$ collisions have a break around $p_T \simeq 2$ GeV/c with a long tail extending to large values of p_T. The hadrons at large p_T are thus not produced by the hadronic piece of the photon. Note that the pointlike diagram in fig. 8.7 predicts the observed p_T^{-4} dependence.

The pointlike diagram leads to jets which are back to back in the $\gamma\gamma$ c.m. system. Due to the Lorentz boost of the $\gamma\gamma$ system, the hadrons will, in general, be focused forward and backward along the jet direction, yielding two non-collinear hadron jets. An example of a $\gamma\gamma \to q\bar{q}$ candidate event is shown in fig. 8.11.

Two procedures have been used to search for acollinear jet events in $\gamma\gamma$ collisions. In the first method, all particles were divided into two groups C_1 and C_2, and a quantity twoplicity T_2 defined:

$$T_2 = \mathrm{Max}\left[\left(\left|\sum_{i \in C_1} P_i\right| + \left|\sum_{i \in C_2} P_2\right|\right)\bigg/ \sum_{\text{all } i} |P_i|\right], \tag{8.8}$$

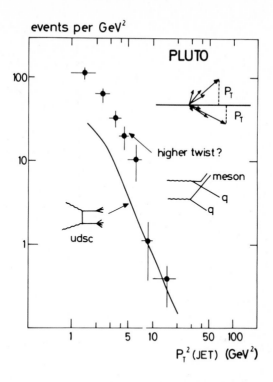

Fig. 8.12a. The transverse momentum distribution of the jets as observed by PLUTO. The prediction based on $\sigma(\gamma\gamma \to q\bar{q}) = R_{\gamma\gamma}\sigma(\gamma\gamma \to \mu^+\mu^-)$ with $R_{\gamma\gamma} = 3\Sigma_{u,d,s,c}e^4 = 34/27$ is also shown. Figures 8.12b and c (below) show the same for TASSO and JADE, respectively.

i.e. the thrust of the event is maximized using two independent axes. This method [284] has been used by PLUTO and by TASSO.

In the second method, jets are defined as particle clusters. Particles spaced within 30° were combined into preclusters and preclusters spaced within 45° were combined into clusters. Clusters which consist of at least two particles and with a total energy of 2 GeV or more are called jets. The events are classified as one-jet, two-jet or more than-two-jet events. This method has been used by JADE [285].

The PLUTO and TASSO groups have evaluated the mean p_T of the hadrons with respect to the jet axis as defined above. Both groups found a value of 300 MeV/c consistent with the value measured in $e^+e^- \to q\bar{q} \to$ hadrons.

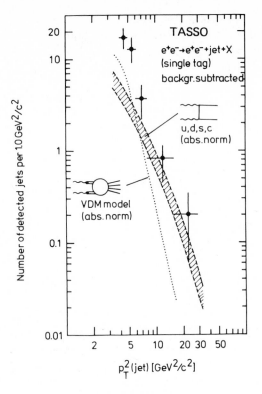

Fig. 8.12b.

The transverse momentum distribution of jets with respect to the beam axis (PLUTO and TASSO) or the direction of the c.m. system (JADE) is plotted in fig. 8.12.

The masses of the light quarks are not important at large p_T, and we expect

$$R_{\gamma\gamma} = \frac{\sigma(\gamma\gamma \to q\bar{q})}{\sigma(\gamma\gamma \to \mu^+\mu^-)} = 3\sum_i e_i^4 = 34/27 \quad \text{at large } p_T. \quad (8.9)$$

The cross section based on this prediction, including u, d, s and c quarks is also shown in fig. 8.12. The data seem to approach the predicted cross section from above and are consistent with the prediction at large values of p_T.

The data at large p_T are indeed consistent with resulting from a pointlike component of the photon. However, at such low energies

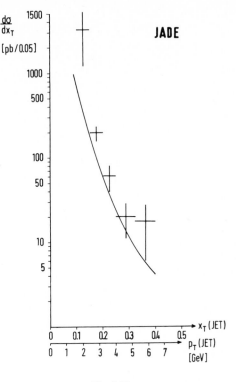

Fig. 8.12c.

hadron jets are not well defined on an event to event basis. Also, the processes [280] shown in fig. 8.13 can contribute to large p_T jets. However, the first process is expected to be smaller, and the second process will lead to a p_T^{-6} behaviour, which seems excluded as the sole source of large p_T $\gamma\gamma$ events.

8.5. Electron scattering on a photon target

The PLUTO group has now completed [286] the analysis of their large Q^2 data on inelastic electron–photon scattering, $e\gamma \rightarrow e'$ hadrons. The process is shown in fig. 8.14. In this reaction they measure the direction and the energy of one of the outgoing leptons, thus defining Q^2 and ν of the virtual photon. The second electron is untagged, resulting in a nearly real transversely polarized target photon. The data were selected using

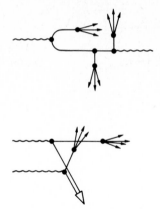

Fig. 8.13. Two possible processes resulting in large p_T jets.

the following cuts: the tagged electron (positron) should scatter between 100 mrad and 250 mrad and deposit at least 8 GeV in the tagging counter. The corresponding Q^2 value should be greater than 1 $(GeV/c)^2$. Less than 4 GeV should be deposited in the tagging counters on the opposite side of the interaction point. At least three hadrons with a visible invariant mass $W_{vis} > 0.75$ GeV should be observed in the central detector. These cuts resulted in 117 events for an integrated luminosity of 2500 nb^{-1} at an average beam energy of 15.5 GeV. Beam–gas scattering, annihilation events and inelastic Compton scattering ($e\gamma \to eX$) contrib-

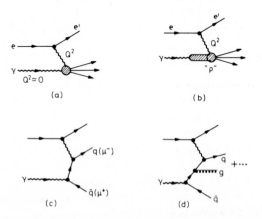

Fig. 8.14. Diagrams contributing to $e\gamma \to e'$ + hadrons.

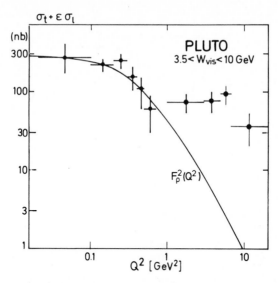

Fig. 8.15. The cross section for $\gamma\gamma$ as a function of Q^2 for a fixed cut on visible energy. The data are from the PLUTO collaboration. The solid line represents the ρ formfactor, $m_\rho^2/(m_\rho^2 + Q^2)$.

ute an estimated background of 6 events resulting in a final data sample of 111 events. The data have not been radiatively corrected. However, this correction is expected to be small.

The value of $\sigma_t + \varepsilon\sigma_\ell$ extracted from these data are plotted versus Q^2 in fig. 8.15 together with data at lower values of Q^2 discussed above. The solid line is the vector dominance prediction, using a simple ρ pole:

$$\sigma_t\left(Q^2, W_{vis}\right) = \sigma\left(W_{vis}\right)/\left(1 + Q^2/m_\rho^2\right)^2. \qquad (8.10)$$

The data are in agreement with the VDM model for Q^2 values below 1 $(GeV/c)^2$, but the observed cross section is well above the prediction at higher values of Q^2, indicating the presence of a hard component.

The data for $Q^2 > 1$ $(GeV/c)^2$ were analyzed in terms of deep inelastic electron–photon scattering [287]. Deep inelastic electron–photon scattering can be parametrized in terms of three structure functions. $F_L(x, Q^2)$, $F_T(x, Q^2)$ and $F_3(x, Q^2)$, where $F_L(x, Q^2)$ and $F_T(x, Q^2)$ correspond to the longitudinal and transverse polarization vector of the virtual photon, $F_3(x, Q^2)$ is the transverse polarization vector of the target photon in the scattering plane, and $x = Q^2/(Q^2 + W^2)$. $F_3(x, Q^2)$ will average to zero since the scattering plane was not determined. The cross section for $e\gamma \rightarrow eX$, with x and $y = 1 - (E'_1/E)\cos^2\theta/2$ as variables, is given by:

$$\frac{d\sigma(e\gamma \rightarrow eX)}{dx\,dy} = \frac{16\pi\alpha^2 EE_\gamma}{Q^4}\left[(1-y)F_2(x,Q^2) + xy^2F_1(x,Q^2)\right],$$

(8.11)

with $F_2(x,Q^2) = 2xF_T(x,Q^2) + F_L(x,Q^2)$. The data can be analyzed in terms of $F_2(x,Q^2)$ only, since x^2y averaged over the acceptance is rather small.

$$\frac{d\sigma(e\gamma \rightarrow eX)}{dx\,dy} = \frac{16\pi\alpha^2 EE}{Q^2}(1-y)F_2(x,Q^2).$$

(8.12)

It is necessary to convert the measured quantities x_{vis} and W_{vis} into x and W in order to extract the experimental value of $F_2(x,Q^2)$ from the data. This can only be done assuming a model for hadron production in $e\gamma \rightarrow eX$. Various models give consistent results and the measured value of $F_2(x,Q^2)$, normalized to α is plotted versus Q^2 in fig. 8.16a. $F_2(x,Q^2)/\alpha$ has changed somewhat from the earlier evaluation and is more consistent with a constant value of 0.35. The average Q^2 value is 5 $(GeV/c)^2$ and its x dependence is shown in fig. 8.16b. Both the hadronic and the pointlike part of the photon contribute to $F_2(x,Q^2)$.

In the hadronic part, the photon transforms into a vector meson and the virtual photon interacts with the quarks in the vector meson analogously to lepton–hadron scattering. This contribution cannot be estimated from first principles, however, it will have an x dependence similar to that observed in the structure function of the pion, and its evolution with Q^2 can be predicted. The PLUTO group uses

$$F_2\rho \simeq \frac{\alpha}{f_\rho^2/4\pi}\cdot\tfrac{1}{4}(1-x).$$

(8.13)

This contribution, shown as the dotted curve in fig. 8.16 can be increased by a factor 1.5 by including the contribution from the ϕ. The hadron part of the photons contributes mainly at low x and it is clear that the bulk of the observed formfactor must be of a different origin.

The pointlike contribution can be computed in all orders of perturbative theory. The simple quanta box diagram leads to

$$F_{2,\text{box}} = (\alpha/\pi)\sum_i e_i^4\left\{x\left[x^2 + (1-x)^2\right]\ln W^2/m_q^2 + 4x^2(1-x)\right\}.$$

(8.14)

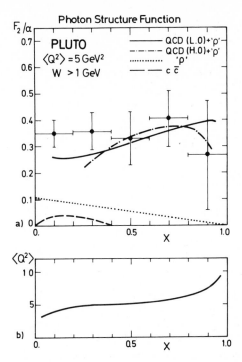

Fig. 8.16. (a) The photon structure function $F_2(x)/\alpha$. The data are averaged over Q^2 values between 1 and 15 GeV^2. The dotted line represents the contribution from the hadronic part of the photon including u, d, and s quarks. The additional contribution from c quarks is shown as the dashed line. The solid line shows the pointlike contribution to $F_2(\alpha)/\alpha$ in leading order QCD with $\Lambda = 200$ MeV (u, d, s quarks). The dashed-dotted curve is the QCD computation including higher order. (b) The correlation between the average value of Q^2 and x.

Gluon corrections modify this to

$$F_{2,\text{LO}} = h(x) \cdot \ln Q^2/\Lambda^2. \tag{8.15}$$

$h(x)$ has been evaluated by many authors. The result for u, d and s quarks with $\Lambda = 0.2$ GeV/c and including the ρ contribution is shown as the solid line in fig. 8.16a. The agreement is good and can be improved by including the contribution from the charmed quarks. This contribution, evaluated using the quark box diagram and $m_c = 1.0$ GeV, is shown as the dashed line.

QCD corrections beyond the leading order have also been calculated [288] and the results are plotted as the dashed–dotted curve in fig. 8.16. These corrections mainly reduce the predicted valued at large and small values of x and change the formfactor but little at medium x values.

Since the QCD higher order contribution is well behaved at medium x values, a measurement of the F_2 formfactor in that region as a function of Q^2 may afford a precise determination of Λ without some of the difficulties which beset its determination in lepton–hadron interactions.

9. Deep inelastic lepton–hadron interactions

The properties of spacelike currents, both electromagnetic and weak, have been determined from measurements of deep inelastic lepton–nucleon or neutrino–electron interactions. The findings, discussed in section 3, are in full agreement with the standard picture. Indeed, the neutral weak current was discovered in such experiments.

The well understood electroweak current is ideally suited to probe the nucleon at small distances, and such studies have resulted in several important discoveries. The existence of pointlike hadronic constituents was first revealed [25,289] in electron–nucleon collisions. Deep inelastic neutrino–hadron experiments [22,290–293] showed that these constituents can be identified with the quarks used to explain the quantum numbers of the hadrons. Measurements of the quark momentum distribution demonstrated that only about one half of the nucleon momentum is carried by quarks—the other half by particles which have neither electromagnetic nor weak interactions. This was the first, albeit indirect evidence, for gluons. The variation of the structure functions with Q^2 for fixed x, i.e., the scaling violation, is in agreement with the QCD predictions.

In the first part of this section we will interpret the nucleon structure —as observed in deep inelastic lepton–hadron interactions—in terms of the quark model.

Any field theory of strong interactions predicts that the momentum distributions of the quarks should vary with Q^2 at fixed x.

In the second part we will discuss the observed scaling violation pattern and compare the data to QCD predictions.

These topics have been the subject of numerous review talks. A more complete discussion, including a complete list of references can be found in ref. [294].

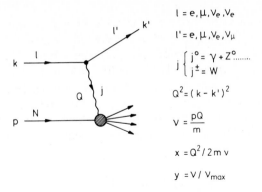

Fig. 9.1. Diagram for lepton–nucleon scattering.

9.1. Cross sections and kinematics

The general form of the lepton–nucleon interactions is shown in fig. 9.1. The incoming lepton ℓ with energy E interacts with the hadron via a neutral $(\gamma + Z^0)$ or charged $(W^{\pm})$ spacelike current. This current has the four-momentum squared

$$Q^2 = -(k - k')^2 \simeq 4E_\ell E_{\ell'} \sin^2\theta/2. \tag{9.1}$$

The energy transfer ν, measured in the rest system of the hadrons is given by:

$$\nu = 2pQ/2m = E_\ell - E_{\ell'} = E_h - m. \tag{9.2}$$

E_ℓ and $E_{\ell'}$ are the energies of the incident and scattered lepton, θ is the scattering angle in the laboratory system, p is the four-momentum of the target nucleon, and m its mass. The mass W of the final hadron system is given by:

$$W^2 = Q^2(2m\nu/Q^2 - 1) + m^2. \tag{9.3}$$

The properties of the current are thus determined by the upper vertex, i.e., by the momentum of the incident lepton and the momentum and angle of the scattered lepton. Such inclusive experiments, summing over all final state hadrons, determine the total absorption cross section for a virtual current of mass Q^2 and energy ν on a nucleon target. This process can be parametrized for an electromagnetic current in terms of two formfactors F_1 and F_2, corresponding to the longitudinal and transverse polarization of the virtual photon. The absorption cross section in the case of a weak charged or neutral current can be parametrized in terms of the formfactors F_1, F_2 and F_3 for the three helicity states of the current. F_3 is the parity violating formfactor, forbidden in purely electromagnetic interactions.

The formfactors F_1, F_2 and F_3 are, in general, functions of two variables, Q^2 and ν. Besides Q^2 and ν also the normalized variables x, y or a combination of Q^2, ν, x or y can be used. The scaled variables x and y are defined as follows:

$$x = Q^2/2pQ = Q^2/2m\nu, \qquad y = 2pQ/s = \nu/E. \qquad (9.4)$$

If the current interacts with a single pointlike constituent—the parton —then the formfactors scale [295], i.e., they are functions of only a single variable x. In any field theory this scaling behaviour will be broken, resulting in formfactors which for fixed x vary only slowly with Q^2. It has therefore become customary to use x and Q^2 as variables.

The differential cross section for deep inelastic electron (muon) scattering on a nucleon target (one-photon exchange) is given by:

$$\frac{d\sigma^{\ell^\pm}}{dx\,dy} = \frac{4\pi\alpha^2 s}{Q^4}\Big[F_2^\ell(x, Q^2)(1 - y + mxy/2E)$$

$$+ 2xF_1(x, Q^2)y^2/2\Big]. \qquad (9.5)$$

In this formula we have neglected the neutral weak current contribution. The weak effects are on the order of $10^{-4}Q^2$ and are in general negligible at present energies. The only exception is the parity violating effect observed in e–d scattering and discussed above.

Neutrinos can interact with matter only through the weak interaction. The different couplings of the two vector bosons, the $W^\pm$ and the Z^0, have already been discussed in section 3. In terms of structure functions, the differential cross section for the reaction

$$\nu(\bar{\nu}) + N \rightarrow \mu^\pm + X$$

can be written as:

$$\frac{d^2\sigma^{\nu,\bar{\nu}}}{dx\,dy}\bigg|_{CC} = \frac{G_F^2 s}{2\pi(1 + Q^2/M_W^2)^2}\Big\{F_2^{\nu,\bar{\nu}}(x, Q^2)(1 - y - mxy/2E_\ell)$$

$$+ 2xF_1^{\nu,\bar{\nu}}(x, Q^2)y^2/2$$

$$\pm \tfrac{1}{2}\big[1 - (1 - y)^2\big]xF_3^{\nu,\bar{\nu}}(x, Q^2)\Big\}. \qquad (9.6)$$

F_i^ν and $F_i^{\bar{\nu}}$ are the formfactors for neutrino and antineutrino induced charged current reactions on a nucleon target respectively. Note that the relative sign of the term proportional to xF_3 changes from neutrinos to antineutrinos. G_F is the Fermi constant.

The cross section $(d^2\sigma^{\nu\bar{\nu}}/dx\,dy)_{NC}$ for the neutral current interaction $^{(\bar{\nu})}+N \rightarrow {}^{(\bar{\nu})} + X$ can be written completely analogously to eq. (9.6).

The processes discussed above are in principle described by 28 different structure functions: Four for the charged leptons (on proton or neutron), six for the neutrinos (charged current), six for the antineutrinos (charged currents) and accordingly twelve for the (anti)neutrino neutral current interactions. However, simple relations between these functions are predicted by the quark–parton model [296] (and the Glashow–Weinberg–Salam model for the neutral currents). These relations reduce the number of independent functions to two.

Let us first assume that the quark with mass m_q carries a fractional momentum μ. In this case:

$$[\mu + (k - k')]^2 = m_q^2,$$

and neglecting all masses:

$$\mu = Q^2/2m\nu = x.$$

The variable x can be identified with the fractional momentum carried by the quarks in the nucleon.

The physical interpretation of the formfactors $2xF_1(x, Q^2)$ and $F_2(x, Q^2)$ can be clarified [294] by evaluating the general cross section [eq. (9.5)] for some simple cases.

The Rutherford cross section is obtained from eq. (9.5) in the limit $\theta \rightarrow 0$ and $y \rightarrow 0$:

$$\frac{d\sigma}{dQ^2} = \frac{4\pi\alpha^2}{Q^4} \int_0^1 \frac{F_2(x)}{x} \, dx. \tag{9.7}$$

A comparison with the Rutherford formula shows that $\int_0^1 F_2(x)/x\,dx$ measures $\Sigma_i e_i^2$, where e_i is the charge of the constituents participating. Rewriting eq. (9.5) in terms of the lepton scattering angle θ yields:

$$\frac{d^2\sigma}{dq^2\,dx} = \frac{4\pi\alpha^2}{q^4} \frac{F_2(x)}{x} \left[\cos^2\theta/2 + \frac{2xF_1(x)}{F_2(x)} \frac{q^2}{2M^2x^2} \sin^2\theta/2 \right]. \tag{9.8}$$

This form can be directly compared to the cross section for $e + \mu \rightarrow e + \mu$:

$$\frac{d\sigma}{dq^2} = \frac{4\pi\alpha^2}{q^4}\frac{E'}{E}\left(\cos^2\theta/2 + \frac{q^2}{2m^2}\sin^2\theta/2\right). \tag{9.9}$$

The comparison yields the Callan–Gross relation for spin-1/2 partons:

$$2xF_1(x) = F_2(x). \tag{9.10}$$

$F_1(x) = 0$ for scalar partons. Thus the spin of the constituents can be determined from a measurement of $2xF_1(x)/F_2(x)$.

To see the significance of $xF_3(x)$, eq. (9.6) is written as:

$$\frac{d\sigma^{\nu,\bar{\nu}}}{dy\,dx} = \frac{G_F^2 s}{2\pi}\left[\frac{F_2(x)\pm F_3(x)}{2} + \frac{F_2(x)\mp F_3(x)}{2}(1-y)^2\right]. \tag{9.11}$$

This equation was obtained from eq. (9.8) using the Callan–Gross relation and neglecting the term $mxy/2E_\ell$.

Eq. (9.11) can then be compared with the cross section for $\nu_e e \rightarrow \nu_e e$ and $\bar{\nu}_e e \rightarrow \bar{\nu}_e e$ discussed in section 3.

$$\frac{d\sigma}{dy}(\nu_e e^- \rightarrow \nu_e e^-) = \frac{d\sigma}{dy}(\bar{\nu}_e e^+ \rightarrow \bar{\nu}_e e^+) = \frac{G_F^2 s}{\pi},$$

$$\frac{d\sigma}{dy}(\bar{\nu}_e e^- \rightarrow \bar{\nu}_e e^-) = \frac{d\sigma}{dy}(\nu_e e^+ \rightarrow \nu_e e^+) = \frac{G_F^2 s}{\pi}(1-y)^2. \tag{9.12}$$

Comparing eqs. (9.11) and (9.12) results in the following relationships:

$$F_2(x) = 2x[q(x) + \bar{q}(x)], \qquad xF_3(x) = 2x[q(x) - \bar{q}(x)], \tag{9.13}$$

where $q(x)$ and $\bar{q}(x)$ denote the parton and antiparton densities as a function of momentum. $xF_3(x)$ thus counts the difference between partons and antipartons in the nucleon.

It is clear from the discussion above that the properties of the constituents and their density distributions can be determined from deep inelastic experiments. We will now consider the data.

9.2. Comparison of the data with the quark–parton model

9.2.1. The total neutrino–hadron cross section

As we have seen above, the formfactors scale if the current interacts directly with pointlike constituents. In this case the total cross section for neutrino–hadron interactions, obtained by integrating eq. (9.6) over x

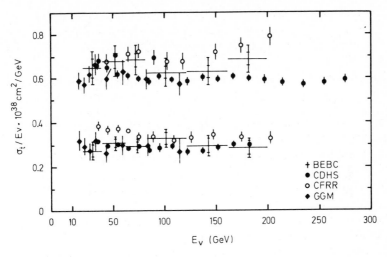

Fig. 9.2. The slopes of the total neutrino and antineutrino cross sections plotted versus incident energy [297].

and y, increases linearly with the incident neutrino energy. The measured cross section, divided by the neutrino energy, is plotted in fig. 9.2 versus neutrino energy for both incident neutrinos and antineutrinos. The slope is clearly constant over a large energy range, providing striking evidence for the pointlike nature of the interactions. The cross section ratio $\sigma(\nu N)/\sigma(\bar{\nu}N)$ is of the order of $1/2$. It follows from eqs. (9.11) and (9.13) that a nucleon made of quarks only yields $\sigma(\bar{\nu}N)/\sigma(\nu N) = 1/3$. The total cross sections for interaction of (anti)neutrinobeams with isoscalar targets, as measured by various groups, are listed in table 9.1.

Table 9.1
Total cross sections on isoscalar targets σ_{tot}/E_ν ($\times 10^{-38}$ cm^2/GeV)
Systematic errors are 4–10%

Experiment	Beam (GeV)	$\bar{\nu}$	ν	$\bar{\nu}/\nu$	Ref.
CITFR	NB 45–225	0.29 ± 0.015	0.61 ± 0.03	–	[298]
CDHS	NB 30–200	0.30 ± 0.02	0.62 ± 0.05	0.48 ± 0.02	[293]
BEBC	NB 20–200	0.305 ± 0.016	0.663 ± 0.032	0.463 ± 0.025	[299]
CHARM	NB 20–200	0.301 ± 0.018	0.604 ± 0.032	0.498 ± 0.019	[63]
CFRR	NB 40–220	0.371 ± 0.020	0.719 ± 0.037	0.517 ± 0.020	[64]
GGM	WB 10–150	0.29 ± 0.04	0.62 ± 0.08	0.47 ± 0.09	[300]

The cross sections measured by CFRR are significantly above the values reported by other groups. This discrepancy is reflected in all structure functions, integrals and ratios. The absolute values of the neutrino structure functions have to be taken with some care as long as the source of the discrepancy remains unknown. Fortunately all groups agree on the value for the ratio of antineutrino to neutrino cross sections. Thus quantities which involve the neutrino and antineutrino data with relative normalizations are not affected.

9.2.2. *The spin of the constituents*

The spin of the constituents can be determined from a measurement of the formfactors $2xF_1(x)$ and $F_2(x)$. Spin-1/2 constituents satisfy the Callan–Gross relation $2xF_1(x) = F_2(x)$, whereas scalar constituents yield $2xF_1(x) = 0$.

Usually the ratio $R = \sigma_L/\sigma_T$ is determined in charged lepton–nucleon interactions. σ_L and σ_T are the cross sections for longitudinal and transversely polarized photons. In the Bjorken limit ($Q^2 \to \infty$, $\nu \to \infty$, $x = Q^2/2m\nu = \text{const.}$, i.e. $Q^2/\nu^2 \to 0$), σ_L vanishes for spin-1/2 constituents and σ_T vanishes for scalar constituents. R can also be written in terms of the formfactors $2xF_1(x)$ and $F_2(x)$ as:

$$R = \sigma_L/\sigma_T = \frac{F_2(x)}{2xF_1(x)}(1 + Q^2/\nu^2) - 1. \tag{9.14}$$

An equivalent definition is often used:

$$R' = \frac{F_2(x) - 2xF_1(x)}{F_2(x)} = R(1 + Q^2/\nu^2) - Q^2/\nu^2. \tag{9.15}$$

The definitions are identical in the scaling limit and lead to $R = R' = 0$ for spin-1/2 constituents, and $R = R' = \infty$ for scalar constituents.

It is rather difficult to determine $R(x, Q^2)$. A measurement of the cross section at a particular value of x, Q^2 gives information only on a combination of the formfactors. To disentangle the formfactors requires cross section measurements at different lepton energies.

Because of its importance, several groups have determined R (or R'). The results are summarized in table 9.2 and some recent data [307] from the CDHS collaboration are plotted in fig. 9.3. The value of R (averaged over Q^2) is measured as a function of ν. The Q^2 values are of the order of 20 GeV$^2/c^2$. R is small and for higher ν clearly compatible with zero. The curve, labelled "QCD" is a QCD calculation and will be discussed below.

Table 9.2
Average values of R and R'

Experiment	Reaction	R	R'	Ref.
SLAC	ep	0.21 ± 0.10	–	[301]
SLAC–MIT	ed	0.17 ± 0.07	–	[302]
CHIO	μp	0.38 ± 0.38	–	[28]
EMC (prelim.)	μp	0.03 ± 0.10	–	[303]
	μN	-0.13 ± 0.19	–	[304]
GGM	νN	–	0.32 ± 0.15	[305]
HPWFOR	νN	–	0.11 ± 0.07^a	[306]
CDHS	νN	0.10 ± 0.07^a	–	[307]
BEBC	νN	–	0.04 ± 0.16	[299]
FIIM	νN	–	0.03 ± 0.12	[308]
CHARM	νN (NC)	–	0.01 ± 0.11^a	[309]

^{a} Radiatively corrected neutrino data.

With the exception of the low Q^2 Gargamelle and SLAC data, all experiments are roughly compatible with $R = 0$, but a careful study of the systematic errors is needed in such an analysis. Note that for the neutrino experiments only the results marked ^{a} have been radiatively corrected. The correction is typically of the order of $\Delta R \simeq -0.05$. The measurements have been done at somewhat different Q^2, typically around

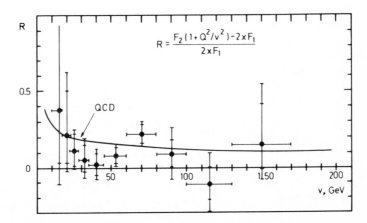

Fig. 9.3. $R = \sigma_L/\sigma_T$ as measured by the CDHS collaboration in neutrino–hadron interactions, plotted versus ν. The QCD prediction is shown by the solid line.

$Q^2 = 25$ GeV$^2/c^2$ with the exception of the quoted two low Q^2 experiments.

The data in table 9.2 show that R is, in general, small and hence the spin of the parton is $1/2$.

9.2.3. Structure functions in the quark – parton model

It is natural of course to identify partons with quarks and in this section we will write the formfactors in terms of the quark (antiquark) distribution functions. $u(x)dx$ is the probability of finding an "up" quark with fractional momentum between x and $x+dx$ in a proton and a similar notation is used for the other flavours. Note that, by isospin invariance, $u(x)dx$ is also the probability of finding a "down" quark in a neutron. Present energies are too small for significant contributions from the "beauty" or "truth" quark and we will hence limit the discussion to u, d, s and c quarks.

The structure functions measured in charged lepton nucleon interactions can be written in terms of the quark distribution functions according to eq. (9.13) as:

$$F_2^{\ell p}(x) = 2xF_1^{\ell p}(x) = \tfrac{4}{9}x(u + \bar{u} + c + \bar{c}) + \tfrac{1}{9}x(d + \bar{d} + s + \bar{s}),$$
$$F_2^{\ell n}(x) = 2xF_1^{\ell n}(x) = \tfrac{4}{9}x(d + \bar{d} + c + \bar{c}) + \tfrac{1}{9}x(u + \bar{u} + s + \bar{s}).$$

$$(9.16)$$

The weak interaction is slightly more complicated since not the quarks but rather some linear combination are the eigenstates of the weak interactions. The mixing angles as discussed in section 3 are generally believed to be of the order of the Cabibbo angle and hence small. To simplify the notation we ignore this effect.

For the charged weak current ($W^{\pm}$ exchange) the structure functions can then be written as;

$$F_2^{\nu p} = 2xF_1^{\nu p} = 2x(d + \bar{u} + s + \bar{c}), \qquad F_3^{\nu p} = 2(d - \bar{u} + s - \bar{c});$$
$$F_2^{\nu n} = 2xF_1^{\nu n} = 2x(u + \bar{d} + s + \bar{c}), \qquad F_3^{\nu n} = 2(u - \bar{d} + s - \bar{c});$$
$$F_2^{\bar{\nu} p} = 2xF_1^{\bar{\nu} p} = 2x(u + \bar{d} + \bar{s} + c), \qquad F_3^{\bar{\nu} p} = 2(u - \bar{d} - \bar{s} + c);$$
$$F_2^{\bar{\nu} n} = 2xF_1^{\bar{\nu} n} = 2x(d + \bar{u} + \bar{s} + c), \qquad F_3^{\bar{\nu} n} = 2(d - \bar{u} - \bar{s} + c).$$

$$(9.17)$$

The factor two enters, because both V and A currents contribute.

The coupling constants of the neutral current have been introduced in section 3. Following the notation of eq. (3.28) and using the abbreviations NC and CC for neutral and charged currents respectively, the structure functions can be written as:

$$F_2(\text{NC}) = F_2(\text{CC})\left(u_L^2 + d_L^2 + u_R^2 + d_R^2\right)$$
$$+ (s + \bar{s} - c - \bar{c})\left[\left(d_L^2 + d_R^2\right) - \left(u_L^2 + u_R^2\right)\right],$$
$$F_3(\text{NC}) = F_3(\text{CC})\left[\left(u_L^2 + d_L^2\right) - \left(u_R^2 + d_R^2\right)\right]. \tag{9.18}$$

The second term in the expression for $F_2(\text{NC})$ is expected to be small. At large values of Q^2, the probability to find a strange or charmed quark is about equal. No results on the neutral current structure functions have been published so far. The experiment is far more difficult than the extraction of the charged current structure functions of the electromagnetic structure functions. Only the momentum or the incident neutrino, and the hadronic energy and its direction of flow can be measured. Both are correlated with the momentum of the struck quark, but in a rather indirect way. The direction of energy flow is not easy to measure, and this results in large uncertainties in Q^2. Rough comparisons [310] of the hadronic energy spectra measured in neutral and charged current reactions respectively, do not reveal striking differences between the structure functions except for a scale factor. It is therefore likely, that the neutral current structure functions are indeed very similar to those seen by charged currents.

The experiments have mostly been performed on isoscalar targets. Before we turn to the data, it is useful to calculate the structure functions for an equal mixture of protons and neutrons in the target by averaging over the proton and neutron structure functions:

$$F_2^{\ell N} = \tfrac{1}{2}\left(F_2^{\ell p} + F_2^{\ell n}\right) = \tfrac{5}{18}x\left(u + \bar{u} + d + \bar{d} + s + \bar{s} + c + \bar{c}\right)$$
$$- \tfrac{1}{6}(s + \bar{s} - c - \bar{c}),$$
$$F_2^{\nu N} = x\left(u + \bar{u} + d + \bar{d} + 2s + 2\bar{c}\right),$$
$$F_2^{\bar{\nu} N} = x\left(u + \bar{u} + d + \bar{d} + 2\bar{s} + 2c\right),$$
$$F_3^{\nu N} = x\left(u - \bar{u} + d - \bar{d} + 2s - 2\bar{c}\right),$$
$$F_3^{\bar{\nu} N} = x\left(u - \bar{u} + d - \bar{d} - 2\bar{s} + 2c\right). \tag{9.19}$$

With the assumption of SU(4) symmetry of quark flavours, expected to hold at high values of Q^2, the sea quark distributions are all equal

($\bar{u} = \bar{d} = s = \bar{s} = c = \bar{c}$). Under these circumstances, or for regions in x where the contribution of the sea is small, one gets:

$$F_2^{\ell N} \approx \tfrac{5}{18} F_2^{\nu N} \approx \tfrac{5}{18} F_2^{\bar{\nu} N}, \qquad F_3^{\nu N} \approx F_3^{\bar{\nu} N}. \tag{9.20}$$

The 24 different structure functions have been reduced to two universal functions. Unfortunately, a considerable fraction of the data is still below the charm threshold. In this case, the reduced couplings due to the Kobayashi–Maskawa matrix have to be taken into account and final differences between the neutrino and antineutrino structure functions remain:

$$F_2^{\nu N} - F_2^{\bar{\nu} N} = \sin^2\theta_c x (\bar{u} + \bar{d} - u - d),$$

$$F_3^{\nu N} - F_3^{\bar{\nu} N} = \sin^2\theta_c (4s - u - d - \bar{u} - \bar{d}). \tag{9.21}$$

9.2.4. Quark distributions

The aim is to measure the x distribution of the formfactors over a wide range in y. The data must be collected at fixed values of Q^2 to exclude effects due to scaling violation. Since $xy = Q^2/s$ this implies that data must be collected over a wide range in lepton energies. The energy is, of course, well defined in the case of charged leptons, whereas the neutrino energy, even in the best case, is rather poorly known.

The kinematical region available in x and $Q^2(y)$ for various experiments is illustrated in fig. 9.4. The upper limit on Q^2 for fixed x is determined by the available lepton energy. The quickly worsening energy resolution at low hadron energies defines the lower limit on Q^2. The x resolution at high x values is mainly given by the momentum measurement of the final state lepton. Bubble chambers tend to be superior in this energy region.

To extract F_2, most experiments assume R to have a small, constant value. In some cases R was assumed to be a function of x with the functional form taken either from theory or from experiments. Fig. 9.5 shows $F_2^{\mu p}$ and F_2^{ep} measured at SLAC [301, 302], FNAL [28] and CERN [303]. The mean Q^2 is of the order of 10 GeV$^2/c^2$ in all four experiments. At high x, where the SLAC data can be compared with the other data the agreement is remarkable. At low x the CERN EMC data, with superior statistics, tend to stay below the FNAL data.

The form factor F_2^ν is determined from the sum of neutrino and antineutrino data, which effectively cancel the xF_3 term in eq. (9.6). Small model-dependent corrections for the differences between the F_2^ν, F_3^ν and

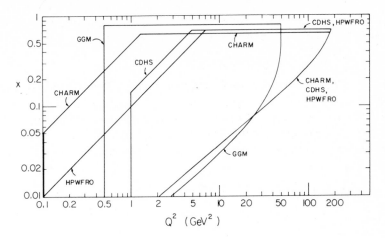

Fig. 9.4. The kinematical region in x and Q^2 available to different experiments.

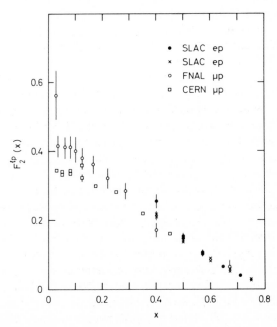

Fig. 9.5. Structure function F_2 for charged lepton scattering on protons. The average Q^2 is around 10 GeV$^2/c^2$. (Data from refs. [28, 301–303].)

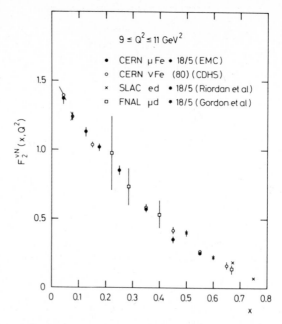

Fig. 9.6. Structure function F_2 for neutrino scattering on nucleons. The data from the charged lepton experiments are scaled by 18/5. (Data from refs. [28, 302, 304, 311].)

$F_2^{\bar{\nu}}$ and $F_3^{\bar{\nu}}$ for the various targets have been made, and the data are radiatively corrected. In fig. 9.6 neutrino data [311] on a Fe target are compared with data from three charged lepton experiments [28, 302, 304]. The charged lepton data have been multiplied by 18/5 following the prescription of eq. (9.20). The structure functions of the various experiments are extracted with slightly different assumption on R, but, nevertheless, the electromagnetic current and the weak charged current data agree remarkably well. To be able to relate two such different processes is a great triumph for the quark–parton model (QPM).

The structure function xF_3 appears only in neutrino scattering. This structure function is determined from the difference between the neutrino and antineutrino cross sections [eqs. (9.11), (9.13)]. Terms containing F_2 and $2xF_1$ cancel (up to small, model-dependent corrections).

The cross section difference is mainly proportional to the average

$$x\bar{F}_3 = \tfrac{1}{2}\left(xF_3^{\nu} + xF_3^{\bar{\nu}}\right). \tag{9.22}$$

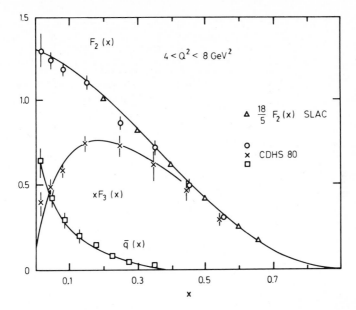

Fig. 9.7. The structure functions F_2, xF_3 and the antiquarks in neutrino–nucleon scattering. The sea contribution is zero beyond $x \simeq 0.4$. (Data from refs. [302, 311].)

This average is, in the QPM, just the sum of the valence quark structure functions

$$\bar{F}_3^{\text{p}} = \bar{F}_3^{\text{n}} = \bar{F}_3^{\text{N}} = 2(u - \bar{u} + d - \bar{d}) + \text{small corrections.} \quad (9.23)$$

Figure 9.7 shows a recent measurement [311] of $F_2^{\nu\text{N}}(x)$ and $xF_3^{\nu\text{N}}(x)$ on iron together with scaled data on F_2^{ed} from SLAC [302]. The resolution in x is just sufficient to demonstrate that xF_3 approaches zero for $x \to 0$, as one would naively guess.

Looking back at eq (9.19), one sees that the average of $F_2^{\nu\text{N}}$ and $F_2^{\bar{\nu}\text{N}}$,

$$\bar{F}_2^{\nu\text{N}} = \tfrac{1}{2}\left(F_2^{\nu\text{N}} + F_2^{\bar{\nu}\text{N}}\right), \quad (9.24)$$

is the momentum distribution function of quarks and antiquarks inside the nucleon. The difference $\bar{F}_2^{\nu\text{N}} - x\bar{F}_3^{\nu\text{N}}$ is therefore the momentum distribution function of the sea quarks. This result is plotted as the squares in fig. 9.7. One notes that the sea quarks ($\bar{q}$) are concentrated at low x values. Above $x = 0.4$ the sea contribution is negligible. The drawn

Table 9.3

Results on the shape of the structure functions

Parameter	Experiments			
	CHARM [309]	CDHS [293]	HPWF [312]	BEBC [22,291]
A	2.63 ± 0.81	–	–	–
a	0.41 ± 0.11	0.5 fixed	–	–
b	2.96 ± 0.31	3.5 ± 0.5	$3.7 \pm 0.1 \pm 0.3$	–
B	0.80 ± 0.09	–	–	–
c	4.93 ± 0.91	6.5 ± 0.5	$4.6 \pm 0.5 \pm 0.6$	$4.9^{+2.7}_{-1.7}$

curves are fits of the empirical form

$$xF_3 = Ax^a(1-x)^b, \qquad 2\bar{q} = B(1-x)^2.$$

The results of several experiments (with slightly different experimental conditions) are summarized in table 9.3.

It will become clear later that the results depend on the Q^2 range. The data listed in table 9.3 all result from experiments done in a similar kinematic region.

Information on the ratio of d to u quark distributions can be obtained by measuring the ratio $F_2^{\ell n}/F_2^{\ell p}$. This ratio can be written [Eq. (9.17)] in the quark–parton model as:

$$F_2^{\ell n}/F_2^{\ell p} = \frac{1+4d/u}{4+d/u},$$

permitting a determination of the ratio d/u. Several predictions for this ratio exist [313–316]. The first measurements were reported by the SLAC MIT group [302]. Figure 9.8 shows these data together with recent measurements by the EMC collaboration [317] in a completely different Q^2 range. The scale on the right hand side gives the ratio d/u. In the region above $x \simeq 0.4$, where the sea quark distribution is practically zero, the u quarks clearly dominate over the d quarks. Similar results are obtained [318] by comparing νp data with ep and μp data. In apparent contradiction to the naive QPM the ratio falls below 0.25 for $x > 0.5$. While the QPM is obviously failing, there are other models:

–Field and Feynman [313] predict a $(1-x)$ behaviour with $d/u = 0$ at $x = 1$.

–Farrar and Jackson [314] predict $d/u = 0.2$ at $x = 1$ (a QCD type model).

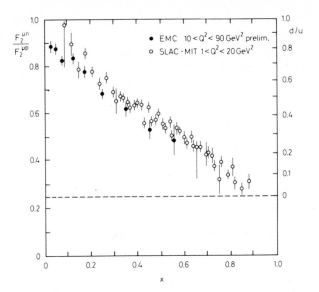

Fig. 9.8. The ratio $F_2^{\mu N}/F_2^{\mu p}$ as a function of x. This corresponds to the ratio of "down" to "up" quarks indicated on the right hand scale. The naive QPM prediction is $d/u = 1/4$ at $x = 0$. (Data from refs. [302, 317].)

–Close and Roberts [316], and Donnachie and Landshoff [315] expect $d/u = \frac{1}{2} \times \frac{3}{7}$ at $x = 1$ in diquark models.

None of these models can be excluded, but the Field and Feynman fit describes the data best.

$\overline{F}_3(x)$ is the probability to find a valence quark with relative momentum x. In the naive quark–parton model $\int_0^1 \overline{F}_3(x)\,\mathrm{d}x = 3$, the number of valence quarks. In practice $x\overline{F}_3$ is measured and the ratio $x\overline{F}_3/x$ is experimentally poorly defined at small x. The procedure adopted by most experiments is to integrate down to a small but variable $x_{\min}$ and extrapolate to zero. The CHARM collaboration uses a slightly different procedure. The result is:

$$\int_0^1 x\overline{F}_3\,\mathrm{d}x = -\begin{cases} 3.2 \pm 0.5, & \text{CDHS [293],} \\ 2.5 \pm 0.5, & \text{BEBC [22, 291],} \\ 2.66 \pm 0.41, & \text{CHARM [309].} \end{cases}$$

It should be emphasized, that scaling violating effects have a strong influence on the result. But nevertheless, the agreement with the naive quark–parton model prediction is striking.

$\bar{F}_2^{\nu}(x)$ describes the momentum distribution of quarks and antiquarks inside the nucleon. The integral $\int_0^1 x\bar{F}_2^{\nu}\,dx$ therefore determines the momentum fraction carried by all objects which couple to the charged current. This sum rule can be rather well determined and the results are:

$$\int_0^1 x\bar{F}_2\,dx = \begin{cases} 0.51 \pm 0.05, & \text{BEBC [319]}, \\ 0.45 \pm 0.02, & \text{CITF [320]}, \\ 0.45 \pm 0.03, & \text{CDHS [293]}, \\ 0.44 \pm 0.02, & \text{CHARM [63]}. \end{cases}$$

Therefore one half of the proton momentum is carried by objects which are invisible to the electromagnetic and the weak current.

The fractional momentum of the valence quarks

$$\int_0^1 x\bar{F}_3\,dx = 0.32 \pm 0.01, \quad \text{CDHS},$$

is about one third of the total momentum. Therefore some $(13 \pm 1)\%$ of the momentum is carried by the sea quarks.

The quark–parton model describes data taken at the same value of Q^2 very well. However, the data vary with Q^2 and this feature cannot be understood in the naive quark–parton model. Furthermore, 50% of the proton momentum is carried by objects different from partons. In the next part we will discuss the Q^2 variation of the data and a possible explanation.

9.3. Scaling violations and QCD

9.3.1. Altarelli – Parisi equations

The resolving power of a spacelike current increases with Q^2 as $(Q^2)^{-1/2}$, i.e. with increasing Q^2 a spacelike current explores clusters of progressively smaller size. It is common to all field theories [12] that a parton can split into smaller clusters of two or more partons sharing the parent momentum. The relevant Feynman graphs, to first order in the strong coupling constant, are shown in fig. 9.9. A comparison of deep inelastic scattering at two different values Q^2 is shown in fig. 9.10. Figures 9.10a, b show that the effect of gluon bremsstrahlung is to decrease the x value of the quark. Figures 9.10c, d illustrate the fact that a gluon which is invisible to the electroweak current at low Q^2 may be resolved into its $Q\bar{Q}$ content by a high Q^2 current, leading to a strong increase of the sea quarks. Therefore the pattern of scaling violations indicated in fig. 9.11,

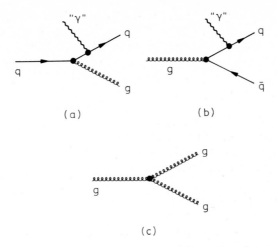

Fig. 9.9. Feynman graphs for: (a) quark gluon bremsstrahlung, (b) splitting of a gluon into a qq̄ pair, and (c) splitting of a gluon into two gluons.

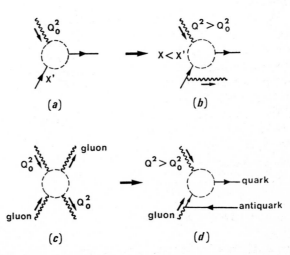

Fig. 9.10. (a) A virtual photon with Q_0^2, striking a quark. (b) A virtual photon of higher Q^2, resolving the quark into a quark and a gluon. (c) A virtual photon with Q_0^2 traversing a gluon without interaction. (d) A virtual photon of higher Q^2 resolving the gluon into a quark–antiquark pair.

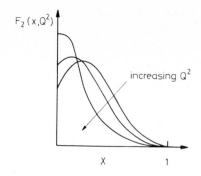

Fig. 9.11. The qualitative change of deep inelastic formfactors with Q^2, expected in field theories of the strong interaction.

where, with increasing values of Q^2, the parton density is enhanced at low x and depleted at large x, is common to all field theories and the Q^2 evolution of the structure function must be studied in detail in order to confront QCD with the data. Note that QCD makes absolute predictions of this evolution.

The cleanest tests [294] on the theory are obtained from measurements of the flavour non-singlet structure functions like for example $xF_3(x, Q^2)$, the valence quark distribution. The reason is that only gluon bremsstrahlung (fig. 9.9a) contributes to the Q^2 evaluation of the non-singlet structure factors, whereas both gluon bremsstrahlung and gluon–quark pair production (fig. 9.9a, b) contribute to the singlet functions like $F_2^{ep}(x, Q^2)$.

The Q^2-evolution [229, 321] of the valence quark distribution $q_V(x, Q^2)$ can be written as:

$$Q^2 \frac{\partial q_V(x, Q^2)}{\partial Q^2} = \frac{\alpha_s(Q^2)}{2\pi} \int_0^1 \frac{dy}{y} P_{qq}(x/y) q_V(y, Q^2), \qquad (9.25)$$

where $\alpha_s(Q^2)$ is the strong coupling constant defined in eq. (5.2) $P_{qq}(z)$ is the probability to find a quark inside the quark with a relative momentum $z = x/y$. The splitting function $P_{qq}(z)$ can be calculated in QCD from the analogy to electron bremsstrahlung. The energy spectrum of the radiated gluon should be proportional to the bremsspectrum of a real photon, i.e.,

$$P_{qg}(z) = \frac{4}{3} \frac{1 + (1 - z)^2}{z}, \qquad (9.26)$$

where z is the energy of the gluon in units of the parent energy and $4/3$ is a colour factor. The probability to find a gluon with relative energy z is equal to the probability of finding a quark with relative momentum $1 - z$. This leads to [229]:

$$P_{qq}(z) = \frac{4}{3}\frac{1+z^2}{1-z}.$$

At $z \to 1$ the quark only contains itself and the complete splitting function can be written as:

$$P_{qq}(z) = \frac{4}{3}\frac{1+z^2}{1-z} + \frac{3}{2}\delta(z-1). \tag{9.27}$$

Similar evolution equations can be written for the singlet quark (anti-quark) distribution function $q(x, Q^2)$ and the gluon distribution function $G(x, Q^2)$:

$$Q^2\frac{\partial q(x,Q^2)}{\partial Q^2} = \frac{\alpha_s}{2\pi}\int_x^1 \frac{\mathrm{d}y}{y}\left[q(y,Q^2)P_{qq}(x/y)\right.$$

$$\left. + G(g,Q^2)P_{gq}(x/y)...\right],$$

$$Q^2\frac{\partial G(x,Q^2)}{\partial Q^2} = \frac{\alpha_s}{2\pi}\int_x^1 \frac{\mathrm{d}y}{y}\left[q(y,Q^2)P_{qq}(x/y)\right.$$

$$\left. + G(g,Q^2)P_{gq}(x/y)...\right]. \tag{9.28}$$

The gluon–quark pair production function $P_{gq}(z)$ and the three gluon vertex function $P_{gg}(z)$ (fig. 9.9c) are given by:

$$P_{gq}(z) = \tfrac{1}{2}\left[z^2 + (1-z)^2\right],$$

$$P_{gg}(z) = 6\left[\frac{z}{(1-z)_+} + \frac{1-z}{z} + z(1-z)\right.$$

$$\left. + (11/12 - N_f/18)\delta(z-1)\right], \tag{9.29}$$

where N_f is the number of flavours participating. It is clear that the singlet functions are more difficult to analyze since they involve the unknown gluon distribution function. In addition there are several theoretical uncertainties [322, 323] which must be evaluated before firm conclusions can be drawn from the data. These involve higher twist corrections, target mass effects and higher orders in QCD.

Two methods are in general use to compare predictions with theory. The first method involves taking the moments [323] of the distributions given by eqs. (9.25) and (9.28), i.e., both sides of the equations are multiplied by x^{N-1} and integrated over x. The results are a set of simple, well defined equations. The non-singlet moments can be written [324] as:

$$M_N^{(3)}(Q^2) = \left[\ln Q^2/\Lambda^2\right]^{-d_N},$$

$$d_N = \frac{4}{33-2N_f}\left[1 - \frac{2}{N(N+1)} + 4\sum_{i=1}^{N} 1/i\right]. \qquad (9.30)$$

The second method is to try to solve the Altarelli–Parisi equations directly. This method [325–327] has the advantage that it only uses measured data, whereas in the moment method, data at all x values are needed to perform the integration. Both analyses however, suffer from uncertainties due to higher order twist or higher order QCD effects.

9.3.2. The value of R

In the naive quark–parton model σ_L approaches 0 in the Bjorken limit. In QCD however, the radiated gluon carries away transverse momentum, leading to a transverse momentum distribution of the quarks and hence to a longitudinal form factor $F_L(x,Q^2)$:

$$F_L(x,Q^2) = F_2(x,Q^2) - 2xF_1(x,Q^2). \qquad (9.31)$$

This formfactor can be written to first order in α_s as [294]:

$$F_L(x,Q^2) = \frac{\alpha_s(Q^2)}{2\pi}\int_x^1 \frac{dy}{y^3}\left[\tfrac{8}{3}F_2(y,Q^2)\right.$$

$$\left. +4A(1-x/y)yG(y,Q^2)\right]. \qquad (9.32)$$

$A = 10/9$ in charged lepton scattering and $A = 4$ in neutrino interactions.

The longitudinal structure function is determined through a measurement of R' (or R):

$$R' = F_L/F_2.$$

At small x $(x < 0.1)$ the main contribution to F_L is from the gluon structure function $G(y,Q^2)$, whereas $F_2(y,Q^2)$ contributes mainly at large x. Thus both the value of α_s and the gluon structure function could be determined from a precise measurement of R'. The results of a recent measurement [307] of R integrated over x are plotted in fig. 9.12 versus Q^2 and compared to the QCD prediction.

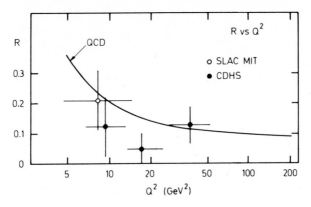

Fig. 9.12. $R = \sigma_L / \sigma_T$ as a function of Q^2. The curve is the QCD prediction.

9.3.3. Moment analysis of $xF_3(x, Q^2)$

The data are, in general, analyzed in terms of the Nachtmann moments [323]. These moments project to definite spin N at all values of Q^2 and they take the finite target mass into account. Using this formalism the non-singlet moment can be written as:

$$M_N^{(3)}(Q^2) = \int_0^1 \frac{dx}{x^3} \xi^{N+1} x F_3(x, Q^2) \frac{1 + (N+1)(1 + 4m^2 x^2/q^2)^{1/2}}{N+2},$$

$$\xi = \frac{2x}{1 + (1 + 4m^2 x^2/Q^2)^{1/2}}. \tag{9.33}$$

The moments at two different values of Q^2 are related in leading order of QCD as:

$$\frac{M_N^{(3)}(Q^2)}{M_N^{(3)}(Q_0^2)} = \left(\frac{\ln(Q_0^2/\Lambda^2)}{\ln(Q^2/\Lambda^2)} \right)^{d_N}, \tag{9.34}$$

d_N is defined in eq. (9.30). The evolution of the valence quark distribution therefore depends on a single constant Λ.

This form suggests to plot the logarithm of the N-th moment of $2xF_3$ versus the logarithm of its M-th moment. Such a plot should yield a straight line with a slope d_N/d_M. The magnitude of the slope is an absolute prediction in QCD. This prediction has been verified by several groups. In fig. 9.13 M_6^3 is plotted versus M_4^3 and M_5^3 is plotted versus M_3^3.

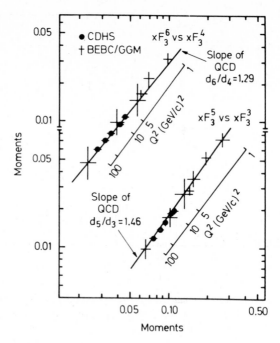

Fig. 9.13. Moments of xF_3 on a double logarithmic plot. The data agree well with the first order QCD predictions.

Data from CDHS [328] and a combination of Gargamelle and BEBC results [291] are shown together with the QCD predictions. Obviously there is excellent agreement between the predicted and the measured value. For d_6/d_4 QCD predicts 1.29 compared to 1.29 ± 0.06 obtained by BEBC/GGM and 1.18 ± 0.09 by CDHS. For d_5/d_3 QCD predicts 1.46 compared to 1.50 ± 0.08 by BEBC/GGM and 1.34 ± 0.12 obtained by CDHS. Note, however, that this excellent agreement was obtained despite the fact that both groups treated these data quite differently. BEBC/GGM included elastic events whereas they were excluded by CDHS. BEBC/GGM included only non-singlet data whereas CDHS include some electron data.

The Q^2 evolution of single moment can be written as:

$$\left[M_n^{(3)}(Q^2) \right]^{-1/d_n} = C\left(\ln Q^2 - \ln \Lambda^2 \right). \tag{9.35}$$

Figure 9.14 shows the bubble chamber data from CERN PS [22] and SPS

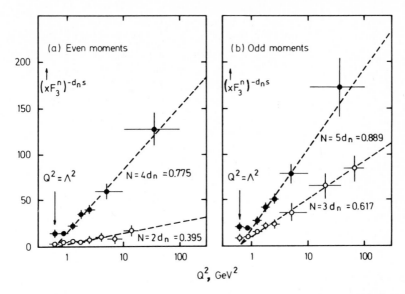

Fig. 9.14. Comparison of the moments of xF_3 with the first order QCD prediction. According to QCD the datapoints are aligned on straight lines with respect to $\ln Q^2$. The lines all intercept the same point $Q^2 = \Lambda^2$.

[291]. The data indeed fall on a straight line, consistent with a logarithmic scaling violation. However, a power series in Q^2, as expected from higher twists, can also mimic a logarithmic Q^2 dependence over the limited Q^2 range investigated.

In spite of the agreement with the predictions there are, however, serious experimental drawbacks in using the moment method. The evaluation of the moments requires the knowledge of the structure function at all x. But kinematical and experimental reasons allow a measurement only in a limited x domain and this domain is different for different values of Q^2 (see fig. 9.3). Thus the data must be extrapolated into unknown regions and this introduces systematic uncertainties. A particularly serious problem arises from the fact, that all moments are derived from the same set of measurements: Different moments are obviously highly correlated. Furthermore target mass effects [324] play an important rôle. Finally, from the theoretical point of view the effects of higher twists could give the same ratios of anomalous dimensions and higher order predictions are missing. The direct solutions of the Altarelli–Parisi equations avoid some of these problems.

9.3.4. *Direct analysis of the formfactors*

The Altarelli–Parisi equations hold at high Q^2, where "high" depends on the unknown parameter Λ and on yet unknown higher order terms in the perturbation expansion. The validity of the equations improves with increasing values of Q^2, however, the bulk of the data is at relatively low Q^2 ($\sim 10 \text{ GeV}^2/c^2$). In addition, corrections due to the finite mass of the target nucleon are necessary. Several methods of solving (or approximately solving) the Altarelli–Parisi equations have been used in analyzing the data: the method of Buras and Gaemers [325], the method of Abbot and Barnett [326] and the method of Gonzales-Arroyo et al. [327].

The valence quark structure function xF_3 can be determined experimentally without assumptions on R or on the gluon structure function. However, target mass effects, the Fermi motion inside the nuclear target, the limited resolutions in x and Q^2, and radiative corrections must be included in the analysis. An analysis of the F_2 structure function must in addition include effects due to R or the gluon structure function. The recent CHARM data [309, 329] are shown in fig. 9.15. The data are corrected for Fermi motion and radiative effects. The resolutions in x and Q^2 are unfolded. $R = 0$ was assumed for the analysis. $F_2(x, Q^2)$ and $xF_3(x, Q^2)$ are plotted versus $\ln Q^2$ for fixed x bin. The data clearly show the scaling violations as expected in a field theory and the solid line represents the results of a Buras–Gaemers-type QCD analysis.

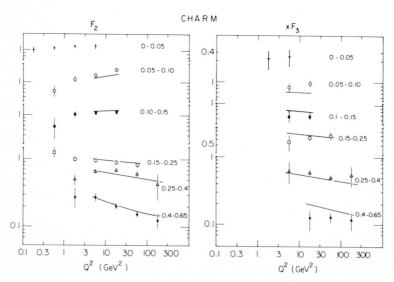

Fig. 9.15. The structure functions F_2 and xF_3 as measured by the CHARM collaboration.

Figure 9.16 shows the recent GGM data [330]. The resolution of this experiment is, in general, superior to the counter experiments. The data are corrected for Fermi motion but not for radiative effects, and agree extremely well with the CHARM data.

The recent CDHS [311] data in fig. 9.17 have by far the highest statistics of all neutrino experiments. The assumption $R = 0.1$ was made for the extraction of F_2. No Fermi motion correction was applied. While both experiments observed the same Q^2 dependence, differences show up in the x dependence. The CDHS structure functions are broader (extend to higher x). The same behaviour can be observed in the HPWFRO data [306] in Fig. 9.18, which were obtained assuming $R = 0.1$ and without

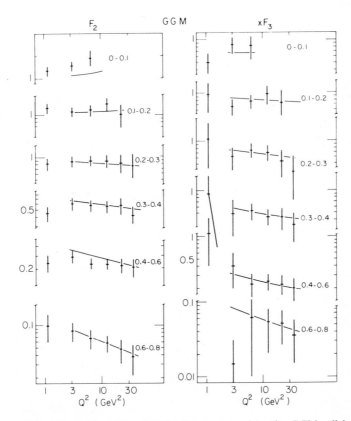

Fig. 9.16. The structure functions F_2 and xF_3 as measured by the GGM collaboration. Curves correspond to a fit to the CHARM data.

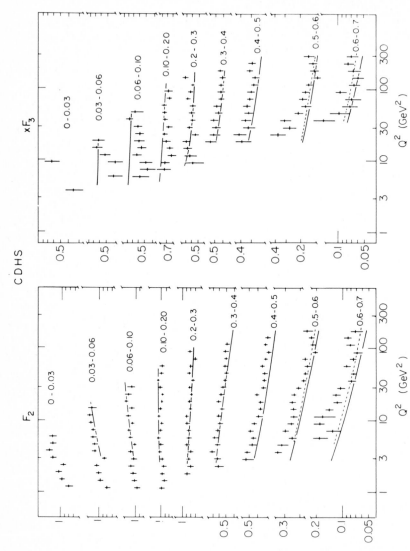

Fig. 9.17. The structure functions F_2 and xF_3 as measured with high statistics by the CDHS collaboration. Curves correspond to a fit to the CHARM data.

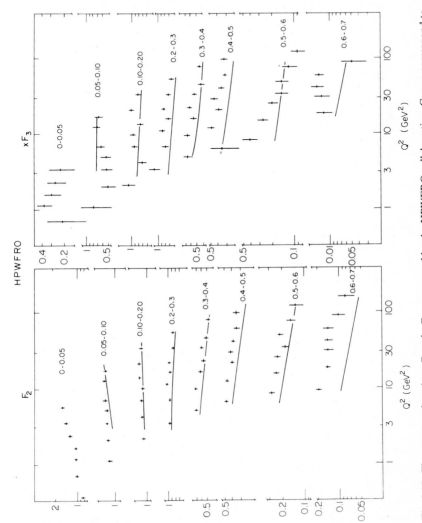

Fig. 9.18. The structure functions F_2 and xF_3 as measured by the HPWFRO collaboration. Curves correspond to a fit to the CHARM data.

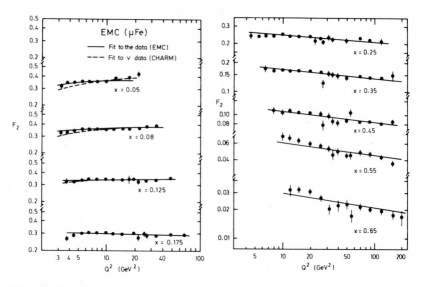

Fig. 9.19. The structure function F_2 as measured by the EMC. The dashed lines correspond to a fit to the CHARM data only. Where the fit to the EMC data (solid lines) agrees with the CHARM fit only a solid line is drawn.

Fermi motion correction. The resolution at high x is dominated by the measurement errors in the muon momentum. On the other hand, the Fermi motion correction is large at large x. It is therefore possible, that the discrepancies are due to the Fermi motion and the different unfolding of the resolution, particularly at high x.

Recent measurements of the F_2 structure function in charged lepton scattering are shown in figs. 9.19 and 9.20. The solid lines represent a fit to the EMC [304] and BCDMS [29] data respectively. The dashed curves are the absolute predictions from the CHARM neutrino data. Where the dashed curve is missing the prediction is actually indistinguishable from the fits to the muon data. The agreement between neutrino data and BCDMS data in fig. 9.20 ($R = 0$) is rather good taking the systematic errors into account. The agreement with the EMC data in fig. 9.19 ($R = 0$) is extremely good.

From the fits mentioned above and those of other experiments, values for Λ can be obtained. Unfortunately, similar scale breaking effects can be simulated by the hadronization of the quark. It is difficult to discriminate between these higher twist effects and QCD effects in the limited Q^2 range presently available.

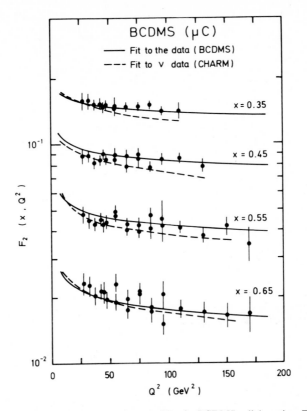

Fig. 9.20. The structure function F_2 as measured by the BCDMS collaboration. The dashed lines correspond to a fit to the CHARM data.

Higher twist effects are in general included in the analysis and they seem to account for roughly 10% of the cross section at high x. The most recent results of Λ_{LO}, evaluated with and without higher twist effects included, are listed in table 9.4.

All experiments agree on a rather low value of Λ despite the different methods used.

Up to now the structure functions F_2, xF_3 and $2xF_1$ have been discussed. It is, however, possible to extract the gluon distribution function from the Q^2 variation of the momentum distributions of light antiquarks. The high statistics of the antineutrino wide band beam exposure made it possible for the CDHS collaboration [331] to extract

Table 9.4
Values for Λ_{LO}

| | No twist | Higher twist | | |
Experiment	Λ_{LO}	Λ_{LO} (MeV)	μ^2 (GeV^2)	Ref.
GGM	190^{+160}_{-120}	700	$0.8\dots0.7$	[330]
BEBC	210 ± 95	–	–	[299]
CHARM	$290\pm120\pm100$	290 ± 120	0.09 ± 0.06	[329]
CDHS	190^{+80}_{-70}	200 ± 20	0.84 ± 0.1	[311]
EMC (Fe)	$122^{+22+114}_{-20-70}$	–	1.16 ± 0.07	[303]
EMC (H$_2$)	$110^{+58+124}_{-46-69}$	–	0.96 ± 0.15	[304]
BCDMS	85^{+60+90}_{-40-70}	–	–	[29]

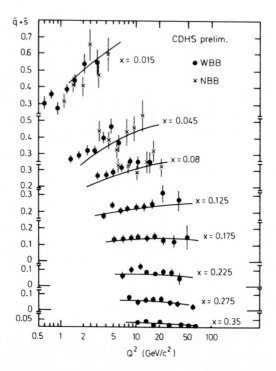

Fig. 9.21. The structure function of the light antiquarks. The curves correspond to a QCD fit to the data obtained by the CDHS collaboration.

the light antiquark momentum distribution as a function of Q^2

$$x\bar{q}\left(x,Q^2\right) = \frac{\pi}{G^2ME}\left(\frac{\mathrm{d}^2\sigma^{\bar{\nu}N}}{\mathrm{d}x\,\mathrm{d}y} - \left(1-y\right)^2\frac{\mathrm{d}^2\sigma^{\nu N}}{\mathrm{d}x\,\mathrm{d}y}\right) + \text{corrections.}$$

(9.36)

The data are shown in fig. 9.21. The distribution is strongly rising with Q^2 for small x and disappears for $x \geqslant 0.4$. The gluon distribution can be extracted from the measured Q^2 evolution of quarks and antiquarks.

The CDHS group has performed [332] an analysis along this line. The gluon distribution at $Q^2 = Q_0^2 = 5$ GeV^2 was parametrized as

$$xG(x) = a(1-x)^p(1+cx),$$

with $a = 2.63$, $p = 5.9 \pm 0.5$ and $c = 3.5 \pm 1.0$. This analysis also confirms the value of Λ and the momentum fraction carried by gluons reported earlier.

10. The search for new particles

Electron–positron annihilation has become a favourite hunting ground for exotic particles, as evidenced by a long list of unsuccessful searches [36,332]. In many cases, however, due to the well defined production mechanism and the high visibility, a negative result can be used to exclude the existence of new particles in the investigated mass range.

10.1. Limits on new sequential leptons

Leptons are pair produced, with the point cross section

$$\sigma = \frac{4\pi\alpha^2}{3s}\frac{\beta(3-\beta^2)}{2}, \quad \beta = P_{\mathrm{L}}/E.$$

(10.1)

The expected decay modes are shown in fig. 10.1. The branching ratios,

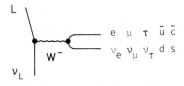

Fig. 10.1. Possible decay modes for a new sequential lepton.

neglecting phase space factors, can be written as:

$$B\left(L \to \nu_L e \bar{\nu}_e\right) = B\left(L \to \nu_L \mu \bar{\nu}_\mu\right) = B\left(\tau \to \nu_L \to \tau \nu_\tau\right)$$

$$\approx \left[(e\nu)+(\mu\nu)+(\tau\nu)+3(d\bar{u})+3(s\bar{c})\right]^{-1} = 11\%,$$

$$B\left(L^- \to \nu_L d\bar{u}\right) \approx B\left(L^- \to \nu_L s\bar{c}\right) \approx 3B\left(L^- \to \nu_L \ell \bar{\nu}_\ell\right) \approx 33\%. \quad (10.2)$$

These decays result in distinct final states:

(i) Approximately 10% of all $L\bar{L}$ pairs result in a final state, $e^+e^- \to e^\pm \mu^\mp$ + neutrinos. The muon and the electron are, in general, acollinear and the total energy visible in the event is less than the available center of mass energy.

(ii) $e^+e^- \to L\bar{L} \to e^\mp (\mu^\mp)$ + (hadrons)$^\pm$ + neutrinos. In this final state the electron (muon) is recoiling against a low-multiplicity hadron jet. The electron (muon) and the jet axis are in general acollinear and some of the available energy is carried off by neutrinos. Roughly 40% of all new sequential lepton pairs will populate this final state. As an example, a tau pair production event observed by TASSO viewed along the beam direction is shown in fig. 10.2.

(iii) $e^+e^- \to L\bar{L} \to$ (hadrons)$^+$ + (hadrons)$^-$ + neutrinos. The two low-multiplicity hadron jets will in general be acollinear and a fraction of the c.m. energy is carried off by neutrinos.

Groups at PETRA and PEP have used these topologies to search for sequential leptons beyond the tau. No evidence has been found, and the resulting mass limits and the method used are listed in table 10.1.

In SU(5) the charged lepton and the charge-1/3 quark within the same generation are degenerate in mass at the unification energy of 10^{15} GeV. This has been used to estimate [336] the mass of the s quark from the muon mass and to predict the mass of the b quark from the tau mass. It seems reasonable to expect that the charged lepton in a new generation of elementary fermions is lighter than the quarks. Indeed, from the present lower limit on the mass of a new lepton, the charge-1/3 quark in a SU(5) model must have a mass above 50 GeV.

10.2. Search for neutral leptons with mass

Heavy neutral leptons [337–339] are not required in the standard theory, however, they are not excluded and indeed they become natural if the weak current should turn out to be a pure vector interaction at high energies. In such models, a heavy neutral lepton E^0 is the partner of the

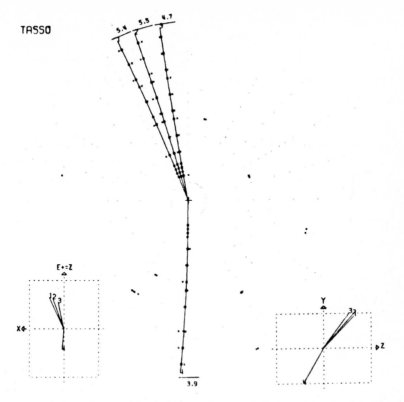

Fig. 10.2. A tau pairproduction event, $e^+e^- \to \tau^+\tau^- \to \mu^- +$ hadrons + neutrinos, observed by TASSO.

Table 10.1
Mass limit on new sequential leptons

Group	Lower limit 95% C.L.	Signature	Ref.
JADE	18.1 GeV	Two acollinear jets	[332]
MARK J	16.0 GeV	Single muon recoiling against many hadrons	[333]
PLUTO	14.5 GeV	Single muon recoiling against many hadrons	[334]
TASSO	15.5 GeV	Single charged particle recoiling against many hadrons	[335]
MAC	14.0 GeV	Acollinear eμ events	[37]
MARK II	13.8 GeV	Acollinear eμ events	[37]

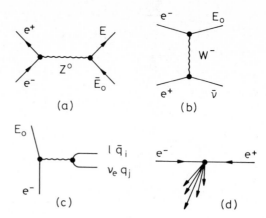

Fig. 10.3. Heavy electronlike neutral lepton. (a, b) Production mechanism via the neutral weak and the charged weak current. (c) Decay modes. (d) Final state in $e^+e^- \to E^0\bar{\nu}$.

electron in a right handed doublet, i.e.,

$$\begin{pmatrix} E^0 \\ e^- \end{pmatrix}_R$$

Possible production and decay mechanisms for an electronlike heavy neutral lepton are shown in fig. 10.3a, b, c. The E^0 can be pair produced via the neutral weak current and singly produced via the charged weak current. Charged current production dominates at present energies, provided the interaction has the usual weak strength. The decay modes $E^0 \to e^- W^+$ can lead to a purely leptonic final state or to a jet consisting of an electron and hadrons.

The production in e^+e^- annihilation of an electronlike heavy neutral lepton has a clean signature; the E^0 and the ν_e are produced back to back, leading to an event with only a single large-angle jet consisting of electrons and hadrons in one hemisphere, and nothing visible in the other hemisphere to balance its transverse momentum (fig. 10.3d).

The JADE collaboration has searched [332] for events with this topology; they select a group of hadrons containing at least one electron. These particles are constrained to be opposite to a cone with an opening angle of 50° which contains neither charged particles nor shower energy.

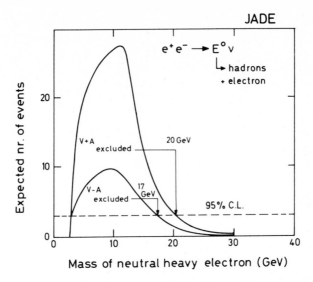

Fig. 10.4. The number of E^0 events in the JADE experiment which satisfy the selection criteria, versus the E^0 mass.

None of the events satisfied the criteria. The expected number of such events resulting from $e^+e^- \to E^0\bar{\nu}_e$ is plotted in fig. 10.4 versus the mass of the heavy lepton. They conclude that electronlike heavy neutral leptons with the properties discussed above do not exist with a mass less than 20 GeV for V + A coupling and less than 17 GeV for a V − A coupling.

10.3. Search for new quarks

All available data [100, 146] are consistent with the assumption that the b quark is member of a doublet, with the 2/3-charge member, the t quark, still missing. However, present theories are not able to constrain the mass of the t quark as evidenced by the flood of theoretical predictions. Production of a new quark has striking signatures: at threshold the normalized hadron cross section will have a step $\Delta R = 3e_i^2$ and the fragmentation of the t$\bar{t}$ quarks will lead to spherical events containing leptons with a high probability. Below threshold there will be narrow 1^{--} t$\bar{t}$ states.

10.3.1. Naked t quark production

The value of $R = \sigma(e^+e^- \to \text{hadrons})/\sigma(e^+e^- \to \mu^+\mu^-)$ is plotted versus the c.m. energy squared in fig. 6.2. It does not show any steps and strongly disfavours the existence of new quarks with charge $\frac{2}{3}e$.

A more sensitive limit on new quark production can be obtained by considering the event shape. Various methods used to determine the event shape have been discussed above. The distribution of events in aplanarity (A) and sphericity (S), as observed [175] by the TASSO collaboration at c.m. energies between 27.4 GeV and 36.6 GeV, is plotted in fig. 10.5. In such a plot, two jet events will cluster at small values of A

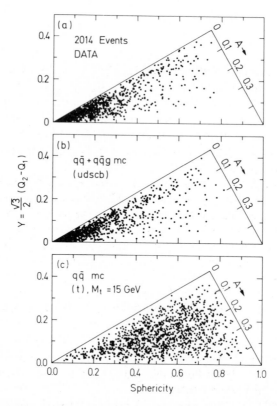

Fig. 10.5. (a) The event distribution in aplanarity and sphericity, observed by the TASSO collaboration between 27.4 GeV and 36.6 GeV in c.m. (b, c) Monte Carlo created events in aplanarity and sphericity for: $e^+e^- \to q\bar{q}g$ with q = u, d, s, c and b, c.m. energy 30 GeV, and (c) $e^+e^- \to t\bar{t}$ with $m_t = 15$ GeV and a c.m. energy of 35 GeV.

Table 10.2
Number of aplanar ($A > 0.15$) multihadron events [332]

Group	Observed number of events	Expected number of events		
		with u,d,s,c,b and QCD	including a new quark with	
			charge $\frac{2}{3}e$	charge $\frac{1}{3}e$
CELLO	9	5.2 ± 1.4	96.3 ± 4.2[a]	31.8[a]
TASSO	12	11 ± 1	138[b]	43[b]

[a] Mass of the new quark 16.0 GeV.
[b] Mass of the new quark 15.0 GeV.

and S, planar events resulting from gluon bremsstrahlung will have small values of A, whereas spherical events will have both A and S large. This is born out by the Monte Carlo results shown in fig. 10.5b, c. The data cluster at small values of S and A, with a long tail of planar events as expected for light quark production including gluon bremsstrahlung. Similar results [332] have also been obtained by CELLO and JADE. The data are listed in table 10.2.

Top quark production is expected to be a prolific source of prompt leptons. The MARK J group select [332] 352 events containing at least one muon. The thrust distribution of these events is plotted in fig. 10.6 and compared with various models. The data are in agreement with the standard model with gluon bremsstrahlung and u, d, s, c, and b quarks. New quarks would lead to an enhancement at low values of thrust, which is not observed in the data. For $T \leqslant 0.75$ they find 14 events to be compared with 13.8 events predicted by the standard model with five quarks and the gluon radiation. A charge-$1/3$ or charge-$2/3$ quark would yield 60 or 163 events, respectively.

A similar analysis is made by the JADE collaboration [332] for c.m. energies between 33 GeV and 36.7 GeV. They select events with one or more muons, in addition the muon must have at least 2 GeV transverse momentum with respect to the thrust axis and the event should have a sphericity $S \geqslant 0.5$. One event satisfies the selection criteria, while they expect 104 such events from a charge-$2/3$ quark.

The lack of a signal in any of these channels excludes the existence of a quark threshold in the mass range up to 36.7 GeV.

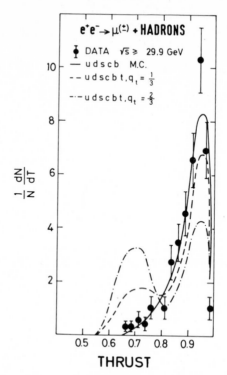

Fig. 10.6. Thrust distribution of multihadron events containing at least one muon, compared with the thrust distribution expected for u, d, s, c, and b quarks, and u, d, s, c, b and t quarks, with the t quark charge $\frac{1}{3}e$ or $\frac{2}{3}e$. The data were obtained by MARK J collaboration.

10.3.2. Search for narrow 1^{--} resonances

Narrow 1^{--} $t\bar{t}$ states are expected to occur about 1 GeV below $t\bar{t}$ threshold. To search for these narrow states the total cross section was measured in steps of 20 MeV in the energy range of 27 GeV to 31.8 GeV and 33 GeV to 37 GeV. The normalized cross section [332] obtained by combining the data from all groups (CELLO, JADE, MARK J and TASSO) is plotted in fig. 10.7 versus c.m. energy. No obvious structure is seen and an upper limit on the width of a resonance is obtained by fitting each measured cross section point to a Gaussian, radiatively corrected, and a constant background term. The width of the Gaussian is determined by the energy resolution of PETRA. Since presumably the width of the resonance is much smaller than the energy resolution of

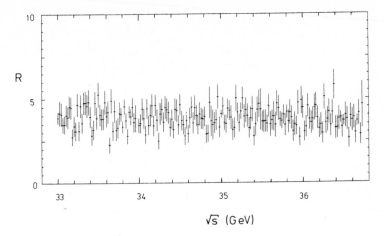

Fig. 10.7. The normalized cross section R, measured in steps of 20 MeV in c.m. The cross section shower was obtained by combining the results from CELLO, JADE, MARK J and TASSO.

PETRA the experiment determines only the area under the resonance

$$\int \sigma_{\text{tot}}(E)\, \mathrm{d}E = \frac{6\pi^2}{M_R} \Gamma_{ee} B_h. \qquad (10.3)$$

Here M_R is the mass, Γ_{ee} the partial width into electrons and B_h the hadronic branching ratio of the resonance.

The upper limits obtained by each group are listed in table 10.3 together with the value obtained by combining the data.

Table 10.3
Upper limits on narrow states

Group	M_R (GeV)	$\Gamma_{ee} B_h$ (keV) (90% upper C.L.)
CELLO	33.52	1.79
JADE	33.34	1.22
MARK J	35.12	0.97
TASSO	33.34	1.33
Combined	33.34	< 0.61

With $B_h = 0.7$ we expect $\Gamma_{ee}B_h = 3.08$ keV for a 1^{--} state made of charge-2/3 quarks and $\Gamma_{ee}B_h = 0.77$ keV for a 1^{--} state made of charge-1/3 quarks. Narrow resonances made of new charge-2/3 quarks are therefore excluded in the mass range investigated.

10.4. Search for free quarks

Free quarks are not in abundance although there is now little doubt that quarks indeed do exist forming colourless hadrons. Conflicting results [340, 341] on the existence of free quarks have been reported from quark searches in bulk material, whereas searches carried out using accelerator beams have not found any evidence of free quarks.

Since in e^+e^- annihilation the initial state consists of back to back quark pairs with large relative momenta one may expect that this reaction is particularly well suited to produce free quarks. Searches for both exclusive quark pair production and for inclusive hadron–quark events have been made at PETRA, PEP and SPEAR. All searches have assumed that a free quark behaves like a pointlike strongly interacting particle with fractional charge.

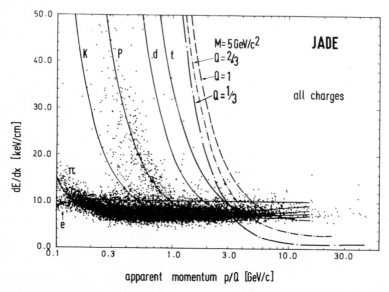

Fig. 10.8. The energy loss dE/dx as a function of apparent momentum p/Q. The predicted energy loss curves for electrons, pions, kaons, protons, deuterons and tritons are shown together with the curves for a hypothetical particle of mass 5 GeV and quark charges 1/3, 2/3 and 1.

The JADE collaboration at PETRA have searched [332, 342] both for inclusive and exclusive quark production using the information from their jet chamber. This detector provides tracking information and measures dE/dx at 48 points along a track. To measure the ionization, tracks must be separated by at least 7 mm. The overall efficiency in the inclusive channel ranges between 0.15 and 0.36 for charge-2/3 quarks and between 0.11 and 0.22 for charge-1/3 quarks, as a function of quark mass. Due to triggering difficulties, only charge-2/3 quarks can be measured in the exclusive channel. The mean energy loss dE/dx is plotted versus apparent momentum in fig. 10.8. The entries cluster along the ionization curves of the known particles. The observed deuterons and tritons result from beam gas or particle–beam-pipe interactions. No quark candidate was found either in the exclusive or in the inclusive channel, in an event sample corresponding to an integrated luminosity of 12 000 nb^{-1} for the quark-2/3 search and 4500 nb^{-1} for the quark-1/3 search. The resulting upper limit cross sections, normalized to the point cross section, are plotted in fig. 10.9 versus quark mass.

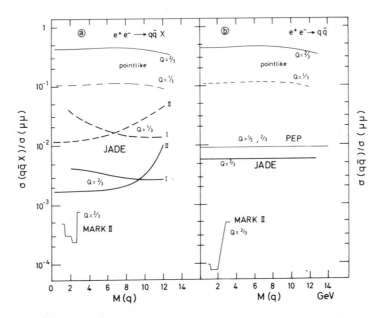

Fig. 10.9. The 90% C.L. upper limit on the cross sections for (a) inclusive quark production $e^{+}e^{-} \rightarrow q\bar{q}X$ and (b) exclusive quark production $e^{+}e^{-} \rightarrow q\bar{q}$ plotted versus quark mass. The curves marked I and II in (a) show the limit obtained by JADE for two different assumptions on the momentum distributions of the quarks [eqs. (10.4) and (10.5), respectively].

The quark mass limit in the inclusive search was extracted from these data for two different momentum distribution of the free quarks:

$$\frac{1}{\sigma(q\bar{q})} E(d^3\sigma/dp^3) = A\exp(-3.5E), \qquad (10.4)$$

$$\frac{1}{\sigma(q\bar{q})} E(d^3\sigma/dp^3) = \text{const.} \qquad (10.5)$$

The resulting 90% C.L. cross sections for inclusive quark production are plotted in fig. 10.9a versus quark mass.

The MARK II collaboration at SPEAR has searched [342] for exclusive pair production of charge-2/3 quarks. They selected two prong events with a collinearity angle of less than 10°, and an apparent momentum p/q greater than half the beam energy. The energy loss normalized to that of a charge-one particle should be between 0.2 and 0.65. The resulting upper limit is plotted in fig. 10.9b.

To search for inclusively produced quarks they evaluated the apparent mass of each track

$$m_t^2 = P_e^2/e^2(1/\beta^2 - 1). \qquad (10.6)$$

A quark candidate must have $m_t^2 > (1.6 \text{ GeV}/c^2)^2$ and an energy loss between 0.2 and 0.65 of that expected for a charge one particle. The resulting upper limit normalized cross section is plotted in fig. 10.9a.

Recently the Northwestern University, Frascati, LBL, Stanford University and the University of Hawaii collaboration have reported [343] first results from a search at PEP for exclusively produced quarks. They used a nonmagnetic detector and the charge of a particle was determined from a measurement of energy loss and velocity. The cross section limit obtained by this experiment is plotted in Fig. 10.9b.

A CERN–Bologna–Frascati–Roma collaboration has searched [344] for fractionally charged quarks in wide-band neutrino and antineutrino beams. The energy loss of the tracks is measured in an array of scintillation counters and in a streamer chamber. Charge-1/3 quarks can be measured for $\beta \geqslant 0.4$ and charge-2/3 quarks for $\beta \geqslant 0.8$. The acceptance was 0.44 with the first and 0.28 with the last assumption. No candidate was found and the experimental limit expressed as the limit of quark candidate events normalized to the number of neutrino interactions is plotted in fig. 10.10 as a function of $\Lambda_{\text{quark}}/\Lambda_{\text{hadron}}$ (Λ is the absorption length). Note that the average Q^2 in neutrino interaction is much lower than the Q^2 investigated in e^+e^- interactions.

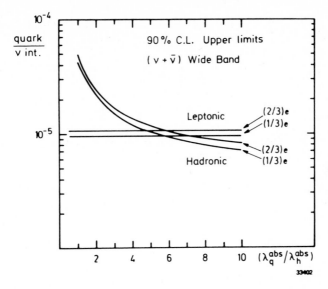

Fig. 10.10. Limit on quark candidate events normalized to the number of neutrino interactions, as a function of the ratio of the absorption lengths $\Lambda_{\text{quark}} / \Lambda_{\text{hadron}}$.

The fact that free quarks have not been observed in accelerator experiments does not necessarily exclude the existence of light free quarks. The colour force of a free quark is not shielded, and this might cause a free quark to have a large size and "eat" normal hadrons. Such quarks could have large interaction cross sections but a small production cross section. It has even been suggested [347] that a quark has an indeterminate mass and expands to a size greater than that of an atom. It is easy to understand why quarks with such properties might have escaped detection.

10.5. Search for excited leptons

If a lepton has a finite size then it would naturally have excited states. Excited states of the leptons can be produced directly in e^+e^- annihilation. An excited electron state will also modify the cross section for $e^+e^- \to \gamma\gamma$.

10.5.1. Excited states of the muon
μ^*, the hypothetical excited state of the muon, can be pairproduced in e^+e^- annihilation with a known cross section. The μ^* decays into a muon

and a photon leading to a final state consisting of two acollinear muons and two photons

$$e^+e^- \to \mu^{*+}\mu^{*-} \to \mu^+\mu^-\gamma\gamma. \tag{10.7}$$

The MARK J collaboration selected [36, 332] events of the type $e^+e^- \to \mu^+\mu^-\gamma\gamma$ where each muon carried at least half of the beam energy and the acoplanarity angle was less than 20°. The number of expected $e^+e^- \to \mu^{*+}\mu^{*-} \to \mu^+\mu^-\gamma\gamma$ events fulfilling the criteria above, is plotted in fig. 10.11 versus the mass of the μ^*. The number of observed events is in agreement with the QED prediction [348] and the resulting upper limit is shown as the dotted line in fig. 10.11. A μ^* with a mass of less than 10 GeV is ruled out.

In principle, single μ^* can be produced in the reaction $e^+e^- \to \mu^{*\mp}\mu^{\pm}$ $\to \mu^{\mp}\mu^{\pm}\gamma$. The differential cross section for a μ^* with spin $1/2$ and mass M can be written as:

$$\frac{d\sigma}{d\Omega} = \Lambda^2\alpha^2 \frac{(s - M^2)^2}{s^3} \left[(s + M^2) - (s - M^2)\cos^2\theta\right]. \tag{10.8}$$

The unknown coupling constant for the $\mu^*\mu\gamma$ vertex is written as $(\alpha\cdot\Lambda)^{1/2}$.

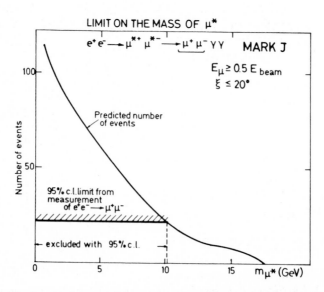

Fig. 10.11. Number of $e^+e^- \to \mu^{*+}\mu^{*-} \to \mu^+\mu^-\gamma\gamma$ events satisfying the selection criteria as a function of μ^* mass. The data were obtained by MARK J.

The JADE collaboration has searched [36, 332] for events of the type $e^+e^- \to \mu^+\mu^-\gamma$ in an event sample representing an integrated luminosity of 10.6 pb^{-1}, collected for c.m. energies between 22 GeV and 37 GeV. They select planar $\mu\mu\gamma$ events with an invariant $\mu^+\mu^-$ mass greater than 1 GeV. The photon should have an energy of at least 1 GeV and the opening angle between the direction of the photon and the direction of any of the muons should be at least 15°. A total of 66 events satisfied the criteria compared to 68 events predicted by QED [349, 350]. The effective $\mu\gamma$ mass distribution obtained by combining either muon in the event with the photon is plotted in fig. 10.12. There is no peak, and the effective $\mu\gamma$ mass distribution is in good agreement with the QED prediction shown as the dotted line. From these data they extract an upper limit on Λ^2 as a function of μ^* mass. This limit is plotted in fig. 10.13. The corresponding cross section limit is $\sigma(e^+e^- \to \mu\mu^*)/\sigma_{\mu\mu} < 0.0075$ for a μ^* mass of 16 GeV.

The MARK J collaboration—in a similar analysis—found 11 $\mu\mu\gamma$ events compared to 12 events predicted for QED. The resulting 95% limit on Λ^2 is also plotted in fig. 10.13 and corresponds to

$$\sigma(e^+e^- \to \mu\mu^*)/\sigma_{\mu\mu} < 0.035 \text{ (95% C.L.)}, \quad \text{for a 16 GeV } \mu^*. \quad (10.9)$$

Similar searches [37] for excited muons have been carried out by the MAC and MARK II collaborations at PEP. The data are in good agreement with the QED prediction, resulting in similar mass limits on a hypothetical μ^* as listed above.

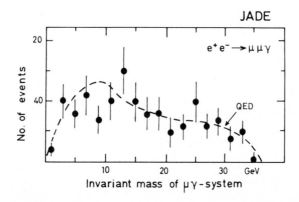

Fig. 10.12. The effective ($\mu\gamma$) mass distribution observed by the JADE collaboration in events of the type $e^+e^- \to \mu^+\mu^-\gamma$. Each event gives two entries. The dashed line represents the QED prediction.

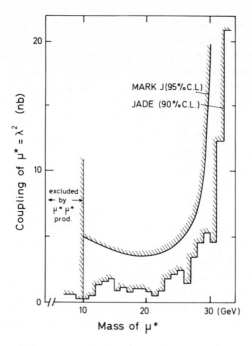

Fig. 10.13. Upper limits to the ($\mu^*\mu\gamma$) coupling constant ($\lambda\alpha$) plotted versus the μ^* mass.

10.5.2. Excited state of the electron

An excited state of the electron with a non-vanishing coupling e*eγ will contribute as a t-channel exchange in $e^+e^- \to \gamma\gamma$ and modify the QED prediction. The data, discussed above in section 3, show that an excited lepton must have a mass greater than 50 GeV if $\alpha^* = \alpha$.

10.6. Search for technipions

It is generally accepted that the gauge symmetry must be spontaneously broken [351] to give mass to the intermediate vector bosons and make the theory renormalizable. However, there is no consensus how this symmetry breakdown is achieved. One mechanism is by introducing [32] $I = 1/2$ fundamental scalar Higgs fields. It has also been proposed [352] that the symmetry breakdown arises dynamically from the gauge interactions themselves. In this model a new set of unbroken non-Abelian gauge interactions with a mass scale on the order of 1 GeV/c^2 is introduced. The interaction gives rise to a complicated spectrum [345] of technicolourless bound states with masses starting around 1 TeV$/c^2$. In addition, technicolour interactions will result in massless pseudoscalar mesons analogous to the pion in QCD. In the extended technicolour scheme which also gives mass to leptons and quarks there are more light pseudoscalars then can be eaten by the longitudinal polarization states of the gauge bosons. In particular we expect charged pointlike pseudoscalar particles with a mass perhaps between 5 and 14 GeV$/c^2$ and decay modes similar to those of the "normal" Higgs meson.

These particles p^+, p^- will be pairproduced in e^+e^- annihilation with cross section for a pointlike scalar particle

$$\sigma(e^+e^- \to p^+ p^-) = \frac{4\pi\alpha^2}{3s} \frac{\beta^3}{4}, \quad \text{with } \beta = p_p/E_B. \tag{10.10}$$

Like the Higgs meson, they couple to mass and $BR(P \to \tau\nu) + BR(P \to \bar{c}s) \approx 1.0$. The relative importance of the two decay modes is model dependent, but it is unlikely that either of them is zero.

The production of technipions leads to distinct final states, however, the β^3 factor suppresses the cross section near threshold. It is therefore necessary to search at energies well above threshold such that the decays produce rather jetlike events and then the topology criteria for new thresholds are not very useful.

The JADE group searched [332] for technipion pairproduction demanding that one of the technipions decays into $\tau\nu$ and the other into $c\bar{s}$. This leads to events where the thrust axis of the $c\bar{s}$ jet does not line up with the thrust axis defined by the τ decay due to the momentum carried off by the neutrinos. Specifically, the JADE collaboration selected events with the thrust axis inside the full detector acceptance. The event was divided into two hemispheres by means of a plane normal to thrust axis. The high-energy hemisphere should contain at least 63% of the beam energy. In the low-energy hemisphere there should be at least one charged particle, and further charged particles should make an angle of at least 70° with respect to the thrust axes. The thrust axes, determined individually for each hemisphere, should be acoplanar with the beam axis to at least 70°. Only one event satisfied the cuts in a sample of approximately 10000 nb^{-1}. The corresponding 90% C.L. mass limit on a technipion as a fraction of $BR(p \to \tau\nu)$ is plotted in fig. 10.14.

10.7. Search for supersymmetric particles

Gauging the isospin led to the successful unification of electromagnetic and weak interactions. In supersymmetric theories [353] the spin is

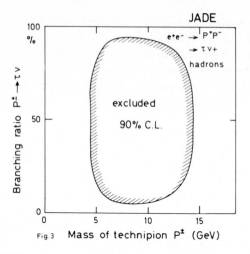

Fig. 10.14. The 90% C.L. mass boundary on charged technipions P is plotted for $B(P \to \tau\nu)$ versus the mass of the technipion.

gauged and this leads to a connection between fermions and bosons. Indeed, the fundamental feature of supersymmetry is that it can generate fermions from bosons and vice versa. Thus for every particle with spin J there will be two new particles with spin $J \pm 1/2$. A partial list of such new particles based on the phenomenology proposed by Farrar and Fayet [354] is given in table 10.4.

Table 10.4
Possible supersymmetric particles

Type of conventional particles	spin		
	1	1/2	0
Matter		quark q leptons ℓ	scalar quarks s_q scalar leptons s_ℓ
Massive gauge bosons	$W^\pm$	supersymmetric heavy leptons $\tilde{W}^\pm, \tilde{Z}^0, \tilde{H}$	Higgs scalar
Massless gauge bosons	g	photino $\tilde{\gamma}^2$ + other nuinos $\tilde{\nu}$ gluinos $\tilde{g}$	

Electron–positron interactions are well suited [345] to search for supersymmetric heavy leptons, scalar quarks and scalar leptons. So far, detailed searches have only been carried out for scalar leptons.

The pair production of scalar leptons in e^+e^- annihilation lead to a distinct final state. According to the possible production graphs and decay modes shown in fig. 10.15 we expect

$$e^+e^- \to \tilde{s}_e\bar{\tilde{s}}_e \to e^+e^- + \text{photinos},$$
$$e^+e^- \to \tilde{s}_\mu\bar{\tilde{s}}_\mu \to \mu^+\mu^- + \text{photinos},$$

i.e., a final state with two leptons acoplanar with respect to the beam axis, and with missing energy.

The CELLO group has carried out a search [355] for scalar leptons at an average c.m. energy of 34.6 GeV. They selected two prong events with an acoplanarity angle with respect to the beam axis of at least 30°. After examining the surrounding shower counter they found that all such events have neutral energy deposited in the shower counters and could be explained as radiative QED events. The number of expected scalar lepton events satisfying their selection criteria is plotted in fig. 10.16 versus the mass of the scalar lepton. The experimental limit is shown as the dashed line. The resulting mass limits are listed in table 10.5, together with earlier results.

10.8. Search for the axion

A possible mechanism to explain [357] the observed *CP* conservation in the strong interaction requires [358] the existence of a light boson, the axion.

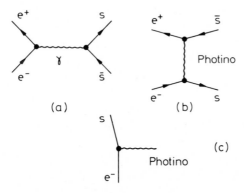

Fig. 10.15. Possible production and decay modes for scalar leptons.

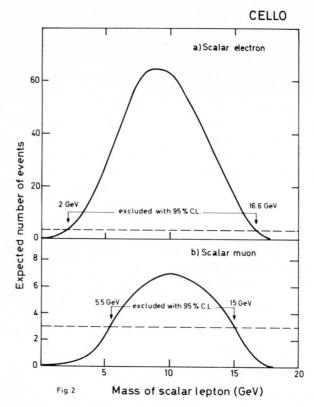

Fig. 10.16. Number of scalar leptons satisfying the selection criteria, plotted versus lepton mass. The 95% C.L. are shown. The data were obtained by the CELLO group.

Table 10.5
Mass limits on scalar leptons

Collaboration	Mass limits in GeV		
	Electron	Muon	Ref.
CELLO	< 5.5, > 15	< 2, > 16.6	[355]
JADE	> 16	–	[175]
MARK J	–	> 15	[333]
PLUTO	> 13	–	

Several beamdump experiments have searched [359] for axions with negative results. Recently, however, evidence for the existence of a light particle decaying into two photons has been presented [360].

The MARK II group at SPEAR has searched [122] for axions in the decay $J/\psi \rightarrow \gamma + axion$. Since the axion does not interact and is assumed to be light, the signature for the decay $J/\psi \rightarrow \gamma + axion$ is simply a monoenergetic photon with an energy of about 1.5 GeV. In a sample of 1.43×10^6 J/ψ decays they found five candidates. These candidates could all be explained as cosmic ray events and they extracted the 90% upper confidence limit of $BR(J/\psi \rightarrow \gamma a) < 3 \times 10^{-5}$. Peccei [360] has estimated an upper limit on the branching ratio of 5×10^{-5}.

References

[1] C. Bernardini, G.F. Corazza, G. Ghigo and B. Touschek, Nuovo Cim. 18 (1960) 1293.

[2] Design study of a 15 to 100 GeV e^+e^- colliding beam machine, CERN/ISR-LEP/78-17 and later updates.

[3] PETRA proposal, DESY Report (1974) and later update.

[4] K. Robinson and G.A. Voss, CEA Report CEA-TM-149 (1965).
K. Steffen, Internal Report DESY M-79/07 (1979).

[5] For a very clear discussion on e^+e^- colliders and related references see:
M. Sands, in: Physics with Intersecting Storage Ring, ed. B. Touschek. (Academic Press, New York, 1971) p. 257.

[6] For a more recent review see:
W. Schnell, CERN N-ISR-RF/81-21 (1981).

[7] S.D. Drell, D.J. Levy and T.M. Yan, Phys. Rev. 187 (1969) 2159; Phys. Rev. D1 (1970) 1617.

[8] N. Cabibbo, G. Parisi and M. Testa, Lett. Nuovo Cim. 4 (1970) 35.

[9] J.D. Bjorken and S.J. Brodsky, Phys. Rev. D1 (1970) 1416.

[10] R.P. Feynman, Photon–Hadron Interactions (Benjamin, Reading, MA, 1972).

[11] The first quantitative discussion on the experimental implications of gluon bremsstrahlung in e^+e^- annihilation was given by: J. Ellis, M.K. Gaillard and G.G. Ross, Nucl. Phys. B111 (1976) 253 [erratum B130 (1977) 516].
See also: T.A. De Grand, Y.J. Ng, and S.H.H. Tye, Phys. Rev. D16 (1977) 3251.

[12] J. Kogut and L. Susskind, Phys. Rev. D9 (1974) 697, 3391.
A.M. Polyakov, in: Proc. 1975 Int. Symp. on Lepton and Photon Interactions at High Energies, Stanford (1975), ed. W.T. Kirk (Stanford Linear Accelerator Center).

[13] T. Appelquist and H.D. Politzer, Phys. Rev. Lett. 34 (1975) 43.

[14] H. Fritzsch, M. Gell-Mann and H. Leutwyler, Phys. Lett. 47B (1973) 365.
D.J. Gross and F. Wilczek, Phys. Rev. Lett. 30 (1973) 1343.
H.D. Politzer, Phys. Rev. Lett. 30 (1973) 1346.
S. Weinberg, Phys. Rev. Lett. 31 (1973) 31.
For recent reviews see:
A.J. Buras, Rev. Mod. Phys. 52 (1980) 199, E. Reya, Phys. Rep. 69 (1981) 195.

[15] E.J. Williams, Kgl. Danske Videnskab. Selskab, Mat. Fys. Medd. 13 (1934) No. 4.
L.D. Landau and E.M. Lifshitz, Phys. Z. Sovjetunion 6 (1934) 244.

A. Jaccarini, N. Arteaga-Romero, J. Parisi and P. Kessler, Compl. Rend. Acad. Sci. Paris 269B (1969) 153, 1129; Nuovo Cim. 4 (1970) 933.

[16] For a review see: H. Terazawa, Rev. Mod. Phys. 45 (1973) 615.

[17] S.M. Berman, J.D. Bjorken and J.B. Kogut, Phys. Rev. D4 (1971) 3388.

[18] JADE: Proposal to construct a detector at PETRA by DESY, Hamburg, Heidelberg, Lancaster, Manchester, Rutherford and Tokyo.

[19] Crystal Ball collaboration: Caltech, Harvard, Princeton, SLAC, Stanford: E.D. Bloom, XIVth Rencontre de Moriond, Les Arcs, France (1979).

[20] J. Steinberger, in: Proc. 1976 CERN School of Phys, CERN 76-20 (1976).
F. W. Büßer, in: Proc. 19th Int. Universitätswochen für Kernphysik, Schladming, Styria, Austria (1980), ed. P. Urban (Springer, Vienna).
R. Turlay, lectures given at the XVI Int. School of Elementary Particle Physics, Kupari, Dubrovnik (1980); and at the XX Cracow School of Theoretical Physics (1981) Department de Physique des particules elementaires (DPh PE 81-02).

[21] F. Sciulli, Hadron structure from lepton beams, in: High Energy Physics 1980, XX Int. Conf., Madison, eds. L. Durand and L.G. Pondrom (American Institute of Physics, New York, 1980).

[22] P.C. Bosetti, et al., Phys. Lett. 70B (1977) 273.

[23] CHDS collaboration: M. Holder et al., Nucl. Instrum. Methods 148 (1978) 235; 151 (1978) 69.

[24] CHARM collaboration: A.N. Diddens et al., Nucl. Instrum. Methods 178 (1980) 27.

[25] W.K.H. Panofsky, Proc. of the XIV Int. Conf. on High Energy Physics, Vienna, Austria (1968), eds. J. Prentki and J. Steinberger (CERN, Geneva).

[26] The SLED Scheme was first suggested by P.B. Wilson, SLAC.

[27] EMC collaboration: O.C. Allkofer et al., Nucl. Instrum. Methods 79 (1981) 445.

[28] CHIO collaboration, B.A. Gordon et al., Phys. Rev. D20 (1979) 2645.

[29] BCDMS collaboration, D. Bollini et al., Phys. Lett. 104B (1981) 403.

[30] S.L. Glashow, Nucl. Phys. 22 (1961) 579.
S. Weinberg, Phys. Rev. Lett. 19 (1967) 1264.
A. Salam, Elementary Particle Theory, ed. N. Svartholm (Almqvist and Wiksells, Stockholm, 1968) p. 367.

[31] R.P. Feynman and M. Gell-Mann, Phys. Rev. 109 (1958) 193.
E.C.G. Sudarshan and R.E. Marshak, Phys. Rev. 104 (1958) 1860.
J.J. Sakurai, Nuovo Cim. 7, (1958) 647.

[32] P.W. Higgs, Phys. Rev. Lett. 13, (1964) 508.

[33] M. Kobayashi and T. Maskawa, Progr. Theor. Phys. 49 (1973) 652.

[34] J.E. Kim, P. Langacker, M. Levine and H.H. Williams, Rev. Mod. Phys. 53 (1980) 211.

[35] For a recent review see:
P. Dittmann and V. Hepp, Z. Phys. C10 (1981) 283.

[36] J. G. Branson, in: Proc. 1981 Int. Symp. on Lepton and Photon Interactions at High Energies, Bonn (1981), ed. W. Pfeil (Univ. of Bonn).

[37] R. Hollebeek, in: Proc. 1981 Int. Symp. on Lepton and Photon Interactions at High Energies, Bonn (1981), ed. W. Pfeil (Univ. of Bonn).

[38] B. Kayser et al., Phys. Rev. D15 (1977) 3407; D20 (1979) 87.

[39] F. Reines et al., Phys. Rev. Lett. 37 (1976) 315.

[40] F.J. Hasert et al., Phys. Lett. 46B (1973) 121.

[41] J. Blietschau et al., Nucl. Phys. B114 (1978) 189; Phys. Lett. 73B (1978) 232.

[42] N. Armenise et al., Phys. Lett. 86B (1979) 225.

[43] D. Bertrand et al., Phys. Lett. 84B (1979) 354.

[44] N. Armenise et al., Phys. Lett. 81B (1979) 385.

[45] A.M. Cnops et al., Phys. Rev. Lett. 41 (1978) 357.

[46] J.P. Berge et al., Phys. Lett. 84B (1979) 357.

[47] H. Faissner et al., Phys. Rev. Lett. 41 (1978) 213.

[48] R.H. Eisterberg et al., Phys. Rev. Lett. 44 (1980) 635.

[49] M. Jonker et al., Phys. Lett. 105B (1981) 242.

[50] R. Budny and A. McDonals, Phys. Lett. 48B (1974) 423.
R. Budny, Phys. Lett. 55B (1975) 227.
J. Ellis and M.K. Gaillard, CERN 76/18 (1976).

[51] F.A. Berends, K.J.F. Gaemers and R. Gastmans, Nucl. Phys. B63 (1973) 381.
F.A. Berends, K.J.F. Gaemers and R. Gastmans, Nucl. Phys. B68 (1974) 541.
F. A. Berends and R. Kleiss, DESY-Report 80-66 (1980).

[52] F.A. Berends and G.J. Komen, Phys. Lett. 63B (1980) 432.

[53] JADE collaboration: W. Bartel et al., DESY 81-072 (1981).

[54] H. Georgi and S. Weinberg, Phys. Rev. D17 (1978) 275.

[55] J.D. Bjorken, Phys. Rev. D19 (1979) 335.

[56] E.H. de Groot, G.J. Gounaris and D. Schildknecht, Phys. Lett. 85B, (1979) 399; Phys. Lett. 90B (1980) 427; Z. Phys. C5 (1980) 127.
E.H. de Groot and D. Schildknecht, Phys. Lett. 95B (1980) 149.

[57] V. Barger, W.Y. Keung and E. Ma, Phys. Rev. Lett. 44 (1980) 1169.

[58] P.A. Hung and J.J. Sakurai, Ann. Rev. Nucl. Part. Sci. 31 (1981) 375.

[59] F. W. Büsser, Neutrino 81, Int. Conf. on Neutrino Physics and Astrophysics 1981, Wailea, Maui, Hawaii (1981), eds. R.J. Cence, E. Ma and A. Roberts (HEP group, Hawaii).

[60] P.Q. Hung and J.J. Sakurai, Phys. Lett. 63B (1976) 295.

[61] E.A. Paschos and L. Wolfenstein, Phys. Rev. D7 (1973) 91.

[62] C. Geweniger, Int. Neutrino Conf., Bergen, Norway (1979), eds. A. Haatuft and C. Jarlskog (Univ. of Bergen).

[63] M. Jonker et al., Phys. Lett. 99B (1981) 265.

[64] M. Shaevitz, Neutrino 81 Int. Conf. on Neutrino Physics and Astrophysics 1981, Wailea, Maui, Hawaii (1981), eds. R.J. Cence, E. Ma and A. Roberts (HEP group, Hawaii).

[65] Illinois–Maryland–Stony Brook–Tohoku–Tufts collaboration: T. Kitagaki et al., Neutrino 81, Int. Conf. on Neutrino Physics and Astrophysics 1981, Wailea, Maui, Hawaii (1981), eds. R.J. Cence, E. Ma and A. Roberts (HEP group, Hawaii).

[66] J. Blietschau et al., Phys. Lett. 88B (1979) 381.

[67] F. A. Harris et al., Phys. Rev. Letters 39 (1977) 437.

[68] M. Derrick et al., Phys. Rev. D18 (1978) 7.

[69] J. Marriner, LBL-6438 (1977) thesis.

[70] V.I. Efremenko et al., Phys. Lett. 84B (1979) 511.

[71] Bari–Birmingham–Brussels–Ecole Polytechnique–Rutherford–Saclay–University College London collaboration: V. Brisson et al., Neutrino 81, Int. Conf. on Neutrino Physics and Astrophysics 1981, Wailea, Maui, Hawaii (1981), eds. R.J. Cence, E. Ma and A. Roberts (HEP group, Hawaii).

[72] J. Okada and S. Pakvasa, Nucl. Phys. B112 (1976) 400.

[73] P.W. Hung, Phys. Lett. 69B (1977) 216.
L.M. Sehgal, Phys. Lett. 71B (1977) 99.

[74] H. Kluttig et al., Phys. Lett. 71B (1977) 446.

[75] L.M. Sehgal, Phys. Lett. 71B (1977) 99.

[76] GGM collaboration: D. Haidt, in: Proc. 1977 Int. Symp. on Lepton and Photon Interactions at High Energies, Hamburg (1977), ed. F. Gutbrod (DESY, Hamburg).

[77] E. Pasierb et al., Phys. Rev. Lett. 43 (1979) 96.

[78] S.L. Glashow, J. Iliopoulos and L. Maiani, Phys. Rev. D2 (1970) 1285.

[79] L. M. Sehgal, Proc. Int. Neutrino Conf. 1978, ed. E.C. Fowler (Purdue Univ.) p. 263.

[80] M. Jonker et al., Phys. Lett. 102B (1981) 67.

[81] CDHS collaboration: Neutrino 81 Int. Conf. on Neutrino Physics and Astrophysics 1981, Wailea, Maui, Hawaii (1981), eds. R.J. Cence, E. Ma and A. Roberts (HEP group, Hawaii).

[82] Y.B. Zel'dovich, JETP 9 (1959) 682.

[83] M.A. Bouchiat and C. Bouchiat, Phys. Lett. 47B (1974) 111.

[84] L.M. Barkov and M.S. Zolotorev, Phys. Lett. 85B (1979) 308.

[85] P.E.G. Baird et al., Phys. Rev. Lett. 39 (1977) 798.

[86] P.E.G. Baird, in: Proc. Int. Workshop on neutral-current interactions in atoms, Cargese, ed. W.L. Williams (Inst. Nat. Phys. Nucl., Paris) p. 77.

[87] Y.V. Bogdanov et al., JETP Lett. 31 (1977) 214.

[88] J.L. Lewis et al., Phys. Rev. Lett. 39 (1977) 795.

[89] J.H. Hollister et al., Phys. Rev. Lett. 46 (1981) 643.

[90] P.H. Bucksbaum et al., Phys. Rev. D24 (1981) 1134; Phys. Rev. Lett. 46 (1981) 640.

[91] V.N. Novikov, O.P. Sustkov and I.B. Khriplovich, JETP 46 (1977) 420.
     P.G.H. Sanders, Phys. Scripta 21 (1980) 289.
     A.M. Märtenson, E.M. Henley and L. Wilets, Univ. of Washington preprint, submitted to Phys. Rev. A.

[92] E.D. Commins and P.H. Bucksbaum, Ann. Rev. Nucl. Part. Sci. 30 (1980) 1.

[93] C. Prescott et al., Phys. Lett. 77B (1978) 347; 84B (1979) 524; 72B (1980) 489.

[94] R.N. Cahn and F.J. Gilman, Phys. Rev. D17 (1978) 1313.

[95] MARK J collaboration: D.P. Barber et al., Phys. Rev. Lett. 64 (1981) 1663.

[96] JADE collaboration: W. Bartel et al., Phys. Lett. 101B (1981) 361.

[97] S.L. Glashow and S. Weinberg, Phys. Rev. D15 (1977) 1958.

[98] G. Goldhaber et al., SLAC-PUB 1973, LBL 6467 (1977).

[99] G. Feldman et al., Phys. Rev. Lett. 38 (1977) 1313.

[100] A. Silverman, in: Proc. 1981 Int. Symp. on Lepton and Photon Interactions at High Energies, Bonn (1981), ed. W. Pfeil (Univ. of Bonn).

[101] For a general review see: J.J. Sakurai, Neutrino 81, Int. Conf. on Neutrino Physics and Astrophysics, Wailea, Maui, Hawaii (1981), eds. R.J. Cence, E. Ma and A. Roberts (HEP group, Hawaii) and J. Meyer, CERN EP, Internal Report 80-09 (1980).

[102] For a review see: F. Scheck, Phys. Rep. C44 (1978) 187.

[103] S. Shrock and L.L. Wang, Phys. Rev. Lett. 41 (1978) 1692.

[104] M. Bourquin et al., Univ. Paris-Sud preprint LAL 81/18 (1981).

[105] R.L. Kingsley et al., Phys. Rev. D10 (1974) 2216.

[106] B. Kayser et al., Phys. Lett. 52B (1974) 385.

[107] M. Jonker et al., Phys. Lett. 86B (1979) 229.

[108] N. Armenise et al., Phys. Lett. 84B (1979) 137.

[109] M. Jonker et al., Phys. Lett. 93B (1980) 203.

[110] J.C. Pati and A. Salam, Phys. Rev. D8 (1973) 1240; Phys. Rev. D10 (1974) 275.

[111] G. Georgi and S.L. Glashow, Phys. Rev. Lett. 32 (1974) 438.

[112] W.J. Marciano and A. Sirlin, Phys. Rev. Lett. 46 (1981) 163.

[113] M. Perl et al., Phys. Rev. Lett. 35 (1975) 1489.

[114] For a general review see: M.L. Perl, Ann. Rev. Nucl. Part. Sci. 30 (1980) 299.

[115] DASP collaboration: R. Brandelik et al., Phys. Lett. 73B (1978) 109.

[116] W. Bacino et al., Phys. Rev. Lett. 41 (1978) 13.

[117] W. Bacino et al., Phys. Rev. Lett. 42 (1979) 749.

[118] H.B. Thacker and J.J. Sakurai, Phys. Lett. 36B (1971) 103.
Y.S. Tsai, Phys. Rev. D4 (1971) 2821.
J.D. Bjorken and C.H. Llewellyn Smith, Phys. Rev. D7 (1973) 887.
For a recent review see: Y.S. Tsai, SLAC-PUB-2105 (1978).

[119] The most recent measurement of $\tau \to \nu_\tau \pi$ is by C.A. Blocker et al., SLAC-PUB-2785 (1981).

[120] G.S. Abrams et al., Phys. Rev. Lett. 43 (1979) 1555.

[121] J.M. Dorfan et al., Phys. Rev. Lett. 46 (1981) 215.

[122] F.C. Porter, SLAC-PUB-2785 (1981).

[123] G.J. Feldman et al., SLAC-PUB-2819 (1981).

[124] For a recent review of $q\bar{q}$ spectroscopy and a complete set of references see: T. Appelquist, R.M. Barnett and K. Lane, Ann. Rev. Nucl. Part. Sci. 28 (1978) 387, and C. Quigg, course 6, this volume.

[125] T. Appelquist and H.D. Politzer, Phys. Rev. D12 (1978) 1404.

[126] R. Van Royen and V.F. Weisskopf, Nuovo Cim. 50A (1967) 617.

[127] L.B. Okun and M. Voloshin, Zh. Eksp. Teor. Fiz. 23 (1976) 369.
C. Rosenzweig, Phys. Rev. Lett. 36 (1976) 697.
M. Bander et al., Phys. Rev. Lett. 36 (1976) 695.
A. De Rújula et al., Phys. Rev. Lett. 38 (1977) 317.

[128] R.C. Giles and S.H.H. Tye, Phys. Rev. Lett. 37 (1976) 1175.
W. Buchmüller and S.H.H. Tye, Phys. Rev. Lett. 44 (1980) 850.
P. Hasenfratz et al., Phys. Lett. 95B (1980) 199.

[129] J.J. Aubert et al., Phys. Rev. Lett. 33 (1974) 1404.
J.E. Augustin et al., Phys. Rev. Lett. 33 (1974) 1406.
G.S. Abrams et al., Phys. Rev. Lett. 33 (1974) 1453.
P.A. Rapidis et al., Phys. Rev. Lett. 39 (1977) 526.
DASP collaboration: R. Brandelik et al., Phys. Lett. 76B (1978) 361.
J. Siegrist et al., Phys. Rev. Lett. 36 (1976) 700.
PLUTO collaboration, J. Burmester et al., Phys. Lett. 66B (1977) 395.
E.D. Bloom, SLAC-PUB-2745 (1981); Proc. XVIth Rencontre de Moriond, Les Arcs, France (1981), ed. J. Tran Thanh Van (Editions Frontières, Dreux, France).
A.A. Zholentz et al., Phys. Lett. 96B (1980) 214.

[130] DASP collaboration, W. Braunschweig et al., Phys. Lett. 57B (1975) 407.
G.J. Feldman et al., Phys. Rev. Lett. 35 (1975) 821.

[131] DASP collaboration, W. Braunschweig et al., Phys. Lett. 67B (1977) 243.

[132] J.S. Whitaker et al., Phys. Rev. Lett. 37 (1976) 1596.

[133] W. Bartel et al., Phys. Rev. Lett. 79B (1978) 492.

[134] D.L. Scharre, in: Proc. 1981 Int. Symp. on Lepton and Photon Interactions at High Energies, Bonn (1981), ed. W. Pfeil (Univ. of Bonn).
D.L. Scharre, Int. Conf. on Physics in Collisions, Blacksburg, VI (1981), eds. W.P. Trower and G. Bellini (Plenum, New York, 1982); SLAC-PUB-2761 (1981).

[135] For a review of the experimental data up to the end of 1978 see:
B.H. Wiik and G. Wolf, Electron–Positron Interactions, Springer Tracts in Modern Physics, vol. 86, ed. G. Höhler (Springer, Berlin, 1979).

[136] D.L. Scharre, SLAC-PUB-2761 (1981).
[137] E. Eichten, K. Gottfried, T. Kinoshita, K.D. Lane and T.M. Yan, Phys. Rev. D21 (1980) 203.
[138] T.M. Himel et al., Phys. Rev. Lett. 34 (1979) 1357.
[139] R. Partridge et al., Phys. Rev. Lett. 45 (1980) 1150.
[140] T.M. Himel et al., Phys. Rev. Lett. 45 (1980) 1146.
[141] S.W. Herb et al., Phys. Rev. Lett. 39 (1977) 252.
       W.R. Innes et al., Phys. Rev. Lett. 39 (1977) 1240.
[142] PLUTO collaboration, Ch. Berger et al., Phys. Lett. 76B (1978) 243.
       DASP2 collaboration, C.W. Darden et al., Phys. Lett. 76B (1978) 246.
[143] J.K. Bienlein et al., Phys. Lett. 78B (1979) 360.
       C.W. Darden et al., Phys. Lett. 78B (1979) 364.
[144] D. Andrews et al., Phys. Rev. Lett. 44 (1980) 1108; 45 (1980) 219.
       T. Böhringer et al., Phys. Rev. Lett. 44 (1980) 1111.
       G. Finocchiaro et al., Phys. Rev. Lett. 45 (1980) 222.
[145] J.L. Rosner, C. Quigg and H.B. Thacker, Phys. Lett. 74B (1978) 350.
[146] D. Schamberger: in: Proc. 1981 Int. Symp. on Lepton and Photon Interactions at High Energies, Bonn (1981), ed. W. Pfeil (Univ. of Bonn).
[147] W. Schmidt-Parzefall, in High Energy Physics 1980, XX Int. Conf., Madison, eds. L. Durand and L.G. Pondrom (American Institute of Physics, New York, 1980).
[148] J.K. Bienlein, in: Proc. 1981 Int. Symp. on Lepton and Photon Interactions at High Energies, Bonn (1981), ed. W. Pfeil (Univ. of Bonn); DESY Report 81/076 (1981).
[149] G. Goldhaber and J.E. Wiss, Ann. Rev. Nucl. Part. Sci. 28 (1978) 387.
       R.H. Schindler et al., Phys. Rev. D21 (1980) 2716.
[150] R.A.J. Lankford, in High Energy Physics 1980, XX Int. Conf., Madison, eds. L. Durand and L.G. Pondrom (American Institute of Physics, New York, 1980).
[151] A. Pais and S.B. Treiman, Phys. Rev. D15 (1977) 2529.
[152] W. Bacino et al., Phys. Rev. Lett. 45 (1980) 329.
[153] For a review of the emulsion data see:
       J. Prentice, Proc. Eur. Phys. Soc. Int. Conf. on High Energy Physics, Lisbon (1981).
[154] F. Foa, in: Proc. 1981 Int. Symp. on Lepton and Photon Interactions at High Energies, Bonn (1981), ed. W. Pfeil (Univ. of Bonn).
[155] N. Cabibbo and L. Maiani, Phys. Lett. 79B (1978) 109.
       N. Cabibbo, G. Corbo and L. Maiani, Nucl. Phys. B155 (1979) 93.
[156] H. Fritsch, in: Proc. 1981 Int. Symp. on Lepton and Photon Interactions at High Energies, Bonn (1981), ed. W. Pfeil (Univ. of Bonn).
[157] G.S. Abrams et al., Phys. Rev. Lett. 43 (1979) 481.
[158] J. Knobloch, Neutrino 81 Int. Conf. on Neutrino Physics and Astrophysics, Wailea, Maui, Hawaii (1981), eds. R.J. Cence, E. Ma and A. Roberts (HEP group, Hawaii).
       J. Allaby, Neutrino 81, Int. Conf. on Neutrino Physics and Astrophysics, Wailea, Maui, Hawaii (1981), eds. R.J. Cence, E. Ma and A. Roberts (HEP group, Hawaii).
[159] S. Pakraso, S.F. Tuan and J.J. Sakurai, Phys. Rev. D23 (1981) 2799.
[160] DASP collaboration, R. Brandelik et al., Phys. Lett. 70B (1977) 132; 80B (1979) 412.
[161] D. Aston et al., Phys. Lett. 100B (1981) 91; preprint CERN-EP/81-47 (1981).
[162] J. Leveille, Univ. of Michigan preprint, U MHE 81-018 (1981).
[163] V. Lüth, in: Proc. 1979 Int. Symp. on Photon and Lepton Interactions at High Energies, Batavia (1979). eds. T.B.W. Kirk and H.D.I. Abarbanel (Fermilab, Batavia).
[164] T. Appelquist and H. Georgi, Phys. Rev. D8 (1973) 4000.

A. Zee, Phys. Rev. D8 (1973) 4038.

G. 't Hooft, Nucl. Phys. B62 (1973) 444.

M. Dine and J. Sapirstein, Phys. Rev. Lett. 43 (1979) 668.

W. Celmaster and R.J. Gonsalves, Phys. Rev. Lett. 44 (1979) 560.

K.G. Chetyrkin, A.L. Kataev and F.V. Tkachov, Phys. Lett. 85B (1979) 277; USSR Academy of Sciences, Inst. of Nuclear Research preprint D-0178 (1980).

[165] R. Felst, in: Proc. 1981 Int. Symp. on Lepton and Photon Interactions at High Energies, Bonn (1981).

[166] A. Quenzer, thesis, Orsay Report LAL 1299 (1977).

A. Cordier et al., Phys. Lett. 81B (1979) 389.

V.A. Sidorov, Proc. XVIII Int. Conf. on High Energy Physics, Tbilisi, USSR, B13 (1976), ed. N.N. Bogolubov (Dubna).

R.F. Schwitters, in: Proc. XVIII Int. Conf. on High Energy Physics, Tbilisi, USSR, (1976), ed. N.N. Bogolubov (Dubna).

J. Perez-Y-Jorba, in: Proc. 19th Int. Conf. on High Energy Physics, Tokyo (1978), eds. S. Homma, M. Kawaguchi, and H. Miyazawa (Physical Society of Japan, Tokyo) p.

PLUTO collaboration: J. Burmester et al., Phys. Lett. 66B (1977) 395.

DASP collaboration: R. Brandelik et al., Phys. Lett. 76B (1978) 361.

[167] J. Siegrist, SLAC-225 (1979).

[168] R.M. Barnett, M. Dine and J. McLerran, Phys. Rev. D22 (1980) 594.

[169] J. Jersak, E. Laermann and P.M. Zerwas, Phys. Lett. 98B (1981) 363.

[170] For a recent review and a complete set of references see:

M.E. Peskin, in: Proc. 1981 Int. Symp. on Lepton and Photon Interactions at High Energies, Bonn (1981).

[171] P. Söding and G. Wolf, DESY 81-013 (1981); Ann. Rev. Nucl. Part. Sci. 31 (1981).

[172] J.D. Bjorken and S.J. Brodsky, Phys. Rev. D1 (1970) 1416.

[173] E. Farhi, Phys. Rev. Lett. 39 (1979) 1587.

S. Brandt and H.D. Dahmen, Z. Phys. C1 (1979) 61.

[174] R.F. Schwitters et al., Phys. Rev. Lett. 35 (1975) 1320.

[175] D. Cords, in: High Energy Physics 1980, XX Int. Conf., Madison, eds. L. Durand and L. G. Pondrom (American Institute of Physics, New York, 1980).

[176] T.F. Walsh and P. Zerwas, Nucl. Phys. B77 (1979) 494.

[177] TASSO collaboration: R. Brandelik et al., Phys. Lett. 100B (1981) 357.

[178] R.D. Field and R.P. Feynman, Nucl. Phys. B136 (1978) 1.

[179] E. Bloom, in: Proc. 1979 Int. Symp. on Lepton and Photon Interactions at High Energies, Batavia (1979) eds. T.B.W. Kirk and H.D.I. Abarbanel (Fermilab, 1979).

[180] J.C. Pati and A. Salam, Nucl. Phys. B144 (1978) 445.

[181] TASSO collaboration: R. Brandelik et al., Phys. Lett. 89B (1980) 418.

PLUTO collaboration: Ch. Berger et al., Phys. Lett. 95B (1980) 313.

[182] C. Bacci et al., Phys. Lett. 86B (1979) 234.

LENA Collaboration: B. Niczyporuk et al., DESY 81-008 (1981).

J.L. Siegrist, SLAC-225 (1979).

[183] A. Bassetto, M. Ciafaloni and G. Marchesini, Phys. Lett. 83B (1978) 207.

W. Furmanski, R. Petronzio and S. Pokorski, Nucl. Phys. B155 (1979) 253.

K. Konishi, Rutherford Lab. preprint RL-79-035 T 241 (1979).

A. H. Mueller, Columbia Univ. preprint CU-TP 197 (1981).

[184] E. Fermi, Prog. Theor. Phys. 5 (1950) 570.

[185] TASSO collaboration: R. Brandelik et al., DESY Report 81/69 (1981).

[186] TASSO collaboration: R. Brandelik et al., Phys. Lett. 94B (1981) 444.

[187] PLUTO collaboration: Ch. Berger et al., Phys. Lett. 104B (1981) 79.

[188] TASSO collaboration: R. Brandelik et al., Phys. Lett. 94B (1981) 91.

[189] TASSO collaboration: R. Brandelik et al., Phys. Lett. 105B (1981) 75.

[190] JADE collaboration: W. Bartel et al., Phys. Lett. 104B (1981) 325.

[191] M. Holder, Eur. Phys. Soc. Int. Conf. on High Energy Physics, Lisbon (1981); Univ. of Hamburg preprint.

[192] B. Anderson et al., Lund Univ. preprint LU TP 81-03 (1981).
E.M. Ilgenfritz, J. Kripfganz and A. Schiller, Acta Phys. Pol. B9 (1978) 881.
A. Casher, H. Neuberger and S. Nussinov, Phys. Rev. D20 (1979) 179.
S. Ritter and J. Ranft, Acta Phys. Pol. B11 (1980) 259.
W. Hofmann, Z. Phys. C10 (1981) 351.
G. Schierholz and M. Tepen, DESY report 81/41 (1981).
M. Bowler, Oxford Univ. report 76-81 (1981).

[193] T. Meyer, DESY report 81/46 (1981).

[194] S.L. Wu, DESY report 81/71 (1981).

[195] S.D. Drell, D. J. Levy and T.M. Yan, Phys. Rev. 187 (1969) 2159; D1 (1970) 1617.

[196] K. Koller and T. F. Walsh, Nucl. Phys. B140 (1978) 449.

[197] PLUTO collaboration: Ch. Berger et al., Phys. Lett. 78B (1978) 176; 82B (1979) 449.

[198] S. Brandt and H.D. Dahmen, Z. Phys. C1 (1979) 61.

[199] T.F. Walsh and P.M. Zerwas, DESY report 80/20 (1920).

[200] LENA collaboration, B. Niczyporuk et al., Z. Phys. C9 (1981) 1.
PLUTO collaboration, Ch. Berger et al., Z. Phys. C8 (1981) 101.

[201] K. Koller and H. Krasemann, Phys. Lett. 88B (1979) 119.

[202] K. Gottfried, Phys. Rev. Lett. 40 (1978) 598.

[203] K. Berkelman, in: High Energy Physics 1980, XX Int. Conf., Madison, eds. L. Durand and L.G. Pondrom (American Institute of Physics, New York).

[204] R. Barbieri et al., Phys. Lett. 57B (1978) 455.
W. Celmaster, Phys. Rev. D19 (1979) 1517.

[205] H.B. Thacker, C. Quigg and J.L. Rosner, Phys. Rev. D18 (1978) 274, 287.
C. Quigg and J.L. Rosner, Phys. Rev. D23 (1981) 2625.

[206] H. Krasemann and S. Ono, Nucl. Phys. B154 (1979) 282.

[207] J.L. Richardson, Phys. Lett. 82B (1979) 272.

[208] A. Martin, Phys. Lett. 93B (1980) 338.

[209] W. Buchmüller, G. Grunberg, and S.H.H. Tye, Phys. Rev. Lett. 45 (1980) 103.

[210] P. B. Mackenzie and G. P. Lepage, CLNS 81/498 (1981), Phys. Rev. Lett. 47 (1981) 1244.

[211] H. Fritzsch and M. Gell-Mann, Proc. XVI Int. Conf. on High Energy Physics, Chicago (1972), eds. J.D. Jackson and A. Roberts (Fermi National Accelerator Laboratory).
H. Fritzsch and P. Minkowski, Nuovo Cim. 30A (1975) 343.
D. Robson, Nucl. Phys. B130 (1977) 328.
K. Koller and T. Walsh, Nucl. Phys. B140 (1978) 449.

[212] G.S. Abrams et al., Phys. Rev. Lett. 44 (1980) 114.
D.L. Scharre et al., Phys. Lett. 97B (1980) 329.

[213] D.L. Scharre, in: Experimental Meson Spectroscopy 1980, eds. S.U. Chung and S.J. Lindenbaum (American Institute of Physics, New York, 1981).

[214] Particle Data Group, Rev. Mod. Phys. 52 (1980) S1.

[215] C. Dionisi et al., Nucl. Phys. B164 (1980) 1.

[216] I. Cohen and H.J. Lipkin, Nucl. Phys. B151 (1979) 16.

[217] B. Berg, Phys. Lett. 97B (1980) 401.
G. Bhanat and C. Rebbi, Nucl. Phys. B189 (1981) 469.
J. Engels, F. Karsch, H. Satz and I. Montvay, Phys. Lett. 101B (1981) 89; 102B (1981) 332.
K. Ishikawa, M. Teper and G. Schierholtz, DESY report 81-089 (1981).

[218] For an up to date review, including references see:
A.J. Buras, in: Proc. 1981 Int. Symp. on Lepton and Photon Interactions at High Energies, Bonn (1981), ed. W. Pfeil (Univ. of Bonn).

[219] B.H. Wiik, Proc. of the Int. Neutrino Conf., Bergen, Norway (1979), eds. A. Haatuft and C. Jarlskog (Univ. of Bergen).
P. Söding, Proc. Eur. Phys. Soc. Int. Conf. on High Energy Physics, Geneva, Switzerland (1979), ed. A. Zichichi (CERN, Geneva).

[220] TASSO collaboration: R. Brandelik et al., Phys. Lett. 83B (1979) 261.

[221] MARK J. collaboration: D.P. Barber et al., Phys. Rev. Lett. 43 (1979) 830.

[222] PLUTO collaboration: Ch. Berger et al., Phys. Lett. 86B (1979) 418.

[223] JADE collaboration: W. Bartel et al., Phys. Lett. 91B (1980) 142.

[224] D. Fournier, in: Proc. 1981 Int. Symp. on Lepton and Photon Interactions at High Energies, Bonn (1981).

[225] P. Hoyer, P. Osland, H.G. Sander, T.F. Walsh and P.M. Zerwas, Nucl. Phys. B161 (1979) 349.

[226] A. Ali, E. Pietarinen, G. Kramer and J. Willrodt, Phys. Lett. 93B (1980) 155.

[227] B. Anderson, C. Gustafson and C. Peterson, Nucl. Phys. B135 (1978) 273.
B. Anderson and G. Gustafson, Z. Phys. C3 (1980) 223.
B. Anderson, G. Gustafson and T. Sjöstrand, Univ. of Lund preprint LU-TP-80-2 (1980).
T. Sjöstrand, Lund Univ. preprint LU-TP-80-3 (1980).

[228] A. Casher, J. Kogut and L. Susskind, Phys. Rev. Lett. 31 (1973) 792.

[229] G. Altarelli and G. Parisi, Nucl. Phys. B126 (1977) 298.

[230] TASSO collaboration: R. Brandelik et al., Phys. Lett. 94B (1980) 437.

[231] G.G. Hanson et al., Phys. Rev. Lett. 35 (1975) 1609.

[232] Physics with high energy electron–positron colliding beams with MARK J detector, D.P. Barber et al., Phys. Rep. 63 (1980) 340.

[233] PLUTO collaboration: Ch. Berger et al., Phys. Lett. 97B (1980) 459.
H.J. Daum, H. Meyer and J. Bürger, DESY Report 80/101 (1980).

[234] J. Dorfan, Z. Phys. C7 (1981) 349.

[235] P. Söding, Particles and Fields Conference, Santa Cruz (1981); DESY preprint in: 81-070 (1981).

[236] W. Braunschweig, in: Proc. 1981 Int. Symp. on Lepton and Photon interactions at High Energies, Bonn (1981), ed. W. Pfeil (Univ. of Bonn).

[237] S.L. Wu and G. Zobernig, Z. Phys. C2 (1979) 207.

[238] H. Newman, in: High Energy Physics 1980, XX Int. Conf., Madison (1980), eds. L. Durand and L.G. Pondrom (American Institute of Physics, New York).
P. Duinker, DESY preprint 81-012 (1981), submitted to Rev. Mod. Phys.

[239] S. Yamada, in: High Energy Physics 1980, XX Int. Conf., Madison (1980), eds. L. Durand and L.G. Pondrom (American Institute of Physics, New York).

R. Marshall, Eur. Phys. Soc. Int. Conf. on High Energy Physics, Lisbon (1981).

[240] JADE collaboration: W. Bartel and A. Petersen, XVth Rencontre de Moriond, Les Arcs, France (1980).

[241] J. Ellis and I. Karliner, Nucl. Phys. B148 (1979) 141.

[242] TASSO collaboration: R. Brandelik et al., Phys. Lett. 94B (1980) 437.

[243] CELLO collaboration: H.J. Behrend et al., DESY report 81/80 (1981).

[244] T.A. De Grand, Y.J. Ng and S.H. Tye, Phys. Rev. D16 (1977) 3251.
J.F. Gunion, III Int. Warsaw Symp. on Elementary Particle Physics, UC Davies preprint (1981).

[245] PLUTO collaboration: Ch. Berger et al., Phys. Lett. 97B (1980) 459.

[246] C. K. Chen, Phys. Rev. D23 (1981) 712.

[247] F.A. Berends and R. Kleiss, DESY reports 80/66 and 80/73 (1980).

[248] G. Sterman and S. Weinberg, Phys. Rev. Lett. 39 (1977) 1436.

[249] K. Fabricius, I. Schmitt, G. Schierholz and G. Kramer, Phys. Lett. 97B (1980) 431; DESY preprint 81-035 (1981).

[250] R.K. Ellis, D.A. Ross and A.E. Terrano, Phys. Rev. Lett. 45 (1980) 1226; Nucl. Phys. B178 (1981) 421.

[251] J.A.M. Vermaseren, K.J.F. Gaemers and S.J. Oldham, Nucl. Phys. B187 (1981) 301.

[252] T.D. Gottschalk, Phys. Lett. 109B (1982) 331.

[253] Z. Kunszt, Phys. Lett. 99B (1981) 429.

[254] CELLO collaboration: H.-J. Behrend et al., DESY preprint 81-080 (1981).

[255] PLUTO collaboration: Ch. Berger et al., DESY preprint 81-054 (1981).

[256] C.L. Basham, L.S. Brown, S.D. Ellis, and T.S. Love, Phys. Rev. Lett. 41 (1978) 1581; Phys. Rev. D17 (1978) 2298; Phys. Rev. D19 (1979) 2018.

[257] G.C. Fox and S. Wolfram, Nucl. Phys. B149 (1979) 413.

[258] PLUTO collaboration: C. Berger et al., Phys. Lett. 99B (1981) 292.

[259] L. Clavelli and D. Wyler, Univ. of Bonn preprint HE-81-3 (1981).

[260] A. Ali, DESY report 81/59 (1981).

[261] K. Konishi, A. Ukawa and G. Veneziano, Nucl. Phys. B157 (1979) 45.

[262] A. Bassetto, M. Ciafaloni and G. Marchesini, Nucl. Phys. B163 (1980) 477.

[263] E.J. Williams, Kgl. Danske Videnskab. Selskab, Mat. Fys. Medd. 13 (1934) No. 4.
L. Landau and E. Lifshitz, Sov. Z. Phys. 6 (1934) 244.

[264] A. Jaccarini, N. Arteaga-Romero, J. Parisi and P. Kessler, Compt. Rend. 269B (1969) 153, 1129; Nuovo Cim. 4 (1970) 933.
V.E. Balakin, V.M. Budnev and I.F. Ginzburg, JETP Lett. 11 (1970) 388.
S.J. Brodsky, T. Kinoshita and H. Terazawa, Phys. Rev. Lett. 25 (1970) 972; Phys. Rev. D4 (1971) 1532.

[265] R.J. Wedemeyer, 1981 Int. Symp. on Lepton and Photon Interactions at High Energies, Bonn (1981), ed. W. Pfeil (Univ. of Bonn).

[266] W.A. Bardeen, 1981 Int. Symp. on Lepton and Photon Interactions at High Energies, Bonn (1981), ed. W. Pfeil (Univ. of Bonn).

[267] F.F. Low, Phys. Rev. 120 (1960) 582.

[268] G.S. Abrams et al., Phys. Rev. Lett. 43 (1979) 477.

[269] PLUTO collaboration: Ch. Berger et al., Phys. Lett. 94B (1980) 259.

[270] TASSO collaboration: R. Brandelik et al., DESY report 81/020 (1981).

[271] A. Roussarie et al., Phys. Lett. 105B (1981) 304.

[272] CELLO collaboration: contribution to the Eur. Phys. Soc. Int. Conf. on High Energy Physics, Lisbon (1981).

[273] D.L. Burke, SLAC-PUB-2745 (1981).

[274] J.A.M. Vermaseren, private communication.
R. Bhattacharya, J. Smith and G. Gramm, Phys. Rev. D15 (1977) 3267.

[275] F. Calogero and C. Zemack, Phys. Rev. 120 (1960) 1860.
S.J. Brodsky and G.P. Lepage, SLAC-PUB-2733 (1981), Phys. Rev. D24 (1981) 1808.

[276] E. Hilger, High Energy Physics, 1980, XX Int. Conf. Madison (1980).
TASSO collaboration: R. Brandelik et al., Phys. Lett. 97B (1980) 448.

[277] D.L. Burke et al., Phys. Lett. 103B (1981) 153.

[278] TASSO collaboration: R. Brandelik et al., DESY preprint 81/058 (1981).

[279] J.J. Sakurai, Ann. Phys. 11 (1960) 1.

[280] S.M. Berman, J.D. Bjorken and J.B. Kogut, Phys. Rev. D4 (1971) 3388.
S. Brodsky, T. De Grant, J. Gunion and J. Weis, Phys. Rev. D19 (1979) 1418.

[281] PLUTO collaboration, Ch. Berger et al., Phys. Lett. 81B (1979) 410; Phys. Lett. 99B (1981) 287.

[282] W. Wagner, in: High Energy Physics 1980, XX Int. Conf., Madison (1980), eds. L. Durand and L.G. Pondrom (American Institute of Physics, New York).

[283] TASSO collaboration: R. Brandelik et al., Phys. Lett. 107B (1981) 290.

[284] D. Cords, 4th Int. Conf. on Photon–Photon Interactions, Paris (1981); DESY Report 81/033 (1981).
H. Spitzer, XVth Rencontre de Moriond, Les Arcs, France (1980), ed. J. Tran Thanh Van (Editions Frontières, Dreux, France).

[285] JADE collaboration: W. Bartel et al., DESY report 81/048 (1981).

[286] PLUTO collaboration, Ch. Berger et al., DESY report 81/051 (1981), Phys. Lett. 107B (1981) 168.

[287] T.F. Walsh, Phys. Lett. 36B (1971) 121.
S.B. Brodsky, T. Kinoshita and H. Terazawa, Phys. Rev. Lett. 27 (1971) 280.
E. Witten, Nucl. Phys. B120 (1972) 189.
C.H. Llewellyn Smith, Phys. Lett. 79B (1979) 83.

[288] W.A. Bardeen and A.J. Buras, Phys. Rev. D20 (1979) 166.
D.W. Duke and J.F. Owens, Phys. Rev. D22 (1980) 2280.

[289] E.D. Bloom et al, Phys. Rev. Lett. 23 (1969) 930.
E.D. Bloom et al., Phys. Rev. Lett. 25 (1970) 1140; Phys. Rev. D4 (1971) 2901.

[290] H. Deden et al., Nucl. Phys. B85 (1975) 269.

[291] P.C. Bosetti et al., Nucl. Phys. B142 (1978) 1.

[292] P. Musset et al., Phys. Rep. 39 (1978) 1.

[293] J.G.H. De Groot et al., Z. Phys. C1 (1979) 143.

[294] J. Ellis, Deep hadronic structure, in: Weak and Electromagnetic Interactions at High Energy, Les Houches (1976) session XXIX, eds. R. Balian and C.H. Llewellyn Smith (North-Holland, Amsterdam).
C.T. Sachrajda, course 2, this volume.
D.H. Perkins, in: NSAI on Techniques and Concepts of High Energy Physics, ed. T. Febel (Plenum, New York, 1981).

[295] J.D. Bjorken, Phys. Rev. 179 (1969) 1547.

[296] R.P. Feynman, Photon–Hadron Interactions (Frontiers in Physics) (Benjamin, Reading, MA, 1972).
J.D. Bjorken and E.A. Paschos, Phys. Rev. 185 (1975) 1969.

[297] J. Wotschack, 1981 Int. Symp. on Lepton and Photon Interactions at High Energies, Bonn (1981).

[298] B.C. Barish et al., Phys. Rev. Lett. 39 (1977) 1595.

[299] H. Deden et al., Contribution presented by P. Fritze to Neutrino 81, Int. Conf. on Neutrino Physics and Astrophysics, Wailea, Maui, Hawaii (1981), eds. R.J. Cence, E. Ma and A. Roberts (HEP group, Hawaii).
[300] J.B. Morfin et al., Univ. of Aachen preprint PITHA 81/14 (1981).
[301] M.D. Mestayer, thesis, SLAC 214 (1978).
[302] A. Bodek et al., Phys. Rev. D20 (1979) 1471.
[303] J.J. Aubert et al., Phys. Lett. 105B (1981) 315.
[304] J.J. Aubert et al., Phys. Lett. 105B (1981) 322.
[305] P. Musset et al., Phys. Rep. 39 (1978) 1.
[306] S.M. Heagy et al., Phys. Rev. D23 (1981) 1045.
[307] H. Abramowicz et al., CERN-EP/81-50 (1981), Phys. Lett. 107B (1981) 141.
[308] V.V. Ammosov et al., Serpuchov, IHEP 81-66 (1981).
[309] J. Panman, thesis, NIKHEF (1981), to be published.
[310] C. Baltay et al., Phys. Rev. Lett. 44 (1980) 916.
[311] H. Wahl et al., talk at the XVIth Rencontre de Moriond (1981), to be published.
[312] B.M. Heagy, Neutrino 79, Int. Conf. on Neutrino Physics and Astrophysics, Bergen, Norway (1979).
[313] R.D. Field and R.P. Feynman, Phys. Rev. D15 (1977) 2590.
[314] G.R. Farrar and D.R. Jackson, Phys. Rev. Lett. 35 (1975) 1416.
[315] A. Donnachie and P.V. Landshoff, Phys. Lett. 95B (1980) 437.
[316] F.E. Close and R.G. Roberts, Rutherford Laboratory preprint RL-80-058 (1980).
[317] J.J. Aubert et al., CERN-EP/83-04.
[318] P. Allen et al., Phys. Lett. 103B (1981) 71.
[319] H. Deden et al., Nucl. Phys. B149 (1979) 1.
[320] F.S. Meritt et al., Phys. Rev. D17 (1978) 2199.
[321] G. Altarelli, Chromodynamics and deep inelastic processes, in: New Phenomena in Lepton–Hadron Physics, eds. E.E.C. Fries and J. Wess (Plenum Press, New York, 1979).
[322] H. Georgi and H.D. Politzer, Phys. Rev. D14 (1976) 1829.
       R. Barbieri, J. Ellis, M.K. Gaillard and G.G. Ross, Nucl. Phys. B117 (1976) 50.
       E.G. Floratos, D.A. Ross and C.T. Sachrajda, W.A. Bardeen,
       A.J. Buras, D.W. Duke, and T. Muta, Phys. Rev. D18 (1978) 3998.
       A.F. Abbott and R.M. Barnett, Ann. Phys. 125 (1980) 276.
[323] J.M. Cornwall and R.E. Norton, Phys. Rev. 177 (1969) 2584.
       O. Nachtmann, Nucl. Phys. B63 (1973) 237; Nucl. Phys. B78 (1974) 455.
[324] D.J. Gross and F.A. Wilczek, Phys. Rev. D8 (1973) 3633; Phys. Rev. D9 (1974) 980.
       H. Georgi and H.D. Politzer, Phys. Rev. D9 (1974) 416.
[325] A.J. Buras and K.F. Gaemers, Nucl. Phys. B132 (1978) 249.
[326] A.F. Abbott and R.M. Barnett, Ann. Phys. 125 (1980) 276.
[327] A. Gonzales-Arroyo et al., Nucl. Phys. B153 (1979) 161; B159 (1979) 512; B166 (1980) 429.
[328] J.G.H. De Groot et al., Phys. Lett. 82B (1979) 292; Phys. Lett. 82B (1979) 456.
[329] M. Jonker et al., CERN EP/81-135 (1981).
[330] J.G. Morfin et al., Phys. Lett. 104B (1981) 235.
[331] J. Steinberger, CERN EP/80-222 (1980).
       F. Eisele, Uni. Dortmund, DO-EXP 6/81 (1981).
[332] For a recent review see: J. Bürger, in: Proc. 1981 Int. Symp. on Lepton and Photon Interactions at High Energies, Bonn (1981), ed. W. Pfeil (Univ. of Bonn).
       P. Duinker, Eur. Phys. Soc. Int. Conf. on High Energy, Physics, Lisbon (1981).
[333] MARK J collaboration: D.P. Barber et al., MIT Report LNS 113 (1980).

[334] PLUTO collaboration: Ch. Berger et al., Phys. Lett. 99B (1981) 489.

[335] TASSO collaboration: R. Brandelik et al., Phys. Lett. 99B (1981) 163.

[336] M.S. Chanowitz, J. Ellis and M.K. Gaillard, Nucl. Phys. B128 (1977) 506.
A.J. Buras, J. Ellis, M.K. Gaillard and D.V. Nanopoulos, Nucl. Phys. B135 (1978) 66.

[337] J.D. Bjorken and C. Llewellyn Smith, Phys. Rev. D7 (1973) 887.

[338] A. Ali, Phys. Rev. D10 (1974) 2801.

[339] F. Bletzacker and H.T. Nieh, Phys. Rev. D16 (1977) 2115.

[340] S. La Rue, D. Phillips and W.M. Fairbank, Phys. Rev. Lett. 46 (1981) 967.

[341] M. Marinelli and G. Morpurgo, Phys. Lett. 94B (1980) 427, 433.

[342] JADE collaboration: W. Bartel et al., Z. Phys. C6 (1980) 295.

[343] J.M. Weiss et al., SLAC-PUB-2671 (1981).
A.M. Litke, Proc. 1981 Int. Symp. on Lepton and Photon Interactions at High Energies, Bonn (1981).

[344] G. Valente, in: High Energy Physics, XX Int. Conf., Madison (1980), eds. L. Durand and L.G. Pondrom (American Institute of Physics, New York).
A. Zichichi, private communication.

[345] For a review including a complete set of references see:
G. Barbiellini et al., Phys. Rep., to be published.

[346] A. De Rújula, R.C. Giles and R.L. Jaffe, Phys. Rev. D17 (1978) 285.

[347] B.M. McCoy and T.T. Wu, Phys. Lett. 72B (1977) 219.

[348] F.A. Berends and R. Kleiss, Nucl. Phys. B178 (1981) 141.

[349] A. Litke, Harvard Univ. thesis (1970).

[350] F.A. Berends et al., Nucl. Phys. B57 (1973) 381.

[351] G. 't Hooft, Nucl. Phys. B35 (1971) 167.

[352] J. Schwinger, Phys. Rev. 125 (1962) 397; Phys. Rev. 128 (1969) 2425
R. Jackiw and K. Johnson, Phys. Rev. D8 (1973) 2286.
J.M. Cornwall and R.E. Norton, Phys. Rev. D8 (1973) 3338.
M.A.B. Bég and A. Sirlin, Ann. Rev. Nucl. Sci. 24 (1974) 379.
S. Weinberg, Phys. Rev. D13 (1976) 974; Phys. Rev. D19 (1979) 1277.
L. Susskind, Phys. Rev. D20 (1979) 2619.

[353] Yu. A. Gol'fand and E.P. Likthman, JETP Lett. 13 (1971) 323.
P.V. Volkov and V.P. Akulow, Phys. Lett. 46B (1973) 109.
J. Wess and B. Zumino, Nucl. Phys. B70 (1974) 39.

[354] For reviews see: F. Fayet, Proc. Orbis Scientiae (Coral Gables, Florida, USA, 1978), eds. A. Perlmutter and L.F. Scott; New Frontiers in High Energy Physics (Plenum, New York, 1978).
G.R. Farrar, in: Proc. Int. School of Subnuclear Physics, Erice, Italy (1978), ed. A. Zichichi (Plenum, New York, 1980).

[355] CELLO collaboration: H.J. Behrend et al., contribution to the 1981 Int. Symp. on Lepton and Photon Interactions at High Energies, Bonn (1981), ed. W. Pfeil (Univ. of Bonn).

[356] H. Spitzer, in: Proc. XVth Rencontre de Moriond, Les Arcs, France (1980), ed. J. Tran Thanh Van (Editions Frontières, Dreux, France); DESY report 80/43 (1980).

[357] R.D. Peccei and H.R. Quinn, Phys. Rev. Lett. 38 (1970) 1440.

[358] S. Weinberg, Phys. Rev. Lett. 40 (1978) 223.
F. Wilczek, Phys. Rev. Lett. 40 (1978) 279.

[359] For a review on axions, including references, see:
J. Ellis, course 3, this volume.

[360] A. Faissner, 1981 Int. Symp. on Lepton and Photon Interactions at High Energies, Bonn (1981), ed. W. Pfeil (Univ. of Bonn).

Seminars related to Courses 7 and 8

1. The CERN P$\overline{\text{P}}$ collider and the UA1 experiment, by Jacques Colas (LAPP).

2. QED corrections to some e^+–e^- processes, by Ronald Kleiss (Leiden).

3. The UA2 and other experiments at the CERN P$\overline{\text{P}}$ collider, by Hartmut Plothow-Besch (Orsay).

4. A-dependence at the ISR, by Roger Rusack (Rockefeller Univ.).

5. The Lund model for jet fragmentation, by Torbjörn Sjöstrand (Lund).

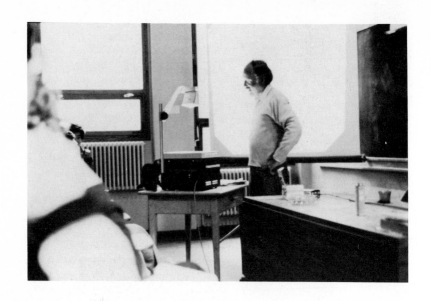